MAN·IN SPACE

CONTRIBUTORS
ANDREW WILSON
CURTIS PEEBLES
H.J.P. ARNOLD

Designed by PHILIP CLUCAS
Commissioning Editor ANDREW PRESTON

This edition published in 1993 by SMITHMARK Publishers Inc.,
16 East 32nd Street, New York, NY 10016.

SMITHMARK books are available for bulk purchase for sales
promotion and premium use. For details write or call the manager
of special sales; SMITHMARK Publishers Inc.,
16 East 32nd Street, New York, NY 10016; (212) 532-6600.

CLB 2594
Produced by CLB Publishing Ltd.,
Godalming Business Centre
Woolsack Way
Godalming, Surrey GU7 1XW
England

ISBN 0-8371-4491-X
Printed in Singapore
10 9 8 7 6 5 4 3 2 1

MAN·IN SPACE

AN ILLUSTRATED HISTORY OF SPACE FLIGHT

EDITED BY
H.J.P. ARNOLD

FOREWORD BY
CAPTAIN EUGENE A. CERNAN

SMITHMARK

CONTENTS

May 1969: Window reflections create mysterious shapes around the Apollo 10 lunar module.

FOREWORD

*I*n December of 1992, the United States, indeed the entire world, recognized the 20th Anniversary of the end of Apollo – the end of a dramatic era in the history of mankind. Those final footsteps of Apollo XVII, not to be retraced for as long as a generation into the future, are now a legacy of a generation from the past.

In reflecting upon events which have now become history, it was once said "...anniversaries somehow become high ground in our journey through life, for they give us an occasion to pause and rethink the past, assess the present and utter cherished hopes for the future." Perhaps this is where the world finds itself today as we stand on the plateau of space looking with uncertainty toward the future – reflecting upon those cherished few moments when we found ourselves standing on the surface of another "planet" – looking back from ¼ million miles away and seeing the Earth in all its splendor and majestic beauty – "one man seeing it for the first time" as only before through the words of poets or perhaps the minds of philosophers. It is now we realize that we were truly on a space voyage as we looked back home at reality from a place where reality itself was almost like a dream. It is now easy to conclude that the world is "just too beautiful to have happened by accident."

Man in Space *gives us the opportunity to relive those moments in history which someday may be looked upon as the greatest human endeavor thus far in the life of mankind. It gives us the opportunity to identify with the dreamers and the pioneers of the past who knew not the meaning of the word impossible – who dared to challenge the concept of those who said "it couldn't be done" – men and women whose commitment and vision now challenge us to reach further than man has ever reached before.*

Captain Eugene A. Cernan
Commander, Apollo XVII

December 7, 1972: launch of Apollo 17 – NASA's sixth and last manned lunar landing mission.

1

DREAMERS, PIONEERS AND ROCKET SOCIETIES

1 *Johannes Kepler (1571-1630) – a brilliant German mathematician who has a hallowed place in the history of astronomy for working out three basic laws of planetary motion. Until that time, explaining the movement of the planets across the sky – so obviously different from the stars – had not been satisfactorily treated even though the views of Claudius Ptolemaeus (Ptolemy) in the second century AD held sway for centuries. Kepler also wrote* Somnium, *a work of combined fantasy and science about a trip to the Moon, which was published in 1634.*

2 *From earliest times there was a belief that the sky and stars were part of a solid dome. This representation of Earth and the Heavens is taken from a medieval woodcut. Even today we still sometimes talk about the "celestial sphere."*

Spaceflight literature is much older than the reality – almost two millenia older in fact – and since the knowledge necessary for the practical accomplishment of spaceflight has been a product only of the last century or so it follows that much of the writing and fiction of previous centuries on the theme is, to say the least, highly speculative and whimsical. For the most part, the techniques of spaceflight were either ignored or solved by a magic formula, and attention was concentrated on a description of the world being visited (almost invariably the Moon) and its strange inhabitants. Over the centuries, the best writing was often concerned either with presenting an enjoyable story or fantasy *per se*, or with using an interplanetary context to satirize aspects (religious and political for instance) of life on Earth. Whatever the form and the precise object of the writing, it constituted a lengthy prelude to the reality – even if there was a major break in the era of the so-called Dark Ages – such that the coming of the reality in the twentieth century certainly did not cause a cultural or intellectual shock in society at large, even if there was some skepticism in what ought to have been better informed circles. The concept of spaceflight, it could be argued, had been well espoused by writers ancient and modern, and was firmly established in the human consciousness.

This outline of the work of the "dreamers" can begin conveniently with the Greeks. Over the centuries of their history, consideration of the nature of the Earth and of the heavens had received much attention. By the time of "The Prince of Astronomers" Claudius Ptolemaeus (Ptolemy), in the second century A.D., notions of a flat Earth had been generally discounted, the size of the Earth had been calculated with reasonable accuracy, the then known planets were recognized as being different bodies from the background stars and the latter had been extensively charted by Hipparchus (second century B.C.) and by Ptolemy himself, who left a list of 48 constellations. Hipparchus had calculated both the size of the Moon and its distance from the Earth with reasonable accuracy, so it was clearly shown to be the nearest celestial body to Earth – and thus the natural target for the heroes of fiction writers.

Ptolemy's name is mainly associated with an astronomical system which rejected movement by the Earth around the Sun and on its axis (both propounded by Aristarchus of Samos in the third century B.C.) in favor of an Earth-centered system with the known planets and the Sun moving around it. Their orbits were taken to be circular, but since this did not fit in with their movements as observed from Earth, Ptolemy supported the concept of each body moving in a small circle or *epicycle*, the center of which moved about the Earth in a perfect circle. The theory was clumsy and contorted but it apparently fitted the then known facts and lasted until the early sixteenth century.

A little before Ptolemy, the Greek writer and philosopher Plutarch had written a treatise *On the Face in the Orb of the Moon* in the form of a dialogue between eight friends. It was partly a consideration of what was then known about the Moon, partly a debate on whether the Moon had inhabitants. (The latter was a question

which greatly intrigued the Roman philosopher and orator Cicero, too.) But the first fictional account of a journey to the Moon was by the Greek satirist Lucian of Samosata, who was born about the time Plutarch died in A.D. 120 and who was therefore a contemporary of Ptolemy. He admitted that his *True History* of voyages to the Moon, Venus and elsewhere was anything but true, but it was certainly fun. In the story, he and fifty companions were aboard a ship which was caught up in a whirlwind that transported them to the Moon. There they found themselves in the middle of a war between the inhabitants of the Sun and Moon over the colonization of Venus. Lucian's imagination worked hard when describing what he found. The lunar inhabitants lived on frogs, were bald but sported beards a little beneath their knees, and when they perspired they exuded honey. There were no women and males up to the age of 25 bore children in their thighs; eyes could be removed and replaced at will; and the creatures had but one finger which grew from their rump. On an extended journey back home, Lucian and his companions "entered the Zodiac and passed the Sun" on the left but did not disembark because the "wind did not allow it."

Famous Astronomers

No less than fourteen hundred years were to pass before the next piece of major spaceflight fiction was written. During that period, the Ptolemaic scheme of an Earth-centered universe held sway, supported on doctrinal grounds by the Church (both Catholic and non-Catholic). Astronomical studies, inextricably combined with a belief in astrology and the casting of horoscopes, largely passed into the keeping of the Arabs, who preserved much of what the Greeks taught. (For example, Ptolemy's major scientific work has only come down to us by means of its translation into Arabic – the *Almagest*.) Beginning in the first half of the sixteenth century, however, the Earth-centered beliefs were challenged by men who were to come to occupy honored places in the history of astronomy – Copernicus, Galileo and Kepler, the last building on the observational genius of Tycho Brahe.

Johannes Kepler was a German mathematician who solved the problem of the observed motions of the planets in three "laws" published in the first part of the seventeenth century. These demonstrated that the planets (including Earth) moved about the Sun in ellipses rather than circles, and also that a planet close in to the Sun moved faster than one further away; a law of great importance not only for understanding the solar system but the behavior of spacecraft and satellites in orbit. In 1687, Isaac Newton's great work *Principia* would provide the detailed mathematical explanation of Kepler's laws and the mechanism – gravity – which controlled the movements of all bodies in motion.

It was this same Kepler who wrote the next piece of spaceflight fiction – *Somnium* (Dream) – which was published in 1634, four years after his death. The story was a mixture of fantasy and science, the latter contained in lengthy footnotes. Kepler's Moon was (accurately) divided into two: Subvolva from which the Earth (Volva) was always visible and Privolva, from which it was never seen. Kepler could not resist the introduction of exotic inhabitants, but one of the most interesting aspects of the story was the means by which the hero, an Icelander called Duracotus (a thinly disguised version of Kepler himself), got to the Moon. His mother contacted a "demon" who had supernatural powers: demons could travel to Levania (the Moon) in about four hours, but the journey was extremely difficult and dangerous for Earth people and could only be undertaken during the period of the waning Moon. The beginning of the journey was the most difficult part for a human because "it is as if he were being catapulted by gunpowder over mountains and oceans." Therefore, the traveler had to be drugged with a potion to prevent his limbs being torn from his body. "Then there is the tremendous cold and breathlessness to be endured: for the latter one has to hold a wet sponge over nose and mouth."

There was in fact much good, perceptive science in *Somnium*; indeed there was more than in any other of the works referred to here. For example, Kepler realized the inherent danger of solar radiation to the space traveler; the need for a traveler to aim at a point in the Moon's orbit where it would be at the appropriate arrival time; and the fact that the Earth's atmosphere did not extend all the way to the Moon. Perhaps most acutely of all, he showed himself to be fully aware (no matter how imperfectly) of the importance of gravity and even of "g" forces as they would be called today. Obviously, knowledge at the time did not permit him to suggest a rational method of traveling to the Moon, and the return to Earth was accomplished simply by having Duracotus wake up from his dream.

In 1638 – four years after the publication of *Somnium* and again posthumously – *Man in the Moone* by Bishop Francis Godwin appeared. He wrote it as a 19-year-old undergraduate at Oxford University during the reign of Elizabeth I, and as his career in the church developed it was possible that he completely forgot the story – though he continued to show a spirited (if non-scientific) disposition throughout his life. The hero of the story (which was enormously popular) was one Domingo **3**

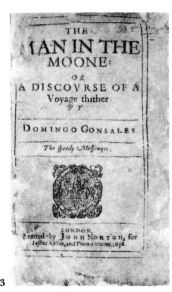

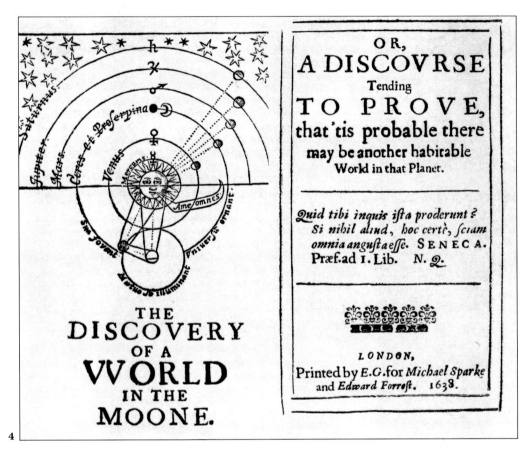

Gonsales, a buccaneering individual who was transported to the Moon by a flock of huge birds called gansas. There he found a delightful Utopia, where there was no crime, all females were beauteous, no wound was incurable and all lived in "love, peace and amity." The book was basically a whimsical and enjoyable story, though there were occasionally more serious insights, as when Gonsales commented (in effect) that the phases of Earth and Moon were opposite: a crescent Moon facing a gibbous or full Earth. Borne by the geese again, he returned to Earth in China and hid the lunar jewels he had been given "never to find them again."

The work of another bishop, John Wilkins of Chester, appeared at about the same time as *Man in the Moone*. Wilkins was a serious scholar and teacher who was to write many scientific works in England during the upheavals of the Civil War. He married Oliver Cromwell's sister and was to become the first Secretary of the Royal Society. *The Discovery of a World in the Moone* fudged the issue of the means of transportation (proffering either wings or a machine "if the motive faculty be properly provided") but as the spirit of colonialism grew, together with friction with the Dutch, Spanish and French, Wilkins

3 *The title page of the book written by Bishop Francis Godwin when an undergraduate at Oxford University and published posthumously in 1638.* Man in the Moone *was an enjoyable work of fiction in which the hero is transported by a flock of huge birds to the Moon, where he finds "love, peace and amity."*

4 The Discovery of a World in the Moone, *by another bishop – John Wilkins of Chester – was not, however, a work of fiction. Wilkins was a scholar and the brother-in-law of Oliver Cromwell as well as the first secretary of the Royal Society. In an age of developing colonialism, he argued that England should not lose the opportunity of adding the Moon to her other possessions.*

1 *By coincidence, a number of the early works, both fictional and (allegedly) factual, about travels to other worlds were published posthumously. This was the case with Cyrano de Bergerac's accounts of voyages to the Moon and to the Sun. This illustration shows one of the ways he proposed that the lunar journey be accomplished: de Bergerac reasoned that, since the Sun draws up dew from grass in the morning, bottles of dew attached to the traveler's body would raise him up too – eventually to the Moon. Even in this work of fancy the attempt was unsuccessful!*

2 *This de Bergerac title page is conventional in style and in showing the heavens containing plants, animals and people. But the Sun's rays fall upon the inveterate traveler in what is recognizably what we would call today a space capsule. The Moon was eventually conquered using a machine powered by a spring and rockets – but even then de Bergerac had to be helped by the Moon drinking the beef marrow with which he had greased himself.*

saw the Moon as an area for expansion. He felt sure the Moon had an atmosphere to breathe and adopted a rigorous attitude to his intended colonisers: "Since our bodies will be devoid of gravity and other impediments of motion, we shall not spend our strength in any labour, and so consequently not much need the reparation of diet." Wilkins presented his work to the Royal Society, confident that England would not ignore the chance of adding the Moon (where Paradise might be situated) to her possessions.

With a greater or lesser scientific content, awareness of astronomical discoveries spread and was incorporated in widely differing literature. The English metaphysical poet John Donne proposed that all Jesuits be transferred to the Moon, where they could establish a new Hell. John Milton sent Satan on a cosmic voyage in *Paradise Lost*. In the second half of the eighteenth century, Voltaire caused an inhabitant of the star Sirius to travel from moon to moon by means of a sunbeam, while other philosophers of the period used similar situations to score witty points against one another. Back in England, Godwin's book was presented as a play and as a comic opera; a poem changed Gonsales' huge birds to tame fleas and lice; a tract had William Pitt and Charles Fox conducting their political feud on the Moon; and a novel means of making the translunar journey – a sort of sled powered by a massive spring – was proposed.

There continued to be published individual works which had a high circulation at the time, and subsequently. Also posthumously published were Cyrano de Bergerac's splendidly written accounts of voyages to the Moon and to the Sun, which appeared in the middle of the seventeenth century. His first attempt at space flight was powered by bottles of dew (to be drawn up by the Sun's power) but this only transported him from France to Canada: he fared better with a machine combining a large spring with rockets clustered around it, but then

only succeeded in the last stages of the flight to the Moon because he had greased himself with beef marrow, which the Moon drank with eagerness. Once arrived, his experiences took the form of a satire on the book of Genesis, although there were some delightful touches such as the statement that the inhabitants of the Moon lived on vapor and odors, and that money took the form of poems.

The tradition of the Moon as the location of satire transposed from Earth was continued by Daniel Defoe in *The Consolidator*, published in 1705. The spacecraft was a machine containing 513 feathers – exactly the number of Members of Parliament at the time. Jonathan Swift in *Voyage to Laputa* had Gulliver climbing a chain to a "vast opaque body" hovering over the Earth in order to escape from an island on which he had been abandoned. In the nineteenth century, growing astronomical knowledge gave a greater impression of reality to the fiction – and on at least one occasion an elaborate hoax was constructed around the serious researches of a prominent astronomer, as when the *New York Sun* in 1835 published a series of despatches describing the alleged discovery of life on the Moon by Sir John Herschel, who was conducting a survey of the southern sky at that time from the observatory at the Cape of Good Hope in South Africa.

In that same year Edgar Allan Poe, critical of a lack of scientific accuracy in the work of other authors, published the *Unparalleled Adventure of One Hans Pfaall*. His hero escaped pressing creditors in Holland by inflating a balloon with a gas of his own invention, which was far less dense than hydrogen. Poe gave a vivid, impressive account of Pfaal's sufferings on his (totally impossible) journey to the Moon by balloon.

Verne and Wells

Far more impressive, however, were Jules Verne's two works *From the Earth to the Moon* and *Round the Moon*, published in 1865 and 1870 respectively. Inevitably there were inaccuracies and frankly impossible situations, perhaps the most obvious of which was the launch of his group of intrepid space travelers in their projectile craft (fitted with shock absorbers, air conditioning and gas lighting!) by being fired from an enormous cannon called Columbiad located in Florida. Verne had correctly adopted a speed of seven miles per second as the necessary escape velocity for a vehicle flying to the Moon, but neither projectile nor its occupants could have withstood instantaneous acceleration to that speed, nor could they have survived a ballistic return to the Earth's surface in the manner described. But this criticism is made with the practical knowledge of over thirty years of spaceflight. The stories were vivid, meticulously worked, persuasive, and make enjoyable reading to this day; the description of the lunar surface from orbit, with its characteristic high contrast and desolate appearance, matching some of the real descriptions of the lunar era of spaceflight.

H.G. Wells, too, was a master story teller, but his *First Men in the Moon*, published in 1901, was far less scientifically based than Verne's work. A scientist and a hard-headed businessman traveled to the Moon using what subsequently came to be known in the science fiction world as an "anti-gravity device." The Moon they discovered had vegetation, breathable air and several forms of life, including an insect-like being that was part human and part robot. With this story the genre reached a stage which might be described as modern, with a proliferation of books and magazines devoted to spaceflight, space adventures and beings from other worlds.

Wells had already written (1898) his Martian classic *War of the Worlds*, which followed on the sensation created some years before by Schiaparelli's report of "canali" on Mars. Life on other planets (but particularly Mars) was a cause energetically adopted in the writings of Camille Flammarion in France over the course of decades in the latter half of the nineteenth century – while life-forms on Mars became an article of faith for Percival Lowell in the USA. As the new century progressed, the Martian novels of Edgar Rice Burroughs enjoyed enormous success – and by then a time had almost been reached which could be regarded as the bridge between fiction and reality.

Perhaps the most apt link between the world of the early literature and that of reality is the fact that, having been the first to send an imaging probe to the far side of the Moon, the Russians named two craters there for Verne and Wells.

As the nineteenth century gave way to the twentieth, what might have appeared to be the stepping stones to space were developed. Balloons carried man into the sky – but only because their gases caused them to be lighter than the air around them: in space there is no air. Propeller-driven aircraft and subsequently jets led to a new age of travel (and warfare), but aircraft need air to provide lift for their wings and oxygen for their engines to function. So they, too, are limited to the lower reaches of the Earth's atmosphere. Sir Isaac Newton had, in fact, pointed the way many years before with his Third Law of Motion – *To every action, there is an equal and opposite reaction.* Consider a compartment containing a chemical or fuel that can be ignited. Provide a means of ignition and one opening through which the hot gases can escape: the pressure at the exit point will be lower than at the opposite, closed end, and the molecules of the hot gases being forced out at high speed (the action) will cause the container to move in the opposite direction to the opening (the reaction). While there were to be innumerable major and minor difficulties to overcome, this was the way to space and the means was the *rocket*, which in basic form had existed for centuries.

The Early History of Rockets

The first rockets were weapons of war, and their invention, using black powder (gunpowder), is usually attributed to the Chinese, with one of the earliest references to the use of "arrows of flying fire" being in 1232, when the Mongols were besieging the city of Kai-Fung-Fu. Subsequently, proposals for the improvement of rockets were published in Europe, and among the most important were those of a Pole, Kasimir Simienowicz, who in 1650 theorized that their accuracy and range could be improved by fitting fins on individual rockets; by arranging the sequential firing (and then discarding) of a number of rockets joined together – a technique that was later to become known as *staging*; and also by clustering a number of rockets together for simultaneous firing.

This treatise was remarkably far-seeing and much studied, but the next significant application of the rocket (again in war) occurred in India. In the last years of the eighteenth century unrest against the rule of the British East India Company had broken out in the southwestern state of Mysore, led by Haidar Ali and his son Tipu Sahib. Haidar Ali formed a rocket force equipped with projectiles weighing 6-11lbs mounted on yard-long bamboo canes and with a range of one mile. The weapons were used against the British troops in at least two major battles, in 1792 and 1799, and this experience led directly to active consideration of the rocket as a weapon by the British themselves – an effort associated with the name of Colonel (later Sir) William Congreve.

Congreve published his *Concise Account on the Origin and Progress of the Rocket System* in 1804 and became an ardent advocate of the rocket, in particular in those situations where the recoil from conventional guns could create severe problems, for example on vessels during a seaborne assault. Congreve's scaled-up rocket took the form of an iron case over 3ft 3in (1m) long, 4in (10cm) in diameter and weighing over 31lb (14kg) that was attached to a wooden pole over 16ft (5m) long. Initially the range was almost 1.2 miles (2km) but this was later improved to somewhat under two miles (3km). During the Napoleonic Wars, British forces bombarded Boulogne, Copenhagen and Danzig with rockets, and before the French Emperor's dreams were finally ended at Waterloo a specialist rocket group was formed in the British army, despite the fact that Lord Wellington was not impressed by the weapon. Congreve rockets were also used during the Anglo-American War of 1812-14, and it was a bombardment by the *Erebus* against Fort McHenry which led directly to the incorporation of the phrase "the rockets' red glare" in the *The Star-spangled Banner*. It was significant as a sign of things to come that Congreve regarded the rocket as an entire system, and that he devoted as much thought to the launcher and its mobility on the battlefield as to the projectile itself.

Space Flight Pioneers

In 1844, some years after Congreve's death, an Englishman, William Hale, developed a spin-stabilised rocket, but while many countries closely studied the potential of the rocket as a weapon, and none more so than Tsarist Russia, where the work of the artillery officers Alexander Zasyadko and Konstantin Konstantinov was important, its technology remained somewhat crude and it lost ground to the evolving sophistication of more conventional forms of artillery. Apart from the ever-popular firework displays, for many years the most valuable use for the rocket was in the breeches buoy system of marine rescue. Tactical rocket weapons reappeared during the Second World War, but by then the role of the rocket in the future of spaceflight was becoming clearer.

Theoretical and practical contributions to the emerging concept of spaceflight were made by workers in many countries, and it was symptomatic that the three men universally regarded as having preeminent places in the formative years of spaceflight came from countries as far apart as Russia (K.E. Tsiolkovskii, 1857-1935), the USA (Robert H. Goddard, 1882-1945), and Transylvania/Rumania (Hermann Oberth, 1894-1990). Each made significantly different contributions but it is Tsiolkovskii who in terms of priority (being first to publish) is justly remembered as the *Father of Astronautics* – the latter term, meaning "sailing among the stars," being first used by a group of enthusiasts in Paris in 1927.

Tsiolkovskii was the son of a forestry worker and amateur inventor, and he became almost totally deaf at the age of nine as a result of complications following scarlet fever. Despite the inevitable learning difficulties, he taught himself from his father's books and, having concentrated on mathematics and physics, studied at a technical institution in Moscow for three years before becoming a tutor. Subsequently he passed the examinations which permitted him to teach. He began serious theoretical research in 1881 and devoted himself to the themes of all-metal dirigibles (airships); aircraft; and to the rocket as a means of interplanetary travel.

He showed he was not lacking in imagination

3 *Konstantin Tsiolkovskii (1857-1935) is widely regarded as the "Father of Astronautics" even though for much of his lifetime his theoretical works were little known outside Russia. A teacher by profession, in the latter part of his life he made far-reaching proposals about rocket flight which were eventually to be put into practice: these included the use of liquid oxygen and hydrogen as propellants and the multi-staging of launch vehicles.*

4 *Robert H. Goddard (1882-1945), unlike Tsiolkovskii, was a practical engineer as well as a theoretical scientist. His prime interest was in research of the upper atmosphere, and rockets were a means of forwarding such research. He experimented initially with solid propellants, but then switched to liquid-fueled rockets and in 1926 achieved the world's first flight by such a vehicle. Little attention was paid to Goddard's achievements during his lifetime, but modern rocketry owes much to his work – the extent of which is demonstrated by the more than 200 patents that were taken out in his name.*

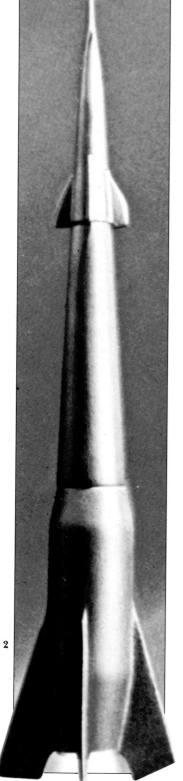

(although far from being a literary giant) by writing two novels in 1893 and 1895, one of which dealt with the creation of an artificial satellite orbiting the Earth at a height of around 200 miles (322km). But his scientific papers were of much greater significance and these began with the publication of *Investigating Space with Reaction Devices* in 1903 and continued until the year of his death in 1935 with a paper on *The Maximum Velocity of a Rocket*.

He dealt with many aspects of the rocket, making theoretical calculations about the behavior of a vehicle in flight, studying the properties of different propellants and making practical suggestions on rocket design, though he was never personally to attempt to put theory into practice. While he devoted considerable attention to what we would now call life support systems in spaceflight – a diagram of a space station concept showed a closed ecological system and garden aboard the vehicle – Tsiolkovskii's most important proposals concerned propellants and the design of the rocket that was needed to achieve flight into Earth orbit and on into interplanetary space.

Thus, as early as in his 1903 paper he proposed the use of liquid oxygen and liquid hydrogen, the latter being the "fuel" and the former the "oxidant" required in the vacuum of space. Liquid propellants were a more powerful and a more controllable source of power than gunpowder and the other solid fuels used hitherto. And in his later papers he elaborated the system of *multi-stage* rockets in which a first stage (the *land rocket* as he called it) would carry an upper stage through the lower and denser layers of the atmosphere before being jettisoned when an upper stage (or stages) would go on into space. Both proposals were a remarkably accurate forecast of what was to happen some decades later when spaceflight was finally achieved.

However, Tsiolkovskii's writings were confined to a relatively small number of colleagues in Russia (and then in the newly-born USSR) and were almost totally unknown in the outside world until the work of Goddard in the USA, and particularly Oberth in Europe, began to give some serious credibility to the idea of spaceflight. The now-retired schoolmaster became a celebrity in the Soviet Union, and accounts of his work penetrated abroad. Since 1958, a monument commemorating his life has stood in the town of Kaluga, which bears the words he wrote in a letter to a colleague in 1911:

"Mankind will not remain forever bound to the Earth. In pursuit of light and space, he will, timidly at first, probe the limits of the atmosphere and afterwards extend his domain to the entire solar system."

Robert Goddard was born in Worcester, Massachusetts and allegedly was inspired after reading Wells' *War of the Worlds*. But his achievements were to be eminently practical, and while others were concerned with dreams and theories, he translated both into reality. Goddard took a science degree in 1908 and obtained his doctorate from Clark University three years later. In his own time, and at his own expense, he began researching ways to improve the performance of rockets, primarily for study of the upper atmosphere. Initially he worked with solid propellants and realized that there was a limit to what could be achieved with these. He worked at refining the design of rockets and, quite independently of Tsiolkovskii, hit upon the idea of step or stage rockets.

During the First World War, Goddard received financial support from the US Government for work on rocket weapons, but while he made progress this was limited by the coming of peace. In 1919, as a Professor at Clark University, he wrote a technical paper for the Smithsonian on *A Method of Reaching Extreme Altitudes*. In one part of the paper he discussed propellant and payload considerations using the then current technology, and suggested that it might be possible to reach the Moon with a rocket carrying a charge of flash-powder which could be seen from Earth upon impact. This suggestion was taken out of context by the press, which tended to portray Goddard as an eccentric, if not mad, professor. The experience was of major importance, for Goddard was by nature a shy man and not a publicity seeker. These characteristics were intensified by his treatment, and thereafter he tended, so far as the press and general public were concerned, to cloak his work in secrecy.

But that work did continue and, having switched to liquid propellants, he made significant progress. In 1926, at a remote farm in Auburn, Massachusetts, he used liquid oxygen and petrol as propellants to accelerate a small rocket to 60mph (100kph) after a burn time of only 2.5 seconds. This was the first flight in the world using liquid propellants and, no matter how limited, can justly be regarded as man's introduction to the space age. In July, 1929, he launched a larger rocket carrying a scientific payload from Auburn, an event that did receive some publicity although the Smithsonian Institution spokesman felt compelled to deny there were any designs on the Moon. Still with limited financial backing, Goddard persisted, and a launch in 1935 achieved a height of 1,000ft (305m) and a maximum speed of 700mph (1,126kph). With the coming of the Second World War he turned his attention to improving fuel pumps and other equipment for the US Navy and, perhaps dreaming of what might have been, died some months after the first Nazi V-2 rockets landed on the south of England in 1944.

Goddard's attitude to publicity was understandable, but his reticence in publishing results, together with the fact that he was not primarily concerned with space travel, enabled the third major figure of this period, Hermann Oberth, to establish a priority by publishing an extensive evaluation of rocketry based on liquid propellants, and to present the possibility of spaceflight as a serious and respectable proposition.

The Importance of Oberth

Oberth was born in Transylvania, which at that time was part of the old Austro-Hungarian Empire, although subsequently he became a German citizen. He was a mathematician and physicist and, like Tsiolkovskii, became a school teacher while devoting much time to research on rocketry. Oberth corresponded with Goddard and became convinced that spaceflight was a practical possibility provided it was comprehensively treated. Tsiolkovskii's work along these lines was totally unknown to Oberth, and in 1923 he published *The Rocket into Interplanetary Space,* which contained much mathematics but also a section on suggestions for a rocket vehicle traveling in space. Importantly (and again like Tsiolkovskii) Oberth dealt with both detailed engineering and wider issues, such as the need for space suits, space food, mission objectives and the numerous requirements to be met in operating a space station or going on interplanetary journeys.

The book was successful, was reprinted, and a second edition followed. More popular works by Willy Ley and Max Valier took the message to the general public. Oberth was invited by the prominent German film producer Fritz Lang to act as technical consultant to his forthcoming production *The Girl in the Moon,* and he moved to Berlin in 1928 to work on the spacecraft designs to be used in the film. Lang also proposed that

Oberth build an actual rocket to be launched on the day that the film was to open. Oberth was not a practical engineer, and while he strove to comply with the request he eventually gave up and returned home. The film opened in 1929 without the launch of an Oberth rocket, but his reputation was little affected by the episode.

Perhaps more than anybody else up to this time, Oberth was able to combine technical knowledge with a wider awareness of the potential of spaceflight that was to inspire others. Those spreading the message in

addition included Valier and Ley, as already mentioned. In the USSR there were Fridrikh Tsander, Yakov Perelman and Nikolai Rynin; and across the Atlantic as this time was Hugo Gernsback, who launched *Amazing Stories* in 1926 and who, while not involved with the practical developments taking place, was a talented promoter of the concept of space travel.

The final steps to space were now ready to be taken, and various societies around the world provided a base for enthusiasts (now beginning increasingly to include engineers at various stages of their careers) to meet, to advance their knowledge and actually to build rockets.

Of the groups formed in the later 1920s and the 1930s, perhaps the most disadvantaged, but certainly among the most durable, was the British Interplanetary Society, which was formed in 1933 on the initiative of a young structural engineer, Philip Cleator. Undaunted by the strictures of the Explosives Act of 1875, which essentially prevented experiments with rocket motors and the launch of rockets, Cleator and his enthusiast friends energetically promoted the concept of space-flight, and by the end of the 1930s they had published in-depth plans for a lunar spaceship; a mission illustrated by the space-art of Ralph Smith, a turbine engineer. The BIS never dropped the adjective "Interplanetary" from its title, which it bears to this day, unlike some groups which considered that such a change created a greater sense of *gravitas*. Out and out romantics rubbed shoulders with those who were, or were to become, involved with astronautics professionally. Before the Second World War put a temporary halt to its activities, the BIS numbered among its members Arthur Cleaver, who was to become a leading British rocket engineer, and a man who was to achieve world fame in the realms of both space reality and fantasy: Arthur C. Clarke.

Enthusiasts in the USA suffered none of the legal restrictions placed upon their British colleagues. The American Interplanetary Society was founded in New York in 1930 with the trusting, albeit over-optimistic, hopes of enthusiasts the world over. In Robert H.

Goddard they had one of the leading figures in astronautics of the time, but while he did become a member he refused to become closely involved in the Society's activities. His attitude was hardly likely to soften when a member of the AIS, following a self-financed visit to experimenters in Germany, claimed in 1931 that no liquid-fuel rocket had ever been flown successfully; a statement that Goddard immediately refuted in the Society's *Bulletin*. In 1935 the AIS became the American Rocket Society, and it thereafter concentrated on rocket motor development. Perhaps the most successful idea was that of James Wyld, who proposed passing the fuel through a vaporising jacket, thereby both cooling the motor and preheating the fuel. Wyld and other members in due course developed their skills in support of the US war effort, but once that war was over the Society advanced powerfully into the new age and in 1963 merged with a science institute to form the American Institute of Aeronautics and Astronautics.

Interplanetary travel and rocketry were ideas that were firmly rooted in Tsiolkovskii's homeland, and they continued to be developed despite the many problems facing the USSR. In 1928 a military organization long involved with war rockets was given expanded facilities and a change of name to the Gas Dynamics Laboratory. In the following year Valentin Glushko, a 21-year-old graduate from Leningrad University, who was subsequently to play a major role in the Soviet space program, suggested that work be started on liquid and electrical rocket engines. His plan was accepted and he was placed in charge. Like von Braun in Germany, Glushko had early dreamed of spaceflight, but also like von Braun was to begin his technical labors in the service of the military.

Korolev's Early Years

In 1931 another organization, the Group for the Study of Reaction Motion, was formed in Moscow. It, too, was ultimately funded by the Red Army, and one of its young members was Sergei Korolev, who was ultimately to

1 *Sergei P Korolev – a picture taken in 1946, but not released until many years later. Once the Soviet developments in space burst upon the world, from 1957 onwards the Communist authorities chose to hide Korolev's identity behind the anonymous title of "Chief Designer." He died in January 1966.*

2 *Just as Korolev's identity for many years was hidden from the world (and Russian people), so detailed information about the Soviet space program was rarely made public. The launch vehicle shown here (which was obviously an artist's impression or model) was simply identified as a "three stage rocket on test with men in it" – which it was obviously not.*

BELIEVES ROCKET CAN REACH MOON

Smithsonian Institution Tells of Prof. Goddard's Invention to Explore Upper Air.

MULTIPLE-CHARGE SYSTEM

Instruments Could Go Up 200 Miles, and Bigger Rocket Might Land on Satellite.

Special to The New York Times.
WASHINGTON, Jan. 11.—Announcement was authorized by the Smithsonian Institution tonight that Professor Robert H. Goddard of Clark College had invented and tested a new type of multiple-charge, high efficiency rocket of entirely new design for exploring the unknown regions of the upper air. The claim is made for the rocket that it will not only be possible to send this apparatus to the higher layers of the air, including those beyond the earth's atmosphere, but possibly even as far as the moon itself.

The highest level so far reached with recording instruments is nineteen miles, accomplished with a free balloon. Professor Goddard believes that his new rocket can be sent through the band of atmosphere around the earth, which he says extends some 200 miles out, and that his new rocket will be of great value to the science of meteorology. To send a rocket beyond the earth influence would require an "initial mass" of 1,274 pounds.

The announcement authorized tonight declares that Professor Goddard is at the present time perfecting the reloading mechanism of his rocket, under a grant from the Smithsonian Institution, and

3 *Robert Goddard at the blackboard of his room in Clark University. His inherently retiring attitude was accentuated in 1919 when his comments about the possibility of reaching the Moon using rockets were ridiculed in some press reports. The cutting reproduced here (**5**) was less sensational.*

4 *The British Interplanetary Society played a prominent role in the popularization and practical development of spaceflight. By the late 1930s, a lunar landing mission had been proposed – illustrated by the space art of Ralph Smith – a member who was a turbine engineer.*

1 *Following his success in launching the world's first liquid-fueled rocket in 1926, Robert Goddard worked steadily at improving and developing rocket technology. By 1935 he was achieving speeds of 700 mph (1,126 kph) and altitudes of 1,000 ft (305 m). This photograph shows a successful launch in 1937 from a site in New Mexico when the rocket reached an altitude of 3,250 ft (1,000 m). By this time vehicle stabilization had also improved significantly.*

become more famous than Glushko, even though the communist authorities for years hid his identity as the architect of Soviet space successes under the anonymous title of *Chief Designer of Carrier Rockets and Spacecraft*. For much of this early period Korolev was concerned with the development of rocket aircraft, and his later work was all the better for his essentially pragmatic approach and organizational skills.

The first Soviet liquid-fuel rocket was launched in 1933, and in that same year most of the development work was gathered together in a new Scientific Research Institute of Jet Propulsion, which was claimed to be the world's first state-owned rocket research establishment. Korolev was a divisional engineer with the effective rank of major-general. However, the years that followed, while much detailed, technical work was accomplished, were years of turmoil because the scientists and

a somewhat doubtful character but, although Valier had been killed by an explosion when experimenting with a rocket powered car, and Oberth returned to Rumania in 1930, development work by the members of the VfR began to accelerate, particularly when it was moved to its own *Raketenflugplatz* (Rocket Flying Place) in Berlin. By the spring of 1932, besides numerous static tests and demonstrations, the VfR had conducted almost ninety flights with its Mirak/Repulsor rockets. This was obviously valuable training and experience for the young engineers and others who were often living on the equivalent of social security. No such condition affected the wealthy and talented young member Wernher von Braun, who had impressed visitors from the German army and who was invited to join their Ordnance Department at the end of 1932. The VfR meantime began to suffer increasingly from the unsound policies and initiatives

2

1

2 *Compared with the scenes at launch sites when the space era had begun, this picture underlines the pioneer status of the new technology. Robert Goddard (second from right) and his assistants were preparing for a launch on October 27, 1931, when the picture was taken at the tower site located 10 miles northwest of Roswell, in New Mexico.*

engineers, as well as their commander-in-chief, Marshal Tukhachevskii, were subjected, like most of Soviet society, to the mass purges with which Stalin cemented his absolute power. The Institute was abolished in 1938 and many of its members were shot or imprisoned. Korolev and Glushko were among those who survived, but it seems certain that the Soviet Union's rocket and spaceflight efforts in the years after the Second World War, despite the achievements, must have been weakened by the purges.

In retrospect, it might be regarded as inevitable that the greatest progress in rocketry was made in Germany, the country whose leaders were to take it into another world war. The VfR (Society for Space Ship Travel) was formed in 1927 in what was then the German town of Breslau. Willy Ley and Max Valier were founder members along with Johannes Winkler, an engineer who was to be the Society's first president. A major aim was to help raise money to support Oberth's rocket experiments.

There was more appearance than substance about the VfR's first two or three years, and it was Winkler acting independently and with the support of a wealthy donor who achieved, in 1931, the first launch in Germany of a liquid-fuel rocket. He had left the VfR some time before, when Oberth had taken over as president.

An increasingly active role in the Society was taken by Rudolf Nebel, who was originally enlisted by Oberth to assist with the abortive scheme to build a rocket as publicity for Fritz Lang's film. Nebel seems to have been

of Nebel, as well as the worsening economic climate, and within a year had collapsed.

Artillery and ordnance officers of the German Army had been researching rocket weapons for some years before Adolf Hitler came to power, because the Versailles Treaty limitations made no mention of rockets. In 1930-31 an artillery captain, Walter R. Dornberger, was given the task of developing a relatively short range, "blanket" bombardment missile and also a liquid-fuel rocket with a range greater than any existing artillery piece. The first task was accomplished quite quickly (the Nebelwerfer rocket system of World War Two) but the second, not surprisingly, proved more difficult. Discussions were held with Nebel and others at the VfR, but these made little progress. However, the contact did enable Dornberger and his colleagues to identify those at the Raketenflugplatz who should be invited to join the army project on a full time basis, and von Braun was but the first.

Developments Under the Nazis

In 1937 the group moved to a new proving ground on the Baltic at Peenemunde, where quite some time later Hermann Oberth, now a German citizen, joined them. But it was the much younger von Braun, a highly talented organizer as well as engineer, who led the rocket development program and who reported directly to Dornberger, now a Major General.

Fortunately for the world the attitude of the Nazi

3 *Two of the leading figures in the early history of spaceflight appear in this group of officials of the US Army's Ballistic Missile Agency photographed at Huntsville, Alabama, in 1956. In the foreground is Hermann Oberth, who from the early 1920s onwards was to be a tireless advocate of spaceflight. Second from the right is Wernher von Braun, creator of the World War II V-2 – but also of the Saturn 5 that was to take men to the Moon.*

4 *Soviet designers and engineers successfully launched their first liquid-fueled rocket on August 17, 1933. This model of the GIRD 09 was displayed at an exhibition in Moscow in 1967.*

4

leadership to the development work at Peenemunde was erratic, with priority status being accorded (with the appropriate funds) and then withdrawn. A major raid by RAF bombers in August 1943 killed many members of the staff and caused considerable damage. Then, some months later, von Braun was among those arrested for a short time on the grounds that he was more interested in spaceflight than in rocket weapons.

Nonetheless, progress was made and the A-4 rocket (later to be called the V-2 – an abbreviation for *Vengeance Weapon 2*) flew successfully in the latter half of 1942, when an explosive warhead weighing one ton (1,016kg) was accelerated to a speed of more than 3,500mph (5,630kph) and impacted 120 miles (193km) away. This range was later improved to 200 miles (322km), with the rocket reaching a height of 60 miles (97km). There was little comparison with the experiments of the early pioneers. The vehicle stood almost 50ft (15m) high and weighed over 28,000lbs (12,700kg) at launch. It burned liquid oxygen mixed with ethyl alcohol, injected into the combustion chamber by pumps powered by a turbine. Guidance and control were obtained by a three-axis gyro acting on exhaust vanes and by movable fin sections.

In 1943 the A-4 went into production at a slave-labor plant near Nordhausen, and a stockpile of almost 2,000 rockets was built up before it went into active service.

Although the weapons were not accurate there was no defence against them, and they hit the ground at many times the speed of sound. The V-2 bombardment was launched in September 1944 against targets in liberated Europe and in southern England. The rockets made the menacing, unpiloted, jet engined V-1 ("buzz-bomb") seem almost archaic in comparison, and by the time Peenemunde fell to the Russians in May 1945 some 1200 V-2s had been fired against England and had killed more than 2500 men, women and children.

The moral issue of scientists working on such a weapons system, and also their awareness or otherwise of the appalling conditions at Nordhausen, is one that cannot be dealt with here. But it was always claimed by von Braun and others that spaceflight was a continuing subject of discussion and planning at Peenemunde. While still a potential weapons system, the A-9/A-10 project designed in 1941-2 married a dart-winged A-4 upper stage with a booster first stage which would have had a launch thrust of 400,000lb (181,000kg). The system was never built but it was the shape of things to come. Even Dornberger allegedly began to think of the wider horizons of spaceflight, and when the A-4 first flew he was reported to have exclaimed to von Braun: "Do you realize what we accomplished today? Today the spaceship was born!"

2

FIRST STEPS INTO SPACE

1 A picture of Sergei Korolev probably dating from about 1960. Released from imprisonment when Joseph Stalin developed an interest in rocketry, Korolev first oversaw the production of captured V-2s in Germany, but toward the end of 1946 the work was moved to the Soviet Union. Within a short period Korolev developed two modified versions of the German rocket.

1

The time had come. By 1945, all the theoretical and technological elements needed to put a satellite into orbit were available. It only remained to take the political decision to develop the scattered elements and put them together. That, too, was now possible. Six years of war had left Europe shattered and the dominant powers were the United States and the Soviet Union. Their differing views of the post-war world quickly broke up the wartime alliance and brought them into conflict. This "Cold War" became increasingly severe during the late 1940s and early 1950s. The years 1948 and 1949 saw the Berlin blockade and then, in August 1949, the Soviets tested their first atom bomb. In response, President Truman ordered the development of the hydrogen bomb. In June 1950, the Korean war began and World War III loomed.

The V-2 in Russia
As the victors in Europe, the US and the Soviet Union inherited Nazi Germany's military technology. The Soviets, like the US, sought V-2 personnel and technology, but all the leading scientists had already fled to the West. The only significant individual they were able to recruit was Helmut Grottrup, former assistant at the Peenemunde Guidance, Control and Telemetry Laboratory, although they did capture technicians and lower-level scientists. Several important facilities were in the Soviet zone: Peenemunde, Nordhausen and the test stands at Lehesten. With no complete V-2s ready to be launched, the Soviets sought to resume production at Nordhausen. By September 1946 some 30 flight-ready V-2s had been built and static tests of engines had begun at Lehesten. Sergei P. Korolev oversaw production while Valentin P. Glushko worked on the V-2 engine systems. (The two scientists had been released from the Gulag after Stalin's interest in rockets had been stimulated by the V-2.) By fall 1946, it was decided to move the effort to Russia: on October 22, Grottrup and about 200 German engineers and scientists, together with their families, were loaded on trains and sent east.

Production facilities were set up at Kaliningrad, near Moscow. Called "Zavod 88," it was very badly equipped during the first months after the Germans' arrival. (One of them had to take an alarm clock apart to use its steel spring). Despite these problems, the first Russian V-2s were ready by the fall of 1947. The flights were made from the Kapustin Yar test range 75 miles (121km) from

Stalingrad. The facilities were very primitive at first and personnel slept in tents or on two trains. The first Russian V-2 was launched on October 30, 1947, and landed near the target due east of Kapustin Yar: a launch the following day, however, failed when the missile began to roll and the fins broke off.

In all, some 20 V-2s were launched during the fall of 1947. Some carried instruments to measure such things as re-entry temperatures, others high-explosive warheads and still others scientific instruments to measure cosmic rays. The launchings were conducted alternately by an all-German and an all-Russian crew, with Korolev directing launch operations.

Just after the war's end, this "collection" of former V-2 personnel and equipment had the highest priority. In 1946 the situation changed, with the major emphasis being shifted to radar and surface-to-air missile technology. This reflected a realization of how destructive US and British bombing had been; improved air defenses were needed immediately to protect the Soviet homeland. (Long-range rockets would take many years to develop).

Early Soviet Lead
Despite this, Stalin approved an ambitious, long-term program of rocket development in 1947 intended to lead to the building of an Intercontinental Ballistic Missile – ICBM. The result was that the Soviets skipped what proved to be the dead end of large, strategic cruise missiles and gained an early lead in rocket technology. It also opened the space frontier. Korolev's first step in this program was a V-2 "clone" – the R-1 (NATO code name SS-1a Scunner). It looked identical to the German V-2 and had a range of about 174 miles (278km). The SS-1a became operational in the late 1940s, and it was followed in 1950 by the R-2 (NATO code name SS-2 Sibling). This was a "stretched" V-2 with a length increased from 46 to 55.5ft (14 to 16.9m). The range was also doubled to 348 miles (556km).

The SS-2 (and possibly the SS-1a) incorporated numerous improvements originally suggested by the German scientists, but their role in the Soviet program was far different from their counterparts in America. Whereas the German scientists in the US were an integral part of both military and industrial development, the Russians kept "their" German scientists at a distance, asking them for ideas and studies but not letting them work on hardware. They were solely a measure of the state of Soviet rocket technology. By 1951 to 1953, they were considered of no further use and were sent home.

The early Soviet start on rocket development could have been due to a different view of the rocket's military role. The Soviets had never conducted strategic bombing during World War II and the Soviet Air Force was limited to ground support. Rockets were regarded as being very long-range artillery, much as the German Army thought of the V-2. As artillery has always played a major role in Russian military thought, this view of rockets would have made them more acceptable, and the SS-1a and SS-2 were operated by the Russian Army, not the Air Force.

The SS-1a and SS-2 were also used for scientific research. The SS-1a (like the V-1 family of sounding rockets) carried two experiment pods on its side and had an elongated nose cone. The SS-2 (called the V-2 or A-4) also carried similar equipment. A common payload was dogs: some were carried in pressurized capsules while others wore space suits and rode in unpressurized capsules, which were then parachuted to a successful landing. Post-flight examination of the dogs indicated no harmful effects.

Sounding rocket flights lasted only a few minutes, and this limited the amount of data that could be collected. An orbiting satellite, on the other hand, could operate indefinitely. In 1947, Professor Gregory A. Tokaty, adviser to the Soviet Air Force, brought up the possibility of satellites in a discussion with Academician S. A. Khristianovitch. This sparked an angry rebuke from Major General Vassilli I. Stalin (Stalin's son) who said: "… others may be wasting their time on abstract projects but you are a service man, and you must be concerned with the practical needs of state defense. Can't you understand that we need jet fighters, not a silly Sputnik!"

Redstone in the US, was derived from V-2 technology. However, it represented an advance in that it was the first nuclear-armed Soviet rocket. Flight tests began in the early 1950s with deployment taking place in 1955. Reflecting its strategic role, the SS-3 was operated by the Soviet Air Force's Long Range Aviation arm.

The scale of Soviet missile development was revealed by their launch activities. Between mid-1953 (roughly the time the SS-3 began flight testing) and late 1957, some 300 ballistic missiles had been launched from Kapustin Yar. The test program was vigorous – as many as 22 missile firings being conducted in one month and up to five in a single 24-hour period. The accuracy was judged to be very good by the CIA.

On June 22, 1957, the Soviets fired their first missile with a range of over 1,000 miles (1,759km), and within two months a total of seven such flights had been made. This was the SS-4 Sandal MRBM. It was an advanced version of the SS-3 and carried a one megaton warhead. Although neither missile could hit the continental US, they did pose a major threat to US forces. By the mid-1950s, the US had a force of some 2,000 nuclear-armed bombers which relied on forward bases in Western Europe, Africa and the Near East. The Soviet missiles could now hit these bases without warning or risk of interception.

Beginning at Tyuratam

Interest in satellites revived in Russia during the mid-1950s. In 1954, Korolev wrote, "It would be timely at this present moment to organize a research division to pioneer a satellite." In 1955, the Soviet Academy of Sciences sent a circular to some one hundred scientists asking them to comment on possible uses for satellites. Some responded positively but others did not: "Not

2 *The Soviet developments of the German V-2s were the SS-1a Scunner and the SS-2 Sibling respectively, to give them their NATO code names. The latter was a stretched version with a range doubled to over 300 miles. This admittedly poor-quality image (it may have been rendered deliberately so) was released in 1964 by TASS with the caption "High combat readiness and practical skills are unthinkable without everyday training and launching of rockets."*

Simply put, the crash nature of the Soviet missile program left little room for satellites. The lack of interest and of vision had different roots from that in the US, but the outcome was the same.

The next step in the Soviet program was development of the SS-3 Shyster. This was a Medium Range Ballistic Missile (MRBM) able to hit targets in Western Europe from the Soviet Union. It had a range of just under 750 miles (1,204km) and, like the shorter range

interested in fantasy; I visualize a space shot only in the year 2000." said one, while another replied: "I don't see of what practical use artificial satellites could be."

During the evening of January 12, 1955, a group of 30 construction workers arrived at the small railstop of Tyuratam in Soviet Central Asia. They were soon followed by a virtual army, who were to build the test site for the Soviet R-7 ICBM (NATO code name SS-6 Sapwood). The launch pad was at the end of a long

3 *This picture – released in 1962 – appeared in the Soviet army newspaper* Krasnaya Zvezda (Red Star). *It shows two Scunner missiles aboard transporters and bears the caption "To the launching stand." Both Scunner and Sibling were also in fact used as sounding rockets in a program of scientific research.*

1 *October 4, 1957, 10:28 PM. Moscow time – and the space age is born as an SS-6 (Sapwood) ICBM lifts off from Tyuratam in Soviet Central Asia with Sputnik 1 aboard. The satellite was to send its radio signals around the world and to administer a fundamental shock to concepts of US technological leadership in the world.*

2 *Sputnik 2, at over 1,100lb (500kg) in weight, was six times heavier than its predecessor when launched early in November 1957. It contained a female husky called Laika which provided valuable biomedical data until life support supplies ran out, when the animal was put down.*

1

3 *Sputnik 3 was not launched until May 1958, but was almost sixteen times heavier than Sputnik 1 and was packed with scientific instruments. Replicas of the spacecraft were shown at exhibitions, and a keyed diagram released very soon afterwards – an original copy being reproduced here. The caption detailing the instruments read: "(1) magnetron to measure gravity; (2) photo multipliers to register corpuscular irradiation of the Sun; (3) solar batteries; (4) apparatus to register photons in cosmic rays; (5) magnetic ionized manometers; (6) ionic trap; (7) electrostatic fluxmeter; (8) mass spectrometer tubes; (9) apparatus to register heavy nuclear effect in cosmic rays; (10) apparatus to measure intensity of primary cosmic irradiation; and (11) micro-meteor recorders." Sputnik 3 remained in orbit for almost two years before re-entering the Earth's atmosphere and burning up.*

rail spur north of Tyuratam. Construction went on around the clock; the earth-moving equipment raising so much dust that lights had to be used constantly. By the end of 1956 the pad was finished and testing of the ground support equipment could begin, using a mockup of an SS-6.

On March 4, 1957, Korolev ordered the checkout of the first flight-ready SS-6 to begin and the missile was rolled out to the pad two months later. The SS-6 was unusual looking: attached to the central core stage were four strap-on stages, giving it the appearance of a fluted column. Each stage had four engines. The vehicle was steered not by gimbaling (moving) the main engines but by using smaller steering rockets: two per strap-on and four on the core stage. Thus, at lift-off, 20 main engines and 12 steering rockets were firing. Designed before the breakthrough in small H-bombs, it carried a 7,000 to 9,000lb (3,175 to 4,082kg) warhead which was originally to be a 500 kiloton A-bomb but

2

this was later changed to a five megaton H-bomb. The range was relatively short – somewhat over 4,000 miles (6,482km) – which meant that from northwest Russia the missile could reach targets in the northeast US. The SS-6 was the ultimate development of the original V-2 technology.

The first SS-6 was ready on May 15, 1957. Lift-off came at seven in the evening Moscow time and the rocket flew for 50 seconds before failing. Two more SS-6 launches took place during the summer and these also failed. The launches attracted CIA attention and reconnaissance U-2 aircraft were sent on over-flights of the suspected launch area; in due course successfully returning with images of the launch pad.

The fourth SS-6 was launched on August 3, 1957. This time everything worked and the dummy warhead impacted near the Kamchatka Peninsula in the Soviet Far East. A second successful SS-6 launch took place on September 7, watched by Communist Party Secretary Nikita Khrushchev.

Korolev now had a golden opportunity. The US Vanguard was still several months from launching a satellite and the Atlas ICBM had failed in both its short-range test flights. The Thor and Jupiter IRBMs were just starting their test programs, and the Eisenhower administration had ordered cutbacks in funding for missile development. The SS-6 was then the only proven large rocket. Moreover, it could launch a larger satellite than any US rocket and could do so with only minor modifications. In 1956 Korolev had proposed the launching of a satellite, but Khrushchev had taken a wait-and-see attitude, probably pending a successful test

of the SS-6. During the summer of 1957, Korolev again raised the idea: this time it was approved.

The designer began working around the clock. It was decided to keep the first satellite simple – an aluminum sphere with four antennas carrying a radio beacon – but it would also provide data on the density of the upper atmosphere. In addition, Korolev had given thought to the satellite's symbolic importance. He said later, "It seems to me the first Sputnik must have a simple and expressive form, close to the shape of natural celestial bodies." This was important, as it would "forever remain in the consciousness of people as a symbol of the dawn of the space age." The radio signals could also be picked up by any amateur "ham" set and the satellite's orbit would take it over both the US and Western Europe. The core stage, which would also go into orbit, would be easily visible to the naked eye.

The "PS" – Russian for Rudimentary Sputnik (sputnik means *satellite*) – weighed 184.3lb (83.6kg), a small fraction of the SS-6's payload capability. (When used as a satellite launcher, the SS-6 was called the A booster in the West). By early September, the satellite was completed and sent to Tyuratam, where it was attached to the booster and checkout began. Early on the morning of October 3, the rocket was rolled out to the pad on its railroad transporter.

Launch of Sputnik 1

By the next morning, October 4, 1957, everything was in readiness. At 10:28 pm Moscow time, the engines roared into life. For a moment the rocket was enveloped in flame, but then it lifted off into the star-filled night sky. Five minutes later the core stage engines shut down, the cone-shaped shroud separated, and Sputnik 1 was pushed free, its radio transmitting a distinct "Beep, beep, beep."

For the Russian people, the news of Sputnik's launch was greeted with a euphoria greater than that following the Nazi surrender. A mood of brooding inferiority was replaced by the feeling that the Soviet Union had overtaken the US in science and technology, and the belief that, because of Sputnik, Russia was now in fact the world's leading power. This was reinforced by political lectures at factories and banners that appeared everywhere literally overnight.

Khrushchev was quick to take advantage of both the Soviet euphoria and US self-examination. At first he had been almost casual: after being told of the launch success, he just went to bed. Within days, however, Khrushchev began his "Sputnik Diplomacy" campaign. Unlike Eisenhower, the Soviet leader understood the political importance of Sputnik. Eisenhower took a narrow view; the satellite was not a weapon and the SS-6 ICBM was still many years away from deployment. In the meantime, US bomber forces were as large and powerful as before. Seen this way, Sputnik had no military importance: like Vanguard, it was a scientific achievement and thus had no role in the Cold War. Khrushchev realized that this did not matter. The Cold War was about symbols: the mushroom-shaped cloud, leadership, diplomatic signals, determination and will. With Sputnik 1, Russia had upstaged the US in one of these symbolic areas – science and technology – and, by implication, military power.

Khrushchev had both foreign and domestic goals for his Sputnik Diplomacy. It created the image of a dynamic, post-Stalinist Russia that was the equal (and soon to be the superior) of the US – an image heightened by the Eisenhower administration's refusal to admit that there was a "Space Race." Khrushchev's policy was meant to lessen respect for the US and to reduce its influence around the world. It also allowed him to overcome opposition within the Communist Party and the military to his policies. The contrast between the shining symbol of Sputnik and the evils of Stalinism put the hardliners on the defensive. By stressing missiles (both ICBMs and SAMs), the huge Soviet Army could be reduced; freeing manpower and investment for economic development.

On November 3, 1957, Sputnik 2 was launched. The 1,120lb (508kg) payload remained attached to the empty core stage and contained a pressurized cabin holding a female husky called Laika. Biomedical data was relayed for a week until life support ran out and, as the satellite could not be recovered, Laika was put to sleep. The achievement went far beyond anything planned for the International Geophysical Year and also underscored the size of the SS-6 ICBM.

Soviet Reaction to Explorer

The first hint that all might not be well with the Soviet space program came with the launch in the US of Explorer 1 on January 31, 1958. The reaction of the Russian people was, in many ways, a parallel to the US response to Sputnik. Within minutes of the announcement being made over Radio Moscow, the mood of cockiness evaporated and was replaced by something like mourning. The official propaganda had led most Russians to believe it would take the US six to ten years to catch up. Some tried to say it was a hoax, paralleling claims in the West following Sputnik 1.

Within three days, however, the propaganda response was ready. According to one account, Khrushchev was being briefed on the US satellite program. When told of the small size of the US payload, he laughed and said, "Oh, but it is no larger than a tiny orange." This was passed on to the propaganda machine and to Moscow's 23 district party secretaries. The word was then sent to the party secretaries of every factory and collective farm, and a similar process spread the theme to every village throughout the USSR. An editorial in a factory newspaper said: "How can such a highly developed country as the United States put into space such a tiny object and want the whole world to believe that it's a real satellite, when it is only make believe?" The message was that even if the US satellite was real, it was an inconsequential challenge to the Soviet space effort.

But there was another, underlying message for the Soviets: being first and staying first are two different things. The next setback was the failure of a satellite launch in early 1958, probably on February 3. Unlike the failure of Vanguard the previous December, no word was made public about the mishap. The CIA detected the failure and was able to identify it as a space launch, but it kept word of the incident secret until 1985.

The Soviets bounced back from this failure with the launch of Sputnik 3 on May 15, 1958. The satellite was a cone-shaped geophysical laboratory weighing 2,925lb (1,327kg), almost three-quarters of which was taken up by scientific instruments to study the Earth's ionosphere, magnetic field, radiation belts and cosmic rays. Sputnik 3 was the most successful of the early Soviet launches in terms of scientific return, and it continued to operate until it re-entered the atmosphere on April 6, 1960, but the large payload was near the maximum for the original SS-6/A booster.

Rockets to the Moon

Following Sputnik 3, the Soviet space program shifted direction. It was now aimed at the Moon, and 3

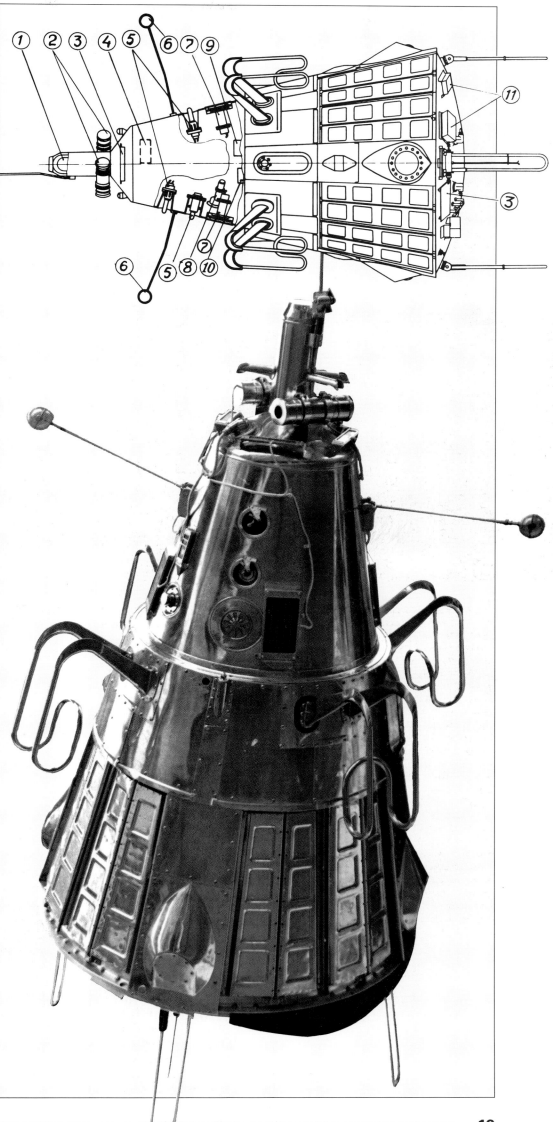

1

a small upper stage was added to the A booster. Called the A-1 booster in the West, it could launch a small payload fast enough to escape the Earth's gravity: around 25,000mph (40,225kph) compared with the 17,500mph (28,157kph) needed to achieve Earth orbit. The first Soviet lunar attempt was made on September 23, 1958, and was unsuccessful. A second failure followed on October 12, and another on December 4. It is probable that all three missions were intended to impact the Moon. This first generation of Soviet lunar probes used a *direct ascent* launch profile, in that the three stages fired one after the other until the third stage reached the pre-set velocity. It was simple but it was sensitive to velocity and/or course errors, which could cause the probe to miss the Moon or fall back to Earth.

At first, US intelligence did not know what to make of the flights. A CIA document, dated December 19, 1958, referred to "attempts to launch space vehicles of unknown nature." This implied the failures came at lift-off or early in the flights, before the nature of the missions would have become apparent.

The first partly successful lunar launch came on January 2, 1959. Luna 1 was a 796lb (361.3kg) sphere carrying a magnetometer on a 3ft (1m) boom, together with radiation and micrometeoroid detectors. It was

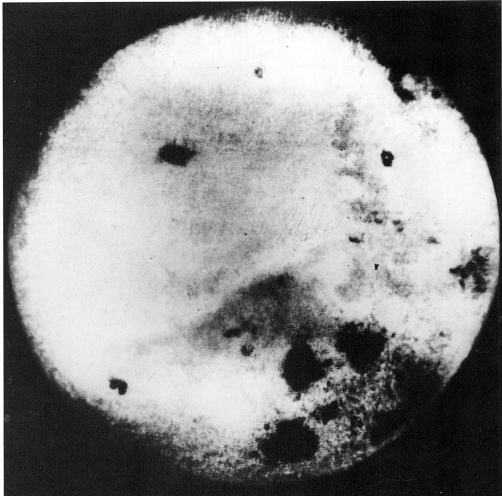

2

1 This picture was released by TASS in 1959 and was identified only as a "cosmic container with scientific and measuring devices." It was almost certainly a Luna (Lunik) spacecraft mounted in a cradle. Luna 1 was only partly successful, but was the first spacecraft to escape the Earth's gravity.

meant to hit the Moon, but a guidance error prevented this and the spacecraft flew past about 34 hours after launch at a distance of 3,100 to 3,700 miles (5,000 to 6,000km). It then continued on into solar orbit. At the time, Luna 1 was known by several names – including *Mechta* (Dream) and *Lunik* – and was the first object to escape the Earth's gravity.

Despite this achievement, the Soviet Moon program again suffered a launch failure during the summer of 1959 (a date of June 18 has been suggested), but the year did conclude with two successful flights. On

September 12, 1959, Luna 2 – which was nearly identical to the first Luna – was launched towards the Moon and as it left the vicinity of Earth its instruments measured the Van Allen radiation belts. Late on the same day, a sodium cloud was released from the third stage to help plot the flight path. The spacecraft was unable to detect any lunar radiation belts or magnetic fields as it neared the Moon, and on September 14 it impacted in the Mare Imbrium, near the crater Archimedes.

The next Soviet lunar flight came on October 4, 1959, the second anniversary of Sputnik 1. Luna 3 was a much more sophisticated spacecraft than earlier lunar missions, either US or Soviet. The barrel-shaped spacecraft carried a camera system equipped with narrow and wide-angle lenses and loaded with 35mm film that had minimum sensitivity to radiation. To point the camera, Luna 3 carried an orientation system that used a Sun and Moon sensor.

The Lunar Far Side

On October 7, the spacecraft passed above the Moon's south pole and then sped on over the far side – an area never seen from Earth. At a distance of 40,522 miles (65,200km), the camera began taking a series of at least 29 photos covering 70 percent of the lunar far side. They also showed part of the visible face, which allowed the location of surface features to be determined. From Luna 3, the Moon appeared to be three degrees across; six times its size from Earth and half again as big as a thumb at arm's length.

Once the 40-minute photo run was completed, the film was developed and dried in a small automatic processor, each frame taking about three minutes to develop. The photos were then scanned by a television camera and the signals transmitted to Earth. The images were not very clear but a number of surface features were visible; subsequently given names such as Tsiolkovsky Crater, the Gulf of Cosmonauts, Sea of Moscow and the Sea of Dreams. The far side was heavily cratered but had none of the large mares that covered much of the visible face of the Moon.

The first period of Soviet lunar exploration ended with two failures in April, 1960. It is believed the launches were of Luna 3-type photographic missions that could have been planned to cover the remaining 30 percent of the far side and/or use a higher resolution system. The A-1 launcher lacked the payload lifting capacity for a lander: a lunar orbit mission may have been too demanding and it would be three years before another Soviet probe to the Moon was launched.

The first generation Lunar missions highlighted the political dimension of the Soviet space program. After Sputnik 3, no more Soviet Earth orbital satellites were launched for two years, and it can be argued that the Soviets had squeezed all the propaganda advantages they could out of such satellites. If they were to maintain their image of technological superiority they would have to beat the US with a Moon probe. And the Soviets were quick to exploit their space accomplishments. Luna 2 was launched only three days before Khrushchev began a tour of the US, but on arrival he presented Eisenhower with a copy of the metal sphere that the spacecraft had carried to the Moon.

The ultimate goal of all this was to create an image of great ICBM strength, using space achievements to back up the claims. The years 1958 and 1959 saw the controversy over the Missile Gap: the difference between the number of ICBMs the US had and estimates of the numbers of Soviet ICBMs deployed. The US Air Force estimated that the Soviet deployment rate could be as high as 500 per year during the early 1960s. Once the

Soviets had a large ICBM force, they would be in a position to attack the US with little warning and could catch US bombers on the ground. The uncertainty that sparked the Missile Gap question was due to the US being unable to discover how many ICBMs the Soviets actually had as opposed to the estimated numbers. Overflights by U-2 aircraft were too few and too limited in area to give an answer, and Khrushchev was able to exploit this uncertainty by pursuing a more aggressive foreign policy. This was aimed at forcing western concessions, particularly on the issue of the divided city of Berlin.

The launch of the Sputniks and Luna missions, as well as Khrushchev's propaganda exploitation of them, were shaping the direction taken by the space programs

in safety over the open sea and, although it was a failure, the space age came to Florida with a first launch in July, 1950.

Visionaries and scientists in the US at this time created a momentum that could only have one outcome. The three issues of *Collier's Magazine* published in 1952, with articles by Wernher von Braun, Willy Ley, Fred Whipple and others, took the concept of spaceflight to the public. By 1953 physicist S. Fred Singer had proposed a Minimum Orbital Unmanned Satellite of the Earth (MOUSE) derived from an earlier study published by the British Interplanetary Society. It was essential to gain government support, and in due course the concept of an International Geophysical Year in 1957-58 came to be centered on the launch of scientific satellites into

2 Another historic Russian achievement was to photograph the far side of the Moon in October 1959. Luna (Lunik) 3 had an onboard camera and film processing system and obtained a sequence of images from a closest distance of about 40,000 miles (65,000 kms). The pictures were crude – but they were the first. Understandably, the features that could be discerned were in many cases given Russian names – such as the Sea of Moscow and Tsiolkovskii Crater.

3   4   5

of both countries. As the 1960s began, it was clear what the next steps in the Space Race would be: to put a man into space and send him to the Moon.

USA

The first postwar attempts on the frontiers of space by the US took place at White Sands in New Mexico. It was from there that the purely American rocket the WAC Corporal was lofted some 40 miles (64km) into the sky, as was the first of the captured V-2s, beginning in March 1946. Designed as a war weapon it was pressed into service for scientific research, and by the time the last V-2 roared upwards in September, 1952, sixty-three had been launched, with a two-thirds success rate. One of the most important developments of the early years took place in February, 1949, when a WAC Corporal was mounted on a V-2 to form a two-step vehicle that achieved an altitude of 250 miles (over 400km) above White Sands before plunging back to Earth. The Aerobee sounding rocket showed promise, as did the Viking from 1949 onwards, and it was not long before engineers and scientists complained that White Sands was too restricting. Following Jules Verne, the new launch site chosen was in Florida. The two sites ran in tandem for a period, but from Cape Canaveral launches of increasingly more powerful rockets could take place

space. Soviet statements on space plans provided a powerful stimulus to the Eisenhower administration and formal support for the US plan to launch satellites in the IGY was announced in July, 1955. Now the question was which route would be taken: via the Viking of the Naval Research Laboratory, the Army's Redstone or the Air Force's Atlas.

Viking Selected

The choice fell on the Viking (the program was shortly renamed Vanguard). As one of the official histories stated, for General Medaris, von Braun and his other colleagues at the Army Ballistic Missile Agency in Huntsville, Alabama the news came as a "shocking blow, [one] difficult to believe. They saw the Vanguard project for what it was, a valiant attempt to build a new and relatively limited launcher under the close to ruinous constraints of an unprecedented development schedule and with underfinancing. They knew their established track record in guiding the evolution of their rapidly maturing ballistic rockets could make the satellite project a routine extension of existing capability .... Yet [the project] was to go in the high risk direction." The reason for the choice was reportedly that the administration wanted the IGY work to have the least impact on the defense ballistic missile programs.

The US made full use of captured V-2 components (and German rocket specialists) in the immediate postwar period, and by September 1952 over sixty had been launched from White Sands in New Mexico, with a two-thirds success rate. Well before then, a WAC Corporal was mounted on a V-2 to form a two-stage vehicle – and subsequently new sounding rockets such as the Aerobee were to be developed.

3 Launch of the first V-2/WAC Corporal vehicle in Project "Bumper" from White Sands on February 24, 1949.

4 The space age came to Florida in July 1950 with the launch of another V-2/WAC Corporal.

5 Technicians prepare for a V-2 launch from White Sands.

1 *Engineers at work on installing a 3.5lb satellite aboard the Vanguard TV-4 launch vehicle at Cape Canaveral. A successful launch took place on March 17, 1958, but the US Army had already orbited America's first satellite by then. However, Vanguard 1 (as it was designated after launch) is still in orbit and is expected to be so until around the year 2258.*

2 *A model of the USAF Pioneer lunar probe that was launched by a three-stage Thor-Able from the Eastern Test Range (Cape Canaveral) on October 11, 1958.*

3 *America's hopes of placing a satellite in orbit under the aegis of the International Geophysical Year initially rested on the US Naval Research Laboratory's Viking program – renamed Vanguard in due course. With Sputniks 1 and 2 already in orbit the first launch attempt of the new three-stage vehicle went disastrously wrong on December 6, 1957, when Vanguard TV-3 (Test Vehicle) rose a few feet and then dropped back in flames.*

4 *TV-3 prior to the launch at Cape Canaveral, Florida.*

From then on the result was perhaps inevitable. Progress in Vanguard was never smooth and development took place in an atmosphere of constant tension. Sputnik 1 won the race and the US hopes of demonstrating that the Soviets' capability was in no way unique suffered a humiliating disappointment on December 6, 1957, when Vanguard TV-3 (Test Vehicle) lifted a few feet off the pad and then dropped back in flames. The army had already received instructions to prepare a modified Jupiter (itself derived from the V-2) as a back-up, and von Braun inaugurated the orbital space age for the US on February 1, 1958, with the launch of Explorer 1. Vanguard 1, after a second failure, followed on March 17 and proved to be a highly-successful and long-lived satellite. But the major prize had gone to the USSR and it would be long haul back for the US.

The shockwaves generated by Sputnik's appearance in October, 1957 provided the impetus for the first phase of US lunar and planetary exploration. Even if the Soviets had orbited the first Earth satellite, the US could try to beat them to the Moon. Proposals for probes had surfaced occasionally throughout the 1950s but it was Sputnik's unheralded debut that provided the focus for a national effort. NASA's creation still lay in the future but the military, then still grappling with the problems of developing long-range rockets, were not averse to expanding their empires to encompass the space arena.

The Department of Defense established its Advanced Research Project Agency in February, 1958 to handle space projects, and the following month authorized the beginning of work on lunar probes. ARPA sifted through some 250 suggestions and finally approved schemes based on the Army's Jupiter and the Air Force's Thor rockets, still under development as IRBMs (Intermediate Range Ballistic Missiles) carrying nuclear weapons. The objective was to fly past the Moon, a simple goal today but a tall order at a time when the US had orbited only the Explorer 1 satellite.

US deep space efforts until the mid-1960s were hampered by a lack of rocket power. The Soviets' SS-6 missile was sized to handle bulky, unsophisticated warheads, whereas American nuclear devices could be carried by smaller vehicles. The SS-6's capacity allowed it to be used as the "workhorse" for launching the first satellites, then deep-space probes and manned spacecraft. In fact, Soviet cosmonauts are still launched by the same basic vehicle. The Air Force's Thor team calculated that, by adding upper stages, their missile could send around 88lb (40kg) to the Moon. Jupiter and its stages was even more severely limited – scraping by with a mere 15.5lb (7kg). When the Soviets' Luna 1 appeared within a year, it tipped the scales at a massive 798lb (362kg).

USAF Pioneer Program

The Air Force went first with three attempts to place a satellite into orbit around the Moon. The aptly-named Pioneer carried a simple infrared camera to transmit pictures of the unseen lunar far side, a microphone for registering micrometeoroid impacts and a magnetic field detector. The plan was for the 84lb (38kg) spacecraft to be dispatched from Earth at six miles per second (9.6km/s) to reach the Moon two-and-a-half-days later, where its small, solid retro-motor would brake it for lunar gravity to capture it into a 40,000 mile (64,000km) orbit. The three-stage Thor-Able lifted off from Florida's Cape Canaveral on the morning of August 17, 1958, marking only the tenth US space launch, but the flight ended abruptly after only 77 sec. The first stage exploded nine miles (15km) up and Pioneer fell into the Atlantic.

Thor was still a development vehicle and such failures were to be expected, but the next attempt proved to be truly frustrating. All three stages seemed to fire perfectly following launch on October 11, 1958, but Pioneer's speed was 534mph (860kph) too low to reach the Moon. Earth's gravity inexorably drew it back from a peak height of 70,200 miles (113,000km) to burn up in the atmosphere some 43 hours later. Analysis showed that stage 2 was carrying more than enough propellant to make up the speed deficit when it was wrongly commanded to shut down by the on board accelerometer. However, Pioneer did penetrate 25 times further than previous satellites, yielding new insights into Earth's encircling radiation belts. The mission was also the first conducted under the authority of the nascent NASA. Although the military conducted the flights, the civilian agency was preparing to take over all deep space programs.

The Air Force's third and final Pioneer took off on November 7, 1958, and this time stage 2's shut down signal was transmitted from ground control. Unfortunately, the solid propellant Altair third stage failed to ignite and the lunar probe burned up in the atmosphere over the Atlantic.

Attempting to fly into lunar orbit was, in retrospect, highly ambitious. In fact, it was not achieved until the Soviet Luna 10, eight years later. As the US Army's two Pioneers had to be small, they were designed solely to fly past the Moon. There was no camera, only a Geiger counter for detecting particle radiation. The Jet Propulsion Laboratory had also provided a similar payload for Explorer, thereby discovering Earth's radiation belts. Jupiter was topped by three stages, powered by eleven, three and one Sergeant solid motors, respectively, spinning at 700rpm for stabilisation. A similar arrangement had been used atop a Redstone for Explorer. This time, following launch on December 5, 1958, the Jupiter first stage shut down four seconds earlier than expected as its propellants ran out. Pioneer was left only 380mph (610kph) short of escape speed and gravity pulled it back from a 63,155 mile (101,700km) peak to burn up over Africa after 38 hours.

Before the final Pioneer could be readied, the Soviets again scooped the headlines by sending Luna 1 past the Moon at a distance of 3,675 miles (5,920km). Although we now know the intention was to hit the surface (as Luna 2 did the following September), it was still a significant achievement and highlighted the superior Soviet lifting power. The last of the five lunar Pioneers authorised by ARPA departed from the Cape in the early hours of March 3, 1959, and came the closest of all to success. Although the upper stages fired slightly off course, it did venture to within 37,282 miles (60,000km) of the Moon, about twice the planned distance, before sailing off to become the first American object in orbit around the Sun. Its tiny, battery-powered transmitter survived for 82 hours, providing contact up to 403,650 miles (650,000km) out.

Early NASA Failures

NASA was already working on its own lunar probes, intending to launch them with the more powerful Atlas ICBM – still a very risky vehicle – in place of the Thor IRBM. That raised the potential lunar orbit mass to about 397lb (180kg) and allowed the scientists more flexibility in designing the payload. There was still no camera, but there was a range of radiation, micrometeoroid and magnetic field detectors packed inside the 275lb/3.3ft diameter (125kg/1m) aluminum sphere. Pioneer's batteries were replaced by paddles of solar cells to provide a longer life. Increasing sophistication was also reflected by the two liquid thrusters for course corrections en route to the Moon, and for braking into

the intended 1,490x2,980-mile (2,400 x 4,800km) lunar orbit. These three Atlas-Able probes flew between November 1959 and December 1960, but none had the chance to succeed: the first vehicle exploded after the protective nose fairing collapsed, the second's stage 2 shut down early, and the third exploded after 68 seconds only 7.5 miles (12km) above the pad. The losses were particularly galling because Luna 2 had already impacted the Moon and Luna 3 had transmitted the first images of the far side.

This was the end of the first wave of American lunar attempts, beginning an interregnum of two years before the next generation Ranger and Luna landers headed towards the Moon on more powerful rockets.

The Atlas-Able orbiters were very much stop-gap craft adopted by a young organization feeling its way. In fact, NASA had already adopted a more sophisticated set of lunar missions in May, 1959, proposed by the California Institute of Technology's Jet Propulsion Laboratory. In the decades since, JPL has become the premier US center for deep space exploration, directing projects such as Mariner, Voyager, Magellan and Galileo. In 1959, JPL offered a five-year menu of unmanned missions from which NASA selected orbital, rough landing and soft landing missions to the Moon. These evolved into the familiar Lunar Orbiter, Ranger and Surveyor projects, respectively, of the 1960s. In the middle of the ill-fated Atlas-Able program, NASA declared that five Ranger spacecraft would fly to the

5

Moon, return increasingly close-up TV pictures before crashing, and release a small instrumented capsule onto the surface.

Ranger benefited from the arrival of Lockheed's Agena B upper stage, designed originally for carrying spy satellites. Combined with the Atlas, it could fire over 790lb (360kg) at the Moon. Not only that, but it would enter an Earth parking orbit before re-starting its engine; ensuring a far more accurate trajectory than Pioneer's direct ascent. Ranger itself was built around an hexagonal aluminum and magnesium bus, carrying the control electronics, nitrogen gas thrusters, two solar panels and a 4ft (1.2m) diameter antenna for transmitting the high resolution pictures to Earth. The Pioneers had been spun in flight for stability, but Ranger was held steady by the thrusters, directing its solar panels towards the Sun.

Turned on by an automatic timer, the TV camera

5 Officials of the US Army Ballistic Missile Agency were convinced that had they been given permission they could have orbited a small satellite before the Soviet Union. With the failure of the Vanguard launch in December 1957 work on a modified Jupiter rocket (derived from the V-2) was forwarded by Huntsville, and on February 1, 1958, America's first satellite Explorer 1 was successfully launched. The spacecraft's greatest achievement was the discovery of the radiation belts named for Dr. James A. Van Allen, who was responsible for placing a geiger counter aboard the spacecraft. In this famous photograph, (right to left) Wernher von Braun, Dr. Van Allen and Dr. William H. Pickering, director of the Jet Propulsion Laboratory in Pasadena, California joyfully celebrate the successful launch.

1 *An artist's impression of NASA's Ranger 1 lunar probe.*

2 *A model of the Ranger 3 spacecraft launched on January 26, 1962, and which missed the Moon by almost 23,000 miles. Failure followed failure in the Ranger program until the summer of 1964.*

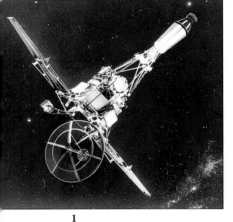

1

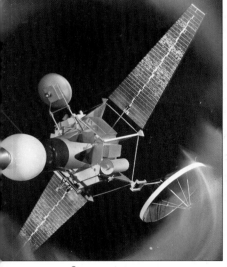

2

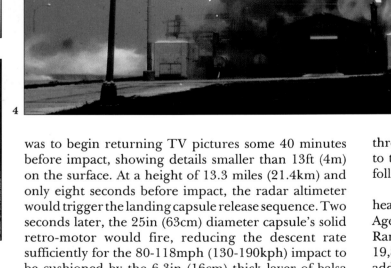

3

3 *Ranger 7 in the laboratory prior to launch on July 28, 1964.*

4 *Success at last! Ranger 7 is lofted towards the Moon by an Atlas-Agena vehicle and before impact three days later transmitted over four thousand pictures of the lunar surface back to California.*

4

was to begin returning TV pictures some 40 minutes before impact, showing details smaller than 13ft (4m) on the surface. At a height of 13.3 miles (21.4km) and only eight seconds before impact, the radar altimeter would trigger the landing capsule release sequence. Two seconds later, the 25in (63cm) diameter capsule's solid retro-motor would fire, reducing the descent rate sufficiently for the 80-118mph (130-190kph) impact to be cushioned by the 6.3in (16cm) thick layer of balsa wood. The seismometer would then spend up to two months returning lunar 'quake data through the battery-powered transmitter.

The first two Rangers, without capsules or TV, were to be fired into high Earth orbits to test the basic spacecraft and its Agena stage. Ranger 1 successfully slotted into the planned parking orbit on August 23, 1961, but Agena's engine failed during the second burn and left the pair stranded. Ranger's battery ran flat after

three days because the solar panels could not lock on to the Sun in the low Earth orbit. Ranger 2's fate the following November 17 was almost identical.

Nevertheless, Ranger 3 and the first lunar payload headed for the Moon on January 26, 1962. Unfortunately, Agena's burn from parking orbit was off target and Ranger was left to sail past the Moon at a distance of 19,872 miles (32,000km). An electronics failure then added insult to injury and the TV screens at JPL remained blank. Unlike modern spacecraft, there were no means for ground controllers to assume command or to re-program the onboard computer. Agena performed well for the next mission, on April 23, 1962, but Ranger's automatic sequencer stopped. As consolation, the inert Ranger 4 became the first US spacecraft to strike the Moon. Ranger 5 brought the series to an ignominious end six months later, when its power supply failed and it missed the Moon by more than 435 miles (700km).

Support for Apollo

Ranger's career might have ended there had it not been for the overriding project of the decade: Apollo. The unmanned lunar programs were devised as scientific enterprises, but as the manned Apollo lunar program gained momentum in the early 1960s the emphasis changed to one of support for their big brother. Scientists of the time could not even agree whether the lunar soil could support a spacecraft, let alone find a suitable landing site. Ranger would be a start, Lunar Orbiter would map potential landing sites around the equator, and Surveyor would demonstrate whether soft landing and survival were possible.

Rangers 6-9 therefore replaced the landing capsule with a 375lb (170kg) battery of four narrow-angle and two wide-angle TV cameras to return pictures of better than 6.5ft (2m) resolution before impact. Ranger had proved itself spectacularly unreliable and Ranger 6 was delayed until January, 1964 while the design was closely scrutinized and subjected to exhaustive tests. Ranger 6 surpassed its predecessors and ground controllers saw the TV cameras switch on to warm up 18 minutes out from impact, but they continued waiting in vain until the luckless craft smashed into the surface in the Sea of Tranquility. The inevitable inquiry pointed to electri-

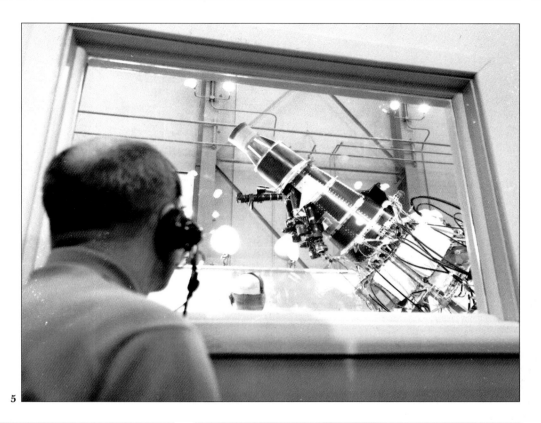

5

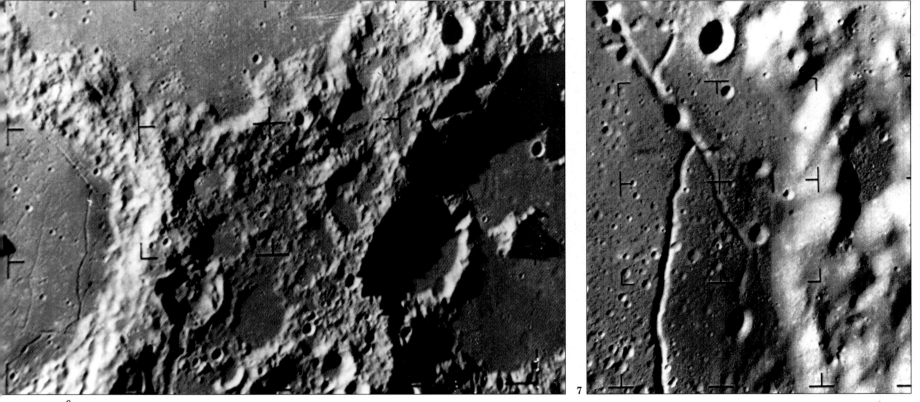

6

cal failures caused by conditions at launch knocking out critical circuits. Ranger was becoming an embarrassment instead of a demonstration of technical superiority over the Soviets.

Ranger 7 proved to be a turning point after thirteen consecutive US lunar failures. Another failure would possibly have killed off the project, but its emphatic success set the pattern for future US deep-space exploration. More than four thousand pictures were transmitted to JPL via the Goldstone antenna in California in late July, 1964 as it homed in on the Sea of Clouds; the last proving a thousand times more detailed than possible from Earth-based telescopes. Apollo planners were delighted with the results: the region appeared generally smooth, with gentle slopes that would make a manned landing easier than previously believed.

Tranquility certainly looked suitable for a manned attempt. Ranger 8 went even better the following February, returning more than 7,000 pictures before hitting the Sea of Tranquility only 15 miles (24km) off target.

The Moon had been revealed in unprecedented clarity without uncovering any showstoppers for Apollo. Ranger 9 was targeted at a more scientifically interesting area only a month later, delivering more than 5,800 images of the crater Alphonsus and bringing the series to a close after costing the equivalent of $1,000m at today's rates. There had been plans for follow-on Rangers with improved cameras and lander capsules, but these were dropped as work intensified on the more complex Lunar Orbiter and Surveyor craft, and as interest grew in the nearer planets.

5 A Ranger spacecraft undergoing tests in a clean room prior to launch.

6 March 1965, and one of the images from Ranger 9 demonstrates the image quality secured by its TV cameras a little over nine minutes from impact at an altitude of 175 miles. Areas of three major craters appear in the picture – Ptolemaeus at top, Alphonsus on the left and Albategnius at right.

7 This Ranger 9 image shows a much smaller area of Alphonsus at higher resolution one minute before impact.

3

GAGARIN AND THE MOON: TRIUMPH AND FAILURE

For the Russian rocket pioneers, manned spaceflight had always been the final goal. Even during the development of Soviet ballistic missiles, Korolev was giving some thought to the possibility. The first proposal was made in the spring of 1956. By this time, several launches of dogs had been made on sounding rockets. Korolev suggested a similar flight be made with a man. At this time, however, construction at the Tyuratam launch complex and planning for the SS-6 missile test flights were underway, so Korolev had little time to devote to the project. After the SS-6 had successfully flown and approval for the Sputnik launches had been given, he abandoned the idea, not taking it up again until May of 1958. This was the time of the Sputnik 3 launch and final preparations for the first (unsuccessful) Luna launches.

1

Mikhail Tikhonravov's design department had proposed a heavy, manned satellite. This was one of three proposals for advanced spacecraft. The second was for a manned, sub-orbital spacecraft that would allow testing of human reactions to space conditions (zero gravity and the G-loading of launch) without committing to orbital flight. The third was by advocates

of unmanned research and involved an advanced automatic satellite.

Initial design work began during the summer, and, in August, 1958, the reports were submitted for consideration. In November, the Council of the chiefs of the design bureaux met to discuss the three projects. The sub-orbital program was rejected. Although easier to accomplish than an orbital flight, it would slow the overall program by spreading the available talent and resources too thinly. The unmanned satellite was rejected as manned flight was the goal. Approval by Khrushchev was given in late 1958.

Development of Vostok

Initial work on the Vostok (East) spacecraft began in early 1959. The design work and construction were done in parallel. This was to speed development – the Mercury program had to be beaten. The initial structural data were ready by March, 1959, while the design information for the electrical system, attitude control, communications, and other systems were ready in May. The two-part design which emerged had several unusual features.

The cosmonaut rode in a large, spherical capsule. This gave the maximum internal volume, and the high-speed aerodynamics of a sphere were already well known. By putting in an off-center weight, the capsule would orient itself under aerodynamic forces during re-entry. Thus no control rockets were needed. Unlike Mercury, Vostok used a normal oxygen/nitrogen atmosphere at sea level pressure and there was enough oxygen, food and water for ten days. This would keep the crewman alive, should the retro-rocket fail, until the spacecraft's orbit decayed naturally. The cosmonaut rode on an ejector seat. This served two functions. In the event of a low-altitude launch failure, the cosmonaut could eject. Secondly, after re-entry, the cosmonaut would eject and parachute to a landing. The capsule would then come down on its own parachute. This reduced the size and weight of the capsule's parachute over that needed to slow the large sphere to a soft landing speed. This would later result in both political and engineering problems. The capsule was attached to a service module. It was a double-cone shape. Around the top end was a ring of bottles containing pressurized N2 for the attitude control jets. The module also carried batteries, communications equipment and the attitude control thrusters. At the base was the liquid fuel retro-rocket.

At the same time, cosmonaut selection was under way. Air Force squadron flight surgeons were asked to recommend physically fit pilots for a "special" training program. Over 3,000 names were submitted. In August, 1959, teams of military doctors visited air bases to interview pilots. The program was initially described as involving "aircraft of a completely new type." One prospect, Georgy Shonin (who was among those later picked) thought they were referring to helicopters and refused. One of the doctors reassured him: "No, no! You don't understand. What we're talking about are long-distance flights, flights on rockets, flights around the Earth." Shonin's jaw dropped in surprise and he volunteered. The 3,000 pilots were quickly cut to 102 who passed both the initial medical tests and showed "a fundamental desire" to fly in space. They underwent rigorous medical and psychological tests at the Central Aviation Research Hospital in Moscow. By February 25, 1960, 20 cosmonauts had been selected.

The group was very different from that of the Mercury

Korabl Sputnik 1 turned so that the retro-rocket was pointed in the wrong direction. When the retro-rocket fired, the spacecraft went into a higher orbit. (It should have been slowed down so it would enter the atmosphere.)

This was the start of a run of bad luck that dogged the Soviet space program for the next year. The next launch was probably of a Vostok B. This was a recoverable version which carried two dogs and other biological test specimens. The launch attempt was made on July, 23, 1960. Soon after lift-off, one of the strap-on rockets failed and the booster, heavy with fuel, came down near the launch pad. If it did carry two dogs, it is not clear if they survived.

The next launch was successful. Korabl Sputnik 2 was launched on August 19. Aboard were two dogs, Strelka and Belka, along with rats, mice, flies, plant seeds, fungi, algae and plant samples. It spent a day in orbit before the retro-rocket was fired. This time everything went as planned. After re-entry, the dogs were ejected from the capsule and landed safely nearby. When the medical data was analyzed, it was found that

One of the major results of the policy of glasnost *(openness) introduced by the former Soviet leader Mikhail Gorbachov was the eventual confirmation that the Soviet government was committed to the race to land men on the Moon before the US – a fact strenuously denied over the twenty years or so that passed after the race was lost. Work on the heavy lift launcher that was intended to take cosmonauts to the Moon began as early as 1962, although numerous design changes were introduced before the N-1 (as it came to be called) took final shape, with a payload capability of 98 tonnes. The first pictures of the N-1 were not released in the West until the early 1990s, and despite their poor quality they are of major historical importance. Four launches of the vehicle took place between February 1969 and November 1972 – all of them unsuccessful.*

2, 3, 4 *The particular N-1 launch is not identified but this sequence of TV images shows respectively the more than 100m-high booster on the launch pad, the ignition of the first stage's 34 engines and the entire vehicle as it lifts away from the launch pad at Baikonur.*
5 *The first N-1 to be launched on the pad shortly before lift off.*

2   3   4

astronauts. All but two were under 30, and only one was over 5ft 7in (1.7m) tall. The biggest difference was in flight experience: none were test pilots. In fact, the most experienced had only 900 hours of flight time; most had approximately 250 hours. Only one had experience in the supersonic MiG 19; the rest flew subsonic MiG 15s and 17s. Their role was also different. Korolev's philosophy stressed automatic systems as the primary control of the spacecraft. The cosmonaut rode as a passenger with little manual control.

The prototype Vostok was completed by the end of 1959. This was the Vostok A version, also called the "electrical analog." It had all the systems of the manned version but lacked the heat shield and parachute. The calculations of heat flow were being checked in wind tunnels and the capsule parachute was still under development. (Landing tests of the capsule began in early 1960.)

The check-out of the electrical analog took several months. It was launched into orbit on May 15 1960, under the name Korabl Sputnik 1. ("Korabl" is Russian for "ship" or "vessel.") Korolev wanted to check the function of its on-board systems during a four-day flight. At the end of the mission, the retro-rocket would be fired. Because Korabl Sputnik 1 lacked a heat shield, it would burn up during re-entry. During the third day of the flight, a problem was detected in the infrared primary orientation system and Korolev was warned a failure might occur. It was suggested that an alternative procedure, orientating the spacecraft using the Sun, be tried. Korolev refused, saying he wanted to follow the original plan. The infrared system did fail and the

5

one of the dogs had gone into convulsions during the fourth orbit. It was recommended, therefore, that the first manned flight should be limited to a single orbit.

The next test launch was made on December 1. Korabl Sputnik 3 was to be a repeat of the previous flight. Aboard were two dogs, Pchelka and Mushka, and the spacecraft was equipped with a new computer

1 Yuri Gagarin with Sergei Korolev in 1961. The cosmonaut was one of the first group of six selected but never flew again in space and was killed in a flying accident in 1968. As the first man in space the political value of Gagarin to the Soviet authorities was greatly enhanced by his personable appearance and character. The system required him to lie on at least one occasion – about his ejection from Vostok 1 and descent by parachute – since there was a risk that his flight would not be recognized, as aircraft flight records required that a pilot land with his aircraft.

2 Gherman Titov flew the Vostok 2 mission on August 6, 1961. He was in space for just over a day and reportedly suffered for most of the time from space sickness – a continuing problem today. Given Titov's experience, it is a little ironic that the official caption to this picture reads "Your blood pressure is excellent!"

control system. The day-long flight was apparently successful up to the re-entry burn. Problems with the retro-rocket caused Korabl Sputnik 3 to go into a too-steep trajectory. The spacecraft was overstressed, broke apart and burned up.

Korolev's problems were not limited to space. He had to cope with a lack of resources and technology, the challenge from the Mercury program and Khrushchev's increasingly capricious demands. Also, Korolev's health had begun to fail. According to one report, he had a heart attack the day following Korabl Sputnik 3's fatal re-entry. At the hospital, it was found that he also suffered from kidney disease, the latter due to his earlier imprisonment. Korolev was told that if he continued his frantic pace, he would die. But he had no choice. Three weeks later he was out of hospital, and facing another failure.

On December 21, 1960, another Korabl Sputnik launch was made. As the third stage began firing, a problem appeared which meant the booster could not develop enough speed to reach orbit. While the third stage continued to burn, the capsule was separated by an emergency command. It made a normal descent, landing near Tura, some 1,800 miles (2,900km) down range from Tyuratam. The biological payload was successfully recovered.

It was a major setback, and Korolev intensified the group's efforts over the next two months. The stakes were high – the first sub-orbital Mercury flight was nearing – and the race to put the first man in space (and thus to maintain the belief in Soviet leadership) was paramount.

The first three of the man-rated "Vostok V" spacecraft were ready. They would conduct two final rehearsals before a manned orbital flight was made. The first, Korabl Sputnik 4, was launched on March 9, 1961. Aboard was a single dog named Chernushka. It completed a single orbit, re-entered and landed safely. A second unmanned flight was prepared.

During this time, the cosmonauts had been undergoing training. This included academic study, simulator training, parachute jumps, centrifuge runs, and spending many days in an isolation chamber. By the summer of 1960, a group of six cosmonauts was selected to make the first space flights. The original group comprised Yuri Gagarin, Anatoli Kartashov, Andrian Nikolayev, Pavel Popovich, Gherman Titov, and Valentin Varlamov. Two were soon eliminated; Kartashov developing hemorrhages during a centrifuge run and Varlamov

fracturing a vertebra in a swimming pool accident. They were replaced by Grigori Nelyubov and Valery Bykovsky.

As launch preparations continued, so did tests and training. On March 23 1961, Valentin Bondarenko had completed a ten-day isolation chamber test. Bondarenko was 24, the youngest of the 20 cosmonauts, and was known as "Valentin Junior" and "Tinkerbell." The chamber was at a reduced pressure, which meant a higher than normal oxygen content. He removed the medical sensors and cleaned off the spots where they had been placed with an alcohol-soaked piece of cotton. When finished, he tossed it aside. It landed on a hot plate and burst into flames, spreading to the interior of the chamber. Bondarenko was severely burned before he could be removed. Gagarin accompanied him to the hospital but nothing could be done. Bondarenko died eight hours later.

The next day, Gagarin and the other five members of the group were flown to Tyuratam for the launch. of Korabl Sputnik 5, which took place on March 25. As before, a single dog, Zvezdochka, was carried and it remained aloft for a single orbit. With its successful recovery, the way was clear for a manned Vostok flight.

Khrushchev was also getting ready. In late March 1961, he met East German leader Walter Ulbricht and told him that they would soon put a man into space. This would give the Soviets such a level of power and prestige that the West would have no choice but to accept Russian demands.

On April 3, the State Commission approved the manned launch. At this time the cosmonaut for the flight had not yet been picked, although it was clear that Gagarin and Titov were the "leaders" of the group. On April 8, the State Commission met to decide who would make the flight, and two days later Gagarin was told he had been selected. Titov was his back-up. The following morning, April 11, the Vostok 1 booster was rolled out to the launch pad. During the day, Gagarin and Titov trained in Vostok 1, the mission spacecraft.

The First Manned Flight

Early on the morning of April 12, 1961, Gagarin and Titov were awakened. They dressed in pressure suits and rode a bus to the launch pad. Gagarin entered the spacecraft and the countdown went smoothly, the only problem being a hatch that was improperly closed and had to be re-sealed. At 9:07 am, Moscow time, the A-1 booster lifted off. Two minutes later, the strap-ons separated while the core stage continued to burn. Five minutes after launch, the core stage shut down. The third stage separated and ignited, sending Vostok 1 towards orbit. Finally, the third stage shut down and Yuri Gagarin was in orbit.

His experiment program was very limited; he was to be launched into space, spend one orbit under weightlessness, then return to Earth – alive. For now, that was enough. Gagarin did eat some food while in orbit. Because of fears of possible adverse effects of weightlessness, the manual controls were locked. But a three-digit code (1-2-5) was set into a "logic clock" and, should Gagarin need to take over control, the combination could either be radioed up or he could open an envelope carried on board. As events turned out, he did not need the code.

Radio Moscow announced the flight at 10:02 am, while it was still under way. At 10:25 am Moscow time, the retro-rocket fired. After re-entry heating was over and the capsule had descended to 4.4 miles (7km), Gagarin ejected. He landed at 10:55 am near the town of Engels.

When Gagarin began his post-flight press conference and tours, it was evident that the parachute landing was a problem. The rules for aircraft flight records required that the pilot land in the aircraft. This was to prevent a record from being set by over stressing the aircraft until it failed and then having the pilot bail out. To prevent Vostok 1's "first" being denied for this reason, the Soviets were evasive. Gagarin was ordered to lie about how he had landed.

Among the Soviet people, the news of Vostok 1's flight was greeted with joy. The propaganda campaign again stressed that Russia was overtaking the US technologically, and emphasized the superiority of the Soviet system over capitalism. Khrushchev was quick to exploit the success. Two weeks later, at the Vienna Summit, he gave President Kennedy an ultimatum: Russia would sign a unilateral peace treaty with East Germany by December 31, 1961, leaving West Berlin isolated.

Korolev had intended to have the second manned flight last three orbits, but it would take some considerable time to do the calculations. However, in the middle of July Khrushchev summoned Korolev and ordered him

Weapons of 20 to 100 megatons would be exploded in the atmosphere. On September 8 and 9, Khrushchev gave his terms for ending the Berlin crisis: Western acceptance of a German peace treaty and of Soviet plans for complete disarmament.

The Vostok 2 flight was thus part of a campaign of psychological pressure over Berlin. It was meant to break Western will by creating the illusion of great Soviet missile power. This illusion was exposed during the fall of 1961 by the US Discoverer photo reconnaissance satellites. They showed the USSR had deployed only a token force of SS-6 ICBMs. The Soviets were forced to back down on their Berlin ultimatum; but if the missile bluff was over, what was later called "The Russian Space Bluff" continued.

The Vostok 2 flight also marked a change in the exploitation of Soviet space accomplishments. Before this, Khrushchev had used missions as they became available. Now the launches were to be made strictly according to his political schedule. The attempt by the Russians to maintain their space leadership was being drawn in two conflicting directions: short-term "stunts" to upstage the Americans, as opposed to the well-thought-

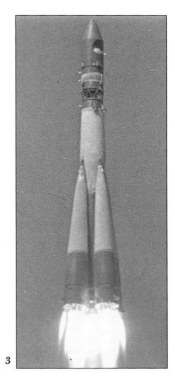

3

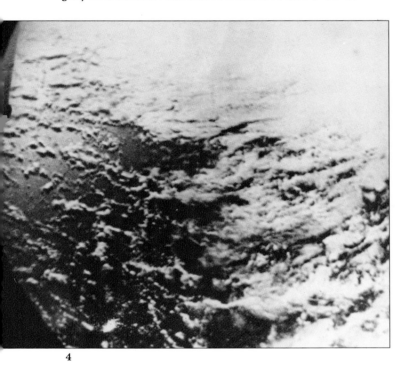

4

5

to conduct the second manned flight by the first half of August. To do this, it was decided to repeat the profile of the Korabl Sputnik 2 and 3 flights. Vostok 2 would orbit for one day and be brought back to Earth. Korolev began an intensive test program to ensure the Vostok systems could operate for a day in space. The Vostok 2 launch was made on August 6, 1961, with Titov as crewman. Soon after launch he suffered from bouts of motion sickness. This limited the experiment program he was able to undertake, but he landed successfully after just over a day in space.

Propaganda Value of Space
At a Kremlin parade on August 9, the link between Soviet achievements in space, Berlin, and Soviet military power was made. Titov said Russia had weapons to "crush an aggressor," while Khrushchev spoke of 100 megaton nuclear weapons. Within a few weeks, the reason for the rush to launch Vostok 2 and the "missile rattling" became clear. Just after midnight on August 13, East German troops began building the Berlin Wall, which sealed off the Western sectors of the city. This was followed on August 30 by the Soviet breaking of a three-year-long moratorium on nuclear testing.

out, step-by-step planning needed to reach the Moon first. With Eisenhower's passive space policy, it was still possible for the Soviets to maintain their lead (since they were racing only themselves). When President Kennedy accepted the Soviets' challenge to a Space Race, the margins became much thinner.

The remaining four Vostok flights were a pair of dual missions. Vostok 3 was launched on August 11, 1962, a full year after Titov's flight. The crewman was Andrian Nikolayev. Once the A-1 booster had placed him into orbit, the pad was quickly prepared and another A-1 was rolled out. As Vostok 3 passed overhead the next day, Vostok 4 lifted off. It was placed into a nearly matching orbit so the two spaceships remained close to each other. The pilot of Vostok 4 was Pavel Popovich. The flight was a long-duration mission; Nikolayev remaining aloft for four days, while Popovich orbited for three days. Both re-entered on August 15. Neither cosmonaut suffered any ill effects from weightlessness; both unstrapped themselves from their seats and floated in the cramped capsules during the flight.

In the Soviet quest for "firsts," this was the first time two manned spacecraft had been in orbit simultaneously, and both long-duration missions were longer than

3 *The launch of an unspecified Vostok mission. It is a testimony to the Soviet design achievement that derivatives of the original SS-6 Sapwood missile (which powered the first manned and unmanned missions) are still used to launch crews into orbit aboard today's Soyuz spacecraft.*

4, 5 *Few actual mission images were released by the Soviet authorities until very recently. These two, poor-quality pictures of clouds over the Earth's surface were taken by Gherman Titov and Andrian Nikolayev respectively. The latter flew the four-day Vostok 3 mission in August 1962 and appears to have used an optical orientation device during photography.*

anything Mercury could achieve. In terms of prestige, the dual mission could be seen as an attempt to restore Russia's status after Khrushchev's withdrawal of the Berlin ultimatum. It was also part of the preparation for shipping ballistic missiles to Cuba, for the Cuban Missile Crisis was only two months away.

The final two Vostok flights were made in June, 1963. Vostok 5 was launched on June 14, and the cosmonaut was Valery Bykovsky. After Vostok 5 had orbited for two days, it was joined by Vostok 6. Its cosmonaut was Valentina Tereshkova – the first (and for two decades the only) woman in space. Western observers expected the two spacecraft to rendezvous and dock. This was beyond Soviet capabilities at the time, however. In fact, Vostok 5 and 6 were in close proximity only twice each orbit. (Vostok 3 and 4 were in nearly identical orbits so remained close together throughout the orbit.) Tereshkova became motion sick at some point in the flight, though this was not confirmed by the Soviets until 1986. It did not seem too serious a problem, however, because her flight was extended from one day to three. (According to some reports, Tereshkova was originally the back-up cosmonaut, but flew when the prime cosmonaut became ill.) Both missions ended on June 19.

Endurance Record

Bykovsky's five days in space was, and still is, the longest duration for a one-man flight. However, this fact was completely overshadowed by Tereshkova. She and four other women, Tatiana Kuznetsova, Valantina Ponomareva, Zhanna Yorkina, and Irina Solovyeva (the back-up Vostok 6 pilot) were selected in March 1962, although the names were kept secret for 25 years. Tereshkova was an ordinary factory worker who had taken up parachute jumping. Yet she spent more time in space than all the Mercury astronauts together. Some in the West condemned NASA for its failure to fly women astronauts. In reality, the Soviet woman cosmonaut program was for propaganda alone; the other women soon left and no more were picked for almost 20 years.

One more Vostok flight was planned. Vostok 7 was to be a week-long biomedical mission. Boris Yegorov, who had an extensive space medicine background, was probably to be the cosmonaut. (Since the Vostok cosmonaut rode the automated spacecraft as a passenger, non-pilots, such as Tereshkova and Yegorov, could also be flown.) The Vostok 7 mission was probably planned for the summer of 1964, and Yegorov trained for six months before the flight was cancelled. Events, in the form of Khrushchev's manipulations, had intervened.

Korolev was also looking ahead: to a Moon flight. The first step was to send a manned spacecraft looping around the Moon. To do this, Korolev chose the Earth orbital assembly approach, which involved launching one or more booster stages, fuel tankers, and the manned spacecraft, then assembling them in orbit. The initial technology was examined in the 1962 Vostok Zh study. ("Zh" is the eighth letter of the Russian alphabet, implying that it was to follow the Vostok 7 flight.) There were three elements: the Vostok Zh, rocket modules, and the payload module. The study's mission profile envisioned the Vostok Zh being launched first. A day later, the rocket module would follow into an orbit three to six miles (5 to 10km) away. The Vostok Zh would then approach under automatic control. When it was within 328 to 656ft (100 to 200m) of the rocket module, the crew (size unspecified) would take over manual control and dock. Additional rocket stages could be launched at daily intervals until the required number had been

assembled. The payload module (nature unspecified) would then be launched and dock with the rocket modules.

The complete assembly had the payload at one end, the Vostok Zh at the other, and the rocket modules in between. Once checkout was completed, the Vostok Zh would undock and re-enter. The rocket modules would then fire one after the other, propelling the payload module deeper into space. This was to demonstrate orbital rendezvous and orbital assembly much in the same way Gemini was used to gain experience for Apollo. It could also use a modified Vostok and existing rockets. The Vostok Zh, with its rendezvous and docking systems, would probably need the A-2 booster (an SS-6 with a larger third stage first used in 1960 and 1961 for planetary launches): the rocket and payload modules could use the A-1 booster.

Moonship Configuration

The actual moonship was called Soyuz (meaning "Union," a common form of referring to the Soviet Union). A 1963-64 study envisioned four elements. The manned spacecraft was the Soyuz A. It had three parts: the Instrument Module with the maneuvering engine and other systems, the bell-shaped Descent Module which carried the two- or three-man crew, and the cylindrical-shaped Orbital Module in which the crew worked. The Soyuz B was the Rocket Block. It was the booster which would send the Soyuz A out of Earth orbit towards the Moon. The stage would be launched "dry" (without fuel). Attached to the rear was an in-orbit maneuvering system called the Rendezvous Block (similar to the Soyuz A propulsion system). The final element was the Soyuz V Tanker Module, which carried a tank of propellant (either LOX or kerosene) for transfer to the Soyuz B Rocket Block. The Soyuz A, B, and V were all launched by A-2 boosters.

The study envisioned the Soyuz B Rocket Block being launched first. This would be followed by four launches of the Soyuz V Tankers at daily intervals. These would dock and fill the Rocket Block's tanks. When the fueling operation was completed, the Soyuz A manned spacecraft would be launched and dock with the Soyuz B Rocket Block. The Rendezvous Block would be discarded and the main rocket engine on the Soyuz B would then fire. Although published Soviet accounts do not specify the mission, Phillip Clark, an English specialist on the Soviet space program, has calculated that the Soyuz B Rocket Block would have been capable of sending the Soyuz A around the Moon.

The Soyuz A-B-V complex had advantages and disadvantages. The primary advantage was that it made a flight around the Moon and back possible using existing rockets. On the other hand, it was extremely complicated. It required six launches, five dockings, four fuel transfers, and the restart of a rocket engine after a week in space. Any failure would break the chain. Although elements of the study later became part of the Soviet space program, it is not clear how much work was done at the time. The dual launch profiles of the Vostok 3-4 and 5-6 missions could have been rehearsals of the precise launch times needed in the Vostok Zh and Soyuz A-B-V studies, for no flight tests were made.

Korolev's Soyuz A-B-V plan faced a competitor, proposed by a rival, Vladimir Chelomei. During the 1940s and 1950s Chelomei had worked on anti-ship cruise missiles, and in the mid-1950s he began work on ICBMs and submarine-launched missiles. Chelomei has often been seen as an opportunist. (He married Khrushchev's daughter and employed his son at his design bureau.) Chelomei's plan envisioned use of the UR-500

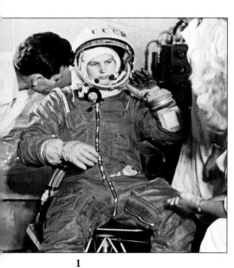

1

1 June 1963 and Valentina Tereshkova is suited up prior to becoming the first woman in space during the three-day flight of Vostok 6. Tereshkova's spacecraft joined that of Valery Bykovsky aboard Vostok 5, which had been launched on June 14. As in the case of the flights of Vostok 3 and 4 in August 1962, the dual flights did not mean that the Soviets had a rendezvous and docking capability – which was the prime aim of NASA's Gemini program. Tereshkova's only experience relevant to her Soyuz flight was that she had taken up parachute jumping, and her time in space was made possible by the fact that the Vostok could fly with no human input. Propaganda was the main purpose of her flight, but this was one "race" that the Soviets would most definitely win since no American women would join the NASA space program as astronauts until the space shuttle was developed.

booster, which was originally developed as an ICBM to carry a 100 megaton bomb. (The UR-500 was later called the "Proton" after its first payload: in the West, it is called the D-1 booster.) The vehicle was large enough to send a manned spacecraft (also of Chelomei's design) directly around the Moon without refueling in Earth orbit. The plan had the advantage of simplicity but it required the development of a new rocket.

Korolev's competitor to the UR-500 (or D-1) was the N-1 (confusingly called the G-1 booster in the West). Work started in 1960. The original N-1 was to have an orbital payload of 40 to 50 tons (compared with the 20 tons of the D-1) and its first flight was planned originally for 1963. The design was reviewed several times, with the planned payload increased each time. But the N-1 also faced internal problems. Valentin Glushko, who designed most of the Soviet's rocket engines, wanted to use exotic propellants for the N-1: nitric acid and dimethyl hydrazine. Korolev's N-1 used LOX and kerosene. There was also disagreement over the design of the first stage engines. Korolev wanted to use 30 medium-thrust engines, while Glushko favored a few, higher-thrust engines. He argued that a large number of engines would be impossible to synchronize. These disagreements led to a split between Korolev and Glushko, and Korolev turned to a design bureau run by N.D. Kuznetsov. It had designed jet engines but had little experience with rockets, and certainly nothing on the scale of the N-1.

Thus, during 1962-63 the Soviet space program was in internal disarray. Chelomei and Korolev were locked in a struggle for power and any attempt to establish an overall plan for landing a Soviet man on the Moon languished. Finally, in early 1964, Khrushchev formally selected Chelomei's Design Bureau to develop the L-1 project: the manned lunar flyby. This was to take place in 1966-67. Korolev was to develop the N-1 superbooster and the L-3 lander, the Soviet Moon landing being made in 1968, well ahead of NASA's schedule. This did not bring peace, however, and Korolev began lobbying for control of the L-1 as well.

Three-Man Mission

At the end of 1963, Khrushchev reportedly summoned Korolev. After asking the designer if he was aware of the Gemini program, Khrushchev set him the task of launching not two but three people into space at the same time before the next anniversary of the Revolution on November 7, 1964. Korolev explained the problems but Khrushchev was not interested.

Korolev arrived back at the Design Bureau and explained the task that had been set to his department heads. He asked for any ideas: none were offered and he seemed dispirited. He said that if anyone had a suggestion to call him and he went home. A similar meeting was held the next day. Korolev began by saying it had been suggested they put three seats in a Vostok and lighten the sphere as much as possible. This was greeted with protests, but Korolev stated that the attempt had to be made.

Khrushchev's demands put incredible pressure on Korolev and the other engineers. Korolev became increasingly short tempered, and it was Konstantin Feoktistov, designer of the descent systems, who found the solution. The ejector seat would be removed, the Vostok interior stripped out and three cosmonauts squeezed in without pressure suits. Korolev reportedly asked, "But who on Earth is going to fly without his spacesuit?" Feoktistov replied: "I will for one."

This raised another problem: the Vostok sphere landed at a speed high enough to injure the cosmonaut

severely if he did not eject. With the three-man design, there was no room for each cosmonaut to have an ejector seat. They would have to land inside the sphere, which meant a major redesign of the parachute system. The drop tests of the new system did not go well. In one case the sphere landed intact, but the three monkeys inside were killed by the impact forces. The system that was ultimately developed used two main parachutes. (Vostok used only one.) To further slow down the sphere, a small, solid-fuel rocket would be fired just before touchdown.

It was fired when a long probe, which hung down from the sphere, touched the ground. This brought the velocity to nearly zero.

Re-entry was another problem. The normal Vostok orbit would decay within the ten days of life-support supplies on board. The A-2 booster would put the revised spacecraft into a much higher orbit; if the retro-rocket failed, the crew would die in space. To prevent this, a second solid-fueled retro-rocket was attached to the top of the sphere.

One danger could not be dealt with under the rushed development program: there was no launch escape system. The Vostok's ejector seat was gone and no escape rocket was added. The crew was betting its life there would not be a problem on the pad or during ascent. Some of the engineers, looking at the risks, said they were working on the perfect "space-grave" for three people.

The modified spacecraft was given the name Voskhod. This was normally translated as "Sunrise" but it can also mean "East" (the same as Vostok, underlining the relationship between the two programs). Because of its political roots and limited capabilities, Voskhod was to undertake three types of mission. The first was a three-man, one-day flight. The second was a two-man

2

mission in which one cosmonaut would make a space walk using an inflatable airlock. The final type would be a 14-day mission with a two-man crew. These would cover all the projected major Gemini "firsts" in the US. The program would fill the gap until tests of the L-1 and L-3 moonships could begin, and also provide experience and biomedical data in preparation for the Moon flight.

The first unmanned Voskhod test flight was ready in early October, 1964. It was launched on October 6, as Cosmos 47. It remained aloft one day before being

2 October 2, 1964: in a picture released in 1971 Sergei Korolev is shown talking to the three-man crew of Voskhod 1: (left to right) Vladimir Komarov, Konstantin Feoktistov and Boris Yegorov. Komarov was the commander of the crew that orbited for one day on October 12, while Feoktistov was an aerospace engineer and Yegorov was a physician with considerable experience of aerospace medicine, who would probably have been the crewman aboard the Vostok 7 mission which was eventually cancelled. The run-up to the first Voskhod flight was characterized by a power struggle between leading figures in the Soviet space program and Khrushchev's political demands. Korolev had for some time been planning for manned missions to the Moon, but the three-man Voskhod missions were reportedly ordered by the Soviet leader to upstage the US two-man Gemini flights. The only way this could be done was for the crewmen to fly without pressure suits in a modified Vostok spacecraft shorn of any ejector seats or any other escape system. The method of landing in the spacecraft also had to be changed drastically, and this involved modifications to the parachute system and the incorporation of a small rocket to slow the descent of the craft just before touchdown. All in all, enormous risks were taken (although an unmanned test flight preceded the Voskhod 1 flight) and the diversion may well have cost the Soviet Union the race to the Moon.

1

2

de-orbited. The way was now clear for the manned launch.

Voskhod 1 was launched on October 12, 1964. The crew was Vladimir Komarov (commander), Konstantin Feoktistov (engineer) and Boris Yegorov (doctor). Yegorov conducted medical tests during the one-day flight, and all three cosmonauts suffered from nausea. During the flight, Khrushchev made the now-traditional phone call to the crew. It was his last public statement: afterwards, he was summoned back to Moscow and told he was being removed from all Party and government posts. When the cosmonauts returned to Moscow for the awards ceremony, they were met by the new Soviet leadership – Aleksei Kosygin and Leonid Brezhnev – making their first public appearance. Despite the government changes, the Russians reveled in the success of Voskhod 1. Pravda said: "Sorry, Apollo! Now such prophecies that the Americans will ever catch up can bring forth only an ironic smile. The gap is not closing, but increasing."

The next Soviet flight was to see the first spacewalk. The Soviets used a different procedure from Gemini. Rather than depressurizing the complete spacecraft, a small inflatable airlock was added to the Voskhod. To compensate for its added weight, one of the three seats was removed. An airlock had advantages; it required far less oxygen and nitrogen than re-pressurizing the sphere, which was much larger and at three times the pressure of the closet-sized Gemini cockpit. Also, Soviet electronics were not designed for exposure to a vacuum; they were air cooled and would soon overheat. Thus, the use of an airlock removed the need to re-qualify the systems.

The airlock and space suit were tested in a vacuum chamber. It was also decided to make an unmanned orbital test. The spacecraft would be launched, then the airlock would be deployed, the load-bearing air frames inflated and the hatch automatically opened. A space suit was placed in the airlock. It would be sealed and inflated. The spacecraft was to make several orbits with the systems being tested in various modes.

Cosmos 57 was launched on February 22, 1965. It reached orbit and the airlock was deployed and inflated. The space suit was also pressurized and the whole sequence took only a few seconds. Then Cosmos 57 fell victim to a bizarre mishap. As the spacecraft passed over Kamchatka, in the Soviet far east, it "seemed to disappear." Analysis showed that ground controllers had sent two commands to the spacecraft simultaneously rather than sequentially. This triggered the spacecraft's self-destruct charge and Cosmos 57 was blown into some 180 pieces.

This was followed by another accident: a Voskhod sphere was dropped from an aircraft to test the airlock ejection and parachute systems. The chutes did not open and the sphere was destroyed on impact. The years of imprisonment, constant work, and now these new problems took their toll on Korolev. One of his engineers recalled he was "a yellowish-green all over" and was having conferences late into the night.

World's First EVA
The decision was made to go ahead with the manned flight because another test mission would have taken a year (and missed the chance to beat Gemini to an EVA). Before the State Commission would approve the flight, KGB Chairman Vladimir Semichastny came to Tyuratam to inspect the equipment. (The KGB apparently thought sabotage may have been the cause of the accidents.) Korolev and Mstislav Keldysh (President of the Soviet Academy of Sciences) were finally able to convince the State Commission the manned flight should go ahead. All this activity was reflected in the delay between the test flights and the manned launches. Six days passed between the Cosmos 47 and Voskhod 1 launches and 24 days between Cosmos 57's mishap and Voskhod 2.

Voskhod 2 went into orbit on March 18, 1965. Its crew was Pavel Belyayev (commander) and Alexei Leonov (pilot). Immediately after reaching orbit preparations for the spacewalk began. As Voskhod 2 crossed over the Pacific the airlock was inflated and Leonov entered. The next several minutes were spent checking out the space suit and backpack. The suit was considerably different from those used on Gemini. The internal pressure of Leonov's suit was higher (this prevented the cosmonaut from getting the bends) and the backpack had its own self-contained oxygen supply. The backpack was needed because the sphere's hatch would be closed during the spacewalk and a Gemini-type hose could not pass through. In fact, Leonov's suit had all the elements of a prototype lunar model.

As Voskhod 2's orbit carried it across Africa, the airlock was depressurized and the outer hatch was opened. Leonov climbed out and floated at the end of a 16ft (5m) tether while television pictures were transmitted to a ground station.

The spacewalk lasted 12 minutes, 9 seconds. When Leonov tried to climb back into the airlock, he began having problems. The space suit had "ballooned" more than anticipated. On the ground, Leonov could climb into the airlock, blind-folded, in a minute, but now it took several attempts, wrestling with the tether, a movie camera and the bulky space suit. Leonov finally reduced the pressure in the suit, running the risk of the bends, and was finally able to squeeze in, close the hatch and repressurize the airlock. He was tired and sweat-covered from his efforts.

Voskhod 2 orbited for a day. As it neared the time for retrofire the automatic orientation system failed, and the spacecraft made another orbit as Belyayev manually oriented it. Great care had to be taken or Voskhod 2 might be sent into a higher orbit, marooning the occupants. During the orientation, the two cosmonauts had to get out of their seats. This altered the spacecraft's balance during retrofire and caused it to come down off target, near the city of Perm. This was a forested region still in the deep grip of the Russian winter. It was far from the waiting recovery forces and no word was received for several hours, during which time Korolev cried openly.

At the landing site, Belyayev and Leonov were in trouble. The sphere had landed in deep snow with its two parachutes caught in tall trees. The hatch could not be opened and a fan inside the sphere could not be shut off, making the cold even worse. Finally they were located, but when warm clothing was dropped the packages caught in the tall trees. Ultimately, a ground team hiked through the forest to reach the crew. The cosmonauts were flown out the day after their landing. They were cold but unharmed, having survived all the flight's mishaps. This was the last good news the Soviet space program was to receive for a very long time.

The next flight, Voskhod 3, was to be a long-duration, two-man flight. Four cosmonauts: Vladimir Shatalov, Georgi Beregovoi, Georgi Shonin and Boris Volynov were selected to make up the prime and back-up crews. The launch was to take place in late 1965/early 1966. Three more Voskhod flights were in the planning stages. One was to be a repeat of the Voskhod 2 spacewalk, but it was to be made by a woman. (Tereshkova is known to have trained in a Voskhod-type space suit.) Another Voskhod flight would have carried a journalist into orbit. Korolev apparently decided such a crewman could convey the experience of space travel in a way understandable to the "man in the street." Two were selected, Yaroslav Golovankov and Yuri Letunov, in July, 1965.

Result of Khrushchev's Fall
The most ambitious Voskhod flight was a second 14-day, two-man biomedical flight. A physician would be matched with a pilot cosmonaut. Also carried on the flight would be a number of rabbits and other experimental animals. Three physicians were selected as possible cosmonauts: Yevgeni Illyin, Yuri Senkevich and Aleksandr Kisilev. They undertook a five-day flight simulation in a Voskhod trainer. They had not yet been paired with pilot cosmonauts. Given their training activities, one suspects this would have been the "Voskhod 4" flight. In all probability the schedule was still tentative. In any event, all the follow-on Voskhod flights were cancelled in late 1965.

To replace the biomedical flights, Cosmos 110 was launched on February 22, 1966. Aboard were two dogs,

1, 2 Cosmonaut Alexei Leonov conducted the world's first spacewalk outside the Voskhod 2 spacecraft on March 18, 1965 – more than two months before America's first planned EVA during the Gemini IV mission. (Leonov was subsequently to assume a leading management position within the Soviet space program.) An inflatable airlock was installed aboard the Voskhod – room being made by taking out the third seat – which had some advantages for the Soviets compared with the NASA system of depressurising the entire spacecraft. Tests of the system scarcely went well and there was some talk of possible sabotage, but it was eventually decided that the mission should go ahead. Leonov's spacesuit – which in essence was required to maintain the desired pressure and supply him with oxygen to breathe – was self-contained, unlike the NASA spacesuit which was supplied with "consumables" from the Gemini craft. With Pavel Belyayev in the commander's seat and the Pacific below, Leonov entered the airlock, which was then depressurised and the outer hatch opened. He was attached to Voskhod by a safety tether and was outside for just over 12 minutes. Because the suit had ballooned more than expected Leonov had difficulty in re-entering and sealing the airlock, but this was eventually accomplished. A mishap caused the crew to land off target and there were fears for their safety for some time. But all was ultimately well and another successful first was recorded. Two movie cameras were used to film the EVA, and these two images are stills taken from the movie record.

1 *While the US Gemini flights paved the way for Apollo, with continued successes during 1965-66, no cosmonauts orbited the Earth. Sergei Korolev died in January 1966, but work on the new Soyuz spacecraft continued and – despite numerous problems during test flights – the first manned mission was fixed for April 23, 1967. The sole occupant of Soyuz 1 was Vladimir Komarov – shown here fully suited up.*

2 *The plan called for a rendezvous and docking to be accomplished following the launch of Soyuz 2 one day after the first launch, but Komarov encountered problems from the beginning and these were serious enough to force cancellation of the second mission. He was ordered to return and was killed when the landing parachute failed to deploy correctly. Here his wife Valentina kisses her husband's portrait at the Kremlin Wall, where his ashes were buried.*

Veterok and Ugolek, who spent 22 days in orbit. An advanced biomedical satellite, based on the Vostok, also began development. A number of reasons can be offered for the sudden end of the Voskhod program. As valuable as the two-week missions would have been, the Soviet space program needed information on rendezvous and docking. Voskhod would have required major modifications and it was doubtful there was that much "stretch" left in the Vostok. The lack of a launch escape system may have come to be seen as unacceptable, too. But the most important reason was a reorganization of the Soviet Moon program.

Khrushchev's fall removed Chelomei's patron, and Korolev intensified his efforts to gain control of the entire Moon effort. He was successful in late 1965. Korolev now had complete control of the Soyuz program (both Earth orbit and lunar). This brought to a close the four "lost years" of the Soviet Moon program: first, the two years of argument between Korolev and Chelomei over the L-1's flight mode, then two more years of political infighting that allowed little work. On top of this had been Khrushchev's demands for flashy, but empty, "firsts."

Then, just when it seemed possible Russia's space program might have a chance to regain the lead lost to Gemini, there was a crushing setback. Following an operation, Korolev died on January 14, 1966. The years in the camps and the unremitting effort as he sought to cope with the American technological juggernaut took their final toll. His abilities would be sorely missed in the years ahead. As an indication of the secrecy that enveloped the Soviet space program, it was only at his death that Korolev's name was made public. Before, he had only been called "the Chief Designer."

Vasily Mishin was named the new director of the design bureau. His first task was to begin work on the Soyuz/L-1 spacecraft. Although approved in late 1963, little work had been done on the project and it was soon in trouble. By early November 1966, both the L-1 and L-3 lander were judged to be unsatisfactory. Orders were sent to improve matters, but as Lev Kamanin, who was on an Air Force committee overseeing the Soviet Moon program, wrote: "But papers and shouts don't help."

It was also in November 1966 that the design of the N-1 booster was finalized. An expert commission headed by Keldysh set the payload at 209,436lb (95,000kg) and flight tests were to begin in the third quarter of 1967. (This was the same time period as the Saturn 5.) After years of delay, the N-1 switched to a "GO-GO" panic mode. One aspect of this was to adopt an "all up" testing program; all the stages would be flown aboard each test rocket. It was quicker but it also ran risks.

The final event of November 1966 was the first unmanned Soyuz test flight. This was Cosmos 133, launched on November 28. It orbited for two days before being de-orbited. During re-entry, the heat shield burned through, thoroughly damaging the Descent Module. The cause was traced to a flaw in the stopper on the forward shield, where it had been mounted on a lathe for machining. This was the first of a series of problems that would end in disaster.

The next Soyuz test was that of Cosmos 140. It was launched on February 7, 1967, and, like its predecessor, remained in orbit for two days. It was followed by two orbital tests of the L-1 spacecraft. Cosmos 146 was launched on March 10, 1967, by a D-1e booster. The following day it maneuvered into a slightly higher orbit and after a further eight or nine days it re-entered. Apparently the booster's e-stage was re-fired to accelerate it for a high-speed re-entry. This would test both the stage's ability to function for over a week in space and the L-1's heat shield. The second L-1 test flight was Cosmos 154, launched on April 8. The spacecraft orbited for 11 days before re-entering. It has been suggested the e-stage firing failed and that a normal-speed re-entry had to be made.

All three flights suffered major problems with on-board systems: a failure of the temperature control system, problems with the attitude control system rockets and burning of the parachute lines. It was clear the Soyuz was not ready for a manned flight, yet the government insisted that the next flight be manned.

The reason was the lost years of the Voskhod program. During 1965-66, ten Gemini missions were flown. These tested rendezvous and docking, navigation, EVAs and long-duration flight. No Soviet manned spaceflight had been made during this time. Clearly, leadership in space had shifted to the US, and the first Soyuz mission was aimed at overtaking the entire Gemini program.

Soyuz 1 Tragedy

The flight plan envisioned the launch of Soyuz 1 with Vladimir Komarov aboard. The following day, Soyuz 2 would be launched with a three-man crew: Valery Bykovsky (commander), Alexei Yeliseyev (flight engineer) and Yevgeny Khrunov (research engineer). Soyuz 1 would then rendezvous and dock with Soyuz 2. Yeliseyev and Khrunov would leave Soyuz 2 and spacewalk over to Soyuz 1. Once the transfer was completed, the two spacecraft would undock. Soyuz 1 would then return to Earth, followed by Soyuz 2 the following day. It was an ambitious plan, with two new, manned spacecraft in orbit at the same time, a docking and an EVA transfer by two crewmen; all in a spacecraft that had shown major flaws in all four of its test flights. It has been reported that Mishin opposed the flight plan and refused to sign the flight endorsement papers.

Soyuz 1 was launched on April 23, 1967, and almost immediately ran into problems. The second half of one of the solar panels did not unfold, thereby cutting the amount of electrical power available. The State Commission met to discuss the status of Soyuz 2. They decided to put off the launch and the crew went back to their quarters. That same evening, the State Commission met again and reversed its decision. After the Soyuzes docked, Yeliseyev and Khrunov would transfer over to Soyuz 1 and manually unfold the solar panel. The Soyuz 2 crew went back to prepare for the launch. When morning came, nobody came to wake them up; the launch had been cancelled.

It appears that Soyuz 1 was also suffering from systems problems. At one point, Komarov said, "Devil-machine, nothing I lay my hands on works." These faults were apparently too serious to permit a second launch and Komarov was ordered to return to Earth. The first attempt was made on the 16th orbit but he was not able to orient the spacecraft for the burn. The next attempt, on the 17th orbit, was also unsuccessful. Komarov was finally able to make the retro-fire on the 18th orbit.

To stabilize the Descent Module during re-entry, Komarov apparently put it into a continuous roll (like a spinning bullet). A 20 to 30rpm spin would cause it to follow a ballistic re-entry. This would double the re-entry G forces from 4 to 5Gs to 8 to 10Gs, and cause it to land some 373 miles (600km) short of the normal landing zone. (The Soyuz Descent Module, like Gemini and Apollo, could develop a small amount of lift: this reduced G forces and allowed a more precise landing.)

launch pads and assembly buildings. On July 26, 1967, NASA Administrator James E. Webb told Congress about the Soviet rocket. It was estimated to have a thrust of between 7.5 and 10m lb (16 to 22m kg). The rocket was known to the US government as the TT-5 (fifth new rocket spotted at Tyuratam). Among the public, it was called the G booster or "Webb's Giant," the latter title being popular among those who suspected it was only a NASA funding ploy.

The Soyuz 1 disaster brought man-related flights to a halt for several months. They resumed with Cosmos 186 on October 27, 1967. This was followed three days later by Cosmos 188. Cosmos 186 then began man-euvering and docked with Cosmos 188 before it had completed its first orbit. (This was the profile for the Soyuz 1/2 mission.)

The first post-Soyuz 1 L-1 test flight was made on November 21, 1967, but the D-1e booster exploded at

3

4

A ballistic re-entry did not require computer control, suggesting this was one of the problems besetting Soyuz 1. The Descent Module came down near Orsk in the Urals, but Komarov was apparently unable to halt the capsule's spin; indeed, he may have blacked out under the high G forces. In any event, when the parachute deployed, its lines were twisted and the capsule impacted the ground at high speed. Komarov was killed, but twelve hours passed before his death was announced.

That same month, the cosmonauts for the L-1 moonflights were selected. The prime crew was Alexei Leonov (commander) and Oleg Makarov (flight engineer). The backup crew was Valery Bykovsky (commander) and Nikolai Rukavishnikov (flight engineer). The support crew was Pavel Popovich (commander) and Vitaly Sevastyanov (flight engineer). It was a mixture of Vostok/Voskhod veterans and new cosmonauts from later selections.

The N-1: "Webb's Giant"
During the late summer of 1966, US intelligence detected the first signs of the N-1 project. Presumably this came from reconnaissance satellite photos of construction on

lift off. Nearby, watching the launch, were Yuri Gagarin and Andrian Nikolayev. They were caught in the fuel cloud and had to run to escape the poisonous nitric acid fumes. Chelomei's D-1 was undergoing a string of launch failures.

In January 1968, a group of 18 cosmonauts was selected to make the Moon landings. They included the L-1 crewmen chosen earlier, along with others such as Georgi Dobrovolski, Yuri Gagarin, Pavel Belyayev, Yevgeny Khrunov and Pyotr Klimuk (commanders) with Yuri Artyukhin and Georgi Grechko (flight engineers). Tragically, Gagarin was killed on March 27, 1968, in a MiG 15 crash during a training flight.

By mid-March 1968 the status of the Soviet Moon program was mixed. The first launch of the N-1 booster had been originally scheduled for the fall of 1967. This had slipped to March of 1968, but by that month there had been another slip to after May at the earliest. In fact, the whole of 1968 was spent dealing with problems. The L-3 lunar lander, too, was not ready: it was sched-uled for tests in Earth orbit during the second half of 1969 with the lunar landing being made in 1970-71.

To put this in perspective, the status of Apollo was

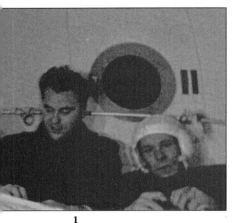

1 *In January 1969 Soyuz 4 and 5 docked, with the transfer of two cosmonauts. Here, two crew members are shown having dinner aboard Soyuz 4 after the transfer.*

2 *The Soyuz 4 spacecraft moves away from Soyuz 5 after undocking. The docking and cosmonaut spacewalks in January 1969 had originally been planned for 1967.*

also mixed at this time. The first Saturn 5 flight had been flawless, but the second revealed problems. The CSM was still being redesigned after the fire. The LM had made its first unmanned test flight in January 1968, but the rest of the year would be spent preparing it for a manned flight. The race to the Moon was still even.

There was one bright spot for the Soviet effort: the L-1 spacecraft was ready. The next L-1 test flight was launched on March 2, 1968. It was given the designation Zond 4 (Russian for "probe"). Actually, the L-1/Zond spacecraft was a stripped-down Soyuz. The Orbital Module was removed and replaced by a collar-like support ring with a dish antenna, while the Descent Module had a strengthened heat shield to withstand the high-speed re-entry. The propulsion system was also based on the Soyuz system. Power was supplied by two solar panels which were smaller than those on Soyuz.

Zond 4 was to be boosted into a highly elliptical orbit that went out to lunar distances, but in a direction opposite to the Moon. This would test communications at long distance. It would then loop back towards Earth and make a high-speed re-entry. The flight was only a partial success and the spacecraft is generally believed to have suffered a communications failure soon after it was boosted out of Earth orbit. Another Zond launch attempt was made on April 22, 1968, but the booster failed.

By late summer, the CIA knew the purpose of Zond's mission and the information was passed to NASA, which decided to change Apollo 8 from an Earth-orbital to a lunar-orbital mission. The final lap in the race to the Moon had begun.

Zond 5 was launched on September 15, 1968, on a trajectory that would send it looping around the Moon. A course correction was made on the way to the Moon and three days later the spacecraft passed 1,212 miles (1,950km) behind the Moon and began the swing back to Earth. As it neared Earth, a second course correction was made to ensure it hit the six to eight mile (10 to 13km) wide re-entry corridor. The angle had to be precise: one degree too steep would cause G forces too high to survive, and one degree too shallow would send the spacecraft skipping back out into space.

Originally, it had been planned for Zond 5 to fly a "double dip" re-entry. It would first hit the atmosphere, losing much of its velocity before skipping back out. A second entry would then be made, with the spacecraft parachuting to a landing. This approach had two advantages; G forces were kept to 4-7 Gs and the Zond could land in Russia.

Before re-entry, a problem appeared in the astro-orientation system. This prevented a skip re-entry and Zond 5 had to fly a ballistic profile. On September 21, it hit the upper atmosphere. The forces were heavy – 10 to 16 Gs – a painfully high level had a human crew been aboard. Zond 5 successfully splashed down in the Indian Ocean, where it was picked up by a Soviet ship for onward transportation to Bombay and then to Russia.

Resumption of Manned Missions
The following month, the Soviets resumed manned orbital flights. The first step was the launch of the unmanned Soyuz 2 spacecraft on October 25, 1968. (This was the same spacecraft that was to have flown on the Soyuz 1/2 linkup.) The following day, Soyuz 3 was launched with Georgi Beregovoi on board. On its first orbit, it maneuvered and closed within a few feet of the other spaceship. Several rendezvous were made but no docking occurred. Beregovoi returned to Earth after four days in orbit.

This was followed on November 10, 1968, by the launch of Zond 6. It made the Moon flyby at a distance of 1,504 miles (2,420km) on November 14. The day before, Kamanin wrote in his diary that two more L-1 test flights were needed before the Soviets could commit to a manned mission. Problems with the astro-orientation and descent control systems made them questionable. Kamanin blamed the over-automating of Soviet spacecraft. He recalled that Korolev came to feel this way shortly before his death, and estimated that an automated flight was ten times more difficult than one (such as Gemini and Apollo) that took advantage of the pilots' skills.

On the way back from the Moon, Zond 6 made two course corrections, the last only eight hours before re-entry. It hit the atmosphere on November 17, and flew the low G, double-dip profile before landing in Russia.

So the Soviets had recovered three Zond spacecraft following high-speed re-entries, two from lunar flybys. On the other hand, of the nine Zond-related flights between 1966 and 1968, several had major problems. For the Soviets, the next Zond launch window opened on December 8, 1968. Apollo 8 would be launched on December 21. After all the work, all the money, all the years, it came down to this.

Soon after Zond 6, there was a D-1 booster failure. According to one account, Alexei Leonov was "spitting fire" when he heard of this, for it would take time to sort out the problems. As Oleg Makarov (flight engineer on Leonov's crew) said years later, "Then the moment came when the big bosses got frightened they would kill people and they cancelled the project."

The same day Apollo 8 left for the Moon, Kamanin wrote in his diary: "For us the holiday is darkened with the realization of lost opportunities and with sadness that today the men flying to the Moon are named Borman, Lovell and Anders, and not Bykovsky, Popovich or Leonov."

Although the Moon flight of Apollo 8 decisively beat Zond, much remained to be done before Americans could stand on the Moon, including two manned tests of the LM. If the two N-1 test flights went well and the L-3 lander was proven, there might still be a chance for a Russian landing in 1970. If Apollo had any setbacks, the Soviets might still be first.

The Soviets' next step began on January 14, 1969, with the launch of Soyuz 4. Aboard was a single cosmonaut, Vladimir Shatalov. The next day, Soyuz 5 was orbited. Its three-man crew was Boris Volynov (commander), Aleksei Yeliseyev (flight engineer) and Yevgeni Khrunov (research engineer). Unlike earlier missions, a full day passed before the docking. Soyuz 4 was the active spacecraft, maneuvering to catch up with the passive Soyuz 5. Once the docking was completed, the Orbital Modules of both spacecraft were depressurized and Yeliseyev and Khrunov climbed out of Soyuz 5. They made their way over handrails to Soyuz 4. Oxygen was supplied from packs strapped to their legs. They climbed through the hatch, closed it and repressurized the Orbital Module. The transfer had taken about an hour. They got out of their spacesuits and joined Shatolov in the Descent Module. Preparations immediately began for undocking. Once this was completed, both spacecraft flew separately. Soyuz 4 returned to Earth on January 17, after three days in orbit and Soyuz 5 came down the following day. The Soviets hailed the flight as the "First Experimental Space Station." In fact, it was the mission intended for the Soyuz 1/2 flight of 1967. Coming after Gemini, it would have been impressive: following Apollo 8 it was an anti-climax.

USSR Fails to Beat the US

At Tyuratam, the first N-1 stood on the pad. It had a tapered first stage and cylindrical upper stages. There were five stages in all – the first three placed it into Earth orbit and such was the high degree of failure tolerance that the N-1 could still reach orbit even if two pairs of first stage engines and one pair of second stage engines failed. Once in orbit, the fourth stage would fire, sending it towards the Moon – which would be reached after a two-day coast. There the fifth stage would fire to place the Soyuz and L-3 lander into lunar orbit.

The mission commander would then don a space suit and space walk from the Soyuz, called the *Lunar Orbital Cabin* (LOK in Russian) over to the L-3 *Lunar Cabin* (LK in Russian), leaving the flight engineer in the Soyuz. He would enter through a hatch in the launch shroud and only then would the LOK and LK separate from the fifth stage.

Mishin described the LK as being very different from the US lunar module. Rather than using a joystick, the Soviet pilot had several large, primitive-looking levers for control. He had only a few seconds before landing to change the touchdown point. The commander would then leave the LK through a hatch in the rear, collect samples, set up experiments and plant the Soviet flag.

At the calculated time, the commander would fire the LK's engine. Unlike the lunar module, the LK used its main engine for both landing and return to lunar orbit. The lower part of the LK, like the lunar module, would be left behind when the spherical cabin lifted off. The LK would then dock with the LOK and the pilot spacewalk back – this time carrying lunar samples. (In retrospect, the cosmonaut transfers planned for Soyuz 1/2 and conducted on Soyuz 4/5 were rehearsals for this part of the flight). Following jettison of the LK, the LOK would fire its own engine for the return to Earth. After two days, the Soyuz would fly a double dip re-entry and land inside Russia. Total flight time would be about eight days.

One problem with Western efforts to reconstruct the history of the Soviet Moon program is that analysts must rely on the twenty-year-old memories of the Soviet participants. This can give rise to errors and incompleteness, for no official Soviet documents have been made available.

But whatever the detailed plan, the first N-1 lifted off, on February 21, 1969. For 70 seconds, it flew normally. Then a fire started in the first stage. All the engines were shut down and the rocket was destroyed. Despite the setback, Soviet cosmonauts were still optimistic about landing on the Moon. On April 9, 1969, Shatalov told a Hungarian news agency correspondent that the Soviets would require "six, seven and perhaps more months" of preparation to make a landing on the Moon. He added: "Who makes the better preparations will get to the Moon first and it is our wish to do so." Leonov was even more explicit about Soviet plans. On June 14 he said: "The Soviet Union also is making preparations for a manned flight to the Moon, like the Apollo program of the United States. The Soviet Union will be able to send men to the Moon this year or in 1970. We are confident that pieces of rocks picked from the surface of the Moon by Soviet cosmonauts will be put on display in the Soviet pavilion during the Japan World Exposition in Osaka in 1970."

It took five months to prepare the second N-1. On July 3, 1969, it was ready for launch. In the final seconds before liftoff, a liquid oxygen turbopump was being spun up to flight speed. Suddenly it disintegrated, sending shrapnel ripping through the fuel tank. Within seconds the spilling fuel ignited. The first stage then exploded. Debris rained down while the pad area was a sea of flame. The smoke cloud was visible on US weather satellite photos. Apollo 11's launch was 13 days away.

Denials of a Race to the Moon

If the Soviets could not beat Apollo 11 to the Moon, they could still try to upstage it. For several years the Soviets had been working on a third generation of unmanned lunar spacecraft. These could soft-land and carry either an ascent stage which could return a small sample of lunar soil to Earth or a small rover. It is believed that three launch attempts were made early in 1969, but all failed before the D-1e booster reached orbit.

However, on July 13, 1969, Luna 15 was successfully launched and sent towards the Moon. On July 17 (the day after Apollo 11's launch), Luna 15 entered orbit around the Moon. In the West, there was speculation that it was intended to land, scoop up some Moon dust and then return to Earth before Apollo 11; the Soviets thus being able to claim the first sampling of the Moon, at lower cost and without risk of life. The US meantime was reassured that Luna 15 would not interfere with Apollo 11, but no information was given by the Soviets about its mission.

4

3 Vladimir Shatalov – who launched alone aboard Soyuz 4 – is shown here receiving letters and newspapers brought by Aleksei Yeliseyev and Yevgeni Khrunov from Soyuz 5. Since Shatalov had only been in space for a couple of days he could have scarcely been homesick! Until the late 1980s and after, Soviet mission photos released were both few and of poor quality.

4 Shatalov during training.

5

5 The Soviet Zond missions tested the procedures demanded by manned flights to the Moon and a safe return to Earth. However, the opportunity was taken during Zond flights in the late 1960s to place a photographic sub-assembly aboard. This used film which was retrieved and processed after the return of the vehicle to Earth. The quality was inevitably much better than any television record. Zond 7 obtained this picture of the Earth from a distance of about 43,000 miles (70,000 kms) in August 1969. Soviet Central Asia is at the center of the disc, with Iran and Saudi Arabia to the south. The bottom of the frame is filled by the African continent, while most of the North Atlantic is covered by extensive cloud.

1

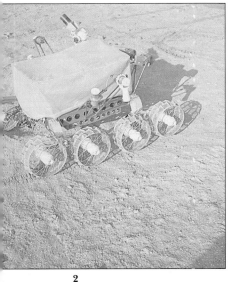

2

3

As Apollo 11 orbited the Moon, Luna 15 maneuvered several times. On July 21, following Armstrong and Aldrin's Moon walk, the Soviet spacecraft fired its engines and began its four-minute descent to the lunar surface of the Mare Crisium, where it crashed.

For the Russian people, the Apollo 11 landing was a tremendous shock. At a vacation home near Moscow, some 270 guests were gathering for breakfast when a downcast director, tears in his eyes, announced the landing. He said a film of the Moon walk would be shown at 10:30 am. As they watched the brief clip, there were more tears. All activities, such as a basketball competition and a dance contest, were cancelled. The Soviet government tried to lessen the impact. Banners lauding Soviet space leadership were removed. There was no live coverage of Apollo 11. Instead, *Through the Eyes of Musical Shows* was shown on Soviet television.

In late October 1969, Mstislav Keldysh announced a shift in Soviet space policy. In Stockholm, he said: "At the moment, we are concentrating wholly on the creation of large [Earth] satellite stations. We no longer have any scheduled plans for manned lunar flights."

In the years following, the Soviets would deny they were ever planning to send men to the Moon. It was a lie to cover the propaganda defeat. In the West, the press, intellectuals and many academics accepted the lie. They had been opposed to Apollo and the Soviet denials fitted their ideological beliefs. Between 1969 and 1974, there were many such statements:

Journey to Tranquility (a book published in 1969): " ... by 1963 it had become clear ... the race for space did not, in fact, exist."

London's *Sunday Times* (1971): "There was never the remotest chance that the Russians would get to the Moon first."

London, the *Guardian* (1971): "Five years ago, some Western observers were arguing that the 'Moon race' was a myth .... This has turned out to be the case."

Walter Cronkite (1974): "It turned out that the Russians were never in the race."

Notwithstanding the Apollo 11 triumph, the Soviet Union continued with a series of unmanned Moon flights. Two missions, Cosmos 300 (September 23, 1969) and Cosmos 305 (October 22, 1969) never progressed beyond Earth orbit. Luna 16 (September 12, 1970) was the first successful lunar sample return mission and Luna 17 (November 10, 1970) carried Lunokhod 1, a small Moon rover that operated under remote control for several months. Two more missions followed in September, 1971: Luna 18, a sample return mission that crashed, and Luna 19, an orbiter mission.

On February 14, 1972, Luna 20 was launched on a sample return mission. A year later, in January 1973, Luna 21 landed Lunokhod 2 on the Moon. The Soviets stressed the low cost and superiority of these missions, giving exaggerated cost savings. The true figure was about $100-120m for a scoop mission versus $450m for an Apollo mission. The unmanned Luna mission brought back about 2lb (1kg) of randomly selected soil compared with over 242lb (110kg) of carefully selected and documented rocks brought back by Apollo 17.

Behind the scenes, the Soviets were continuing development flights in their manned lunar program. There was a flurry of launches during 1969-71. The first was Zond 7, launched on August 7, 1969. It looped around the Moon and landed in Russia. On November 28, 1969, the first unmanned test of the L-3 lunar lander was attempted, but the D-1 booster failed before reaching orbit. The next Soviet lunar-related launch was made a year later. The series began with the launch of Zond 8 on October 20, 1970. What proved to be the last of the series looped around the Moon and splashed down in the Indian Ocean.

The following month, on November 24, 1970, Cosmos 379 was boosted into orbit by an A-2 and is thought to have been a test of the L-3 ascent stage. The full L-3 spacecraft was flown on Cosmos 382 (December 2, 1970), followed almost three months later by Cosmos 398, another L-3 ascent stage test. All the flights conducted orbital maneuvers typical of a lunar spacecraft.

Glushko and Mishin

For the past two years, the Soviets had been rebuilding the pad destroyed in the July 1969 blast. By late spring of 1971, the work was completed and the third N-1 stood ready. The launch took place on June 27, 1971. For seven seconds everything worked perfectly and the booster climbed several hundred feet above the pad. Then the rocket began rolling. This became too great for the steering rockets to control and the launcher fell back onto the pad, destroying it a second time.

The last test of the L-3 Ascent Stage was made by Cosmos 434, launched on August 12 1971. As with the other ascent stage tests, the launch vehicle was an A-2. The fourth (and final) N-1 launch was made on November 23, 1972. The first stage completed its 107 second burn and shut down on time. As it coasted before second stage separation and ignition, there was a malfunction and a massive fire started in its tail. For a fourth time there was a huge blast.

Despite the setbacks, Mishin and the others working on the project were still optimistic over the N-1's ultimate success. Four or five tests for a new rocket were normal. (The first three SS-6s blew up, yet it became the workhorse

4

1 *A pre-flight image of the "automatic station" Luna 16. Launched to the Moon in September 1970, this was the first successful lunar sample return mission.*

2 *Luna 17 two months later carried the Lunokhod 1 remote control lander, which operated for several months. The vehicle is shown here operating at a test site on Earth – a "lunodrome."*

3 *Zond 7 imaged the Moon as well as Earth. This picture, taken from a distance of about 6,200 miles (10,000 kms), is of an area on the border between parts of the lunar near side seen from Earth and of the far side never seen. The Ocean of Storms is to the east (right). Just above left of center are the three "old" superimposed craters Russell, Struve and Eddington. Beneath them and slightly to the east are the much smaller, sharply delineated and younger, circular craters Krafft and Cardanus. The craters Hevelius and Cavalerius are just visible on the terminator (line between lunar night and day) at the lower right corner.*

4 *Zond 5 photographed the Earth in September 1968. The Red Sea, Saudi Arabia and the Gulf area are easily seen at top right from the spacecraft distance of around 56,000 miles (90,000 kms), and much of the Mediterranean is clear, although most of Europe is blanketed by a large depression. As in most views of this kind, the shape of the African continent is quite unmistakable.*

of the Soviet space program.) Two more N-1s were built. In early 1974, it was planned to launch the first in August and the second at the end of the year. If they succeeded, the N-1 would be declared operational and Soviet manned lunar flights could begin in 1975.

In 1991, the payloads for the N-1 test launches were revealed. The 1969 launches (called 3L and 5L) each carried an L-1 Zond and a mass model of the LK. The 1971 flight (6L) carried no useful payload. The 1972 7L mission carried a Zond and LK mass model. Had the 1974 8L and 9L missions been flown, they would have carried operational LOK and LK spacecraft.

However, the politics of the Soviet space program intervened. Mishin had undergone several stays in the hospital, and Glushko saw his chance. He harped on the N-1's failure and said policy and personnel had to be changed. In May, 1974, he succeeded. On his first day out of the hospital, Mishin was told he was being "retired" and Glushko, who was named as his replacement at the Korolev bureau, ordered him to turn in his security pass. On his first day Glushko signed a decree cancelling the N-1. Soviet space history was then re-written, eliminating any mention of Mishin, the N-1 and the whole lunar program. Not until Glushko's death in

January, 1989, would the Soviets acknowledge the existence of the N-1.

With the end of the N-1, Soviet unmanned lunar exploration petered out. Luna 22 (an orbiter) was launched on May 29, 1974 and Luna 23 (a sample return mission which crashed) followed on October 28. The final flight, Luna 24 (a sampler), was made two years later on August 9, 1976.

In retrospect, the quest for the Moon stretched Soviet capabilities to breaking point. Had a manned flight been made, there would have been a very high probability the crew would have died. The one-man L-3 lander would have put a tremendously high work load on the pilot. In all areas – boosters, spacecraft, navigation, communications and computers – there was too small a margin to survive an Apollo 13-type failure.

In the end, it came down to this. Khrushchev challenged the US in space: Kennedy accepted and set the Moon as the goal. He was betting US prestige that the Soviets would fall short. Zond, the L-3 and N-1 proved him right. And yet, the race went to the final lap. The US won and the Soviets lost. For the rest of the 1970s and into the 1980s, both sides would have to live with that outcome.

4

MERCURY AND GEMINI – THE ROAD TO APOLLO

1 T. Keith Glennan was NASA's first Administrator – from August 1958 until replaced by a President Kennedy appointee in January 1961. He graduated in electrical engineering from Yale University and worked as a sound engineer in Hollywood before moving on, after war research work for the US Navy, to become President of the Case Institute. He joined many boards and committees in the atomic energy, defense and aerospace industry sectors, and it was his reputation as an administrator that led President Eisenhower to appoint him to the leadership of NASA.

With the launch of the first satellite, it was inevitable that attention would be directed quickly to the possibility of putting man into space. While there would be many problems to overcome, considerable work had already been done and nowhere more so than in aerospace medicine. In high performance, high altitude aircraft; in balloons and in laboratory experiments, man had already demonstrated that weightlessness, the high "G" loads resulting from extreme acceleration and deceleration and living in a controlled atmosphere and pressure environment, for limited periods at least, were not only survivable but need not prevent him from operating efficiently. However, areas of continuing concern for spaceflight included noise and vibration levels at launch; radiation levels above the atmosphere (although Explorer 1 in 1958 indicated that there should be little danger in a low-Earth orbit of around 100 miles high); and the potential physical, and particularly psychological, effects of enforced confinement in a restricted space over a lengthy period, where communication with the outside world was limited.

Much of this work had been done by the military, and the US Air Force was not slow to claim the lead in a unified US space program, although the other major service arms were by no means uninterested. Less than a week after the launch of Sputnik 1, an eventual manned mission to the Moon was recommended by a USAF committee. By the beginning of 1958, the service was proposing a long-term plan which included a "manned capsule test system," then manned space stations and an eventual manned base on the Moon. In March of the same year the USAF published detailed plans for a manned spaceflight to the Moon and back, and this was followed in April by a proposal for a "Man-in-Space-Soonest" project to orbit a ballistic capsule (as distinct from a vehicle which would "skip" or glide in a controlled return through the atmosphere), first carrying instruments, then a primate, and finally man. In the summer of 1958, von Braun's group at the Army's Redstone Arsenal proposed in Project Adam that by the following year a man be launched on a sub-orbital trajectory by a Jupiter IRBM (Intermediate Range Ballistic Missile).

In its plans, the USAF sought the cooperation of NACA – the National Advisory Committee for Aeronautics. Formed by the US Government as long ago as 1915, NACA had established a reputation that was second to none as an aeronautical research organization. While many of its staff of engineers and scientists (numbering 8000 by the late 1950s) had adopted a somewhat skeptical approach to the – as they saw it – new "Buck Rogers" world of astronautics, others embraced it with enthusiasm. This was to be a sign for the future, because in emphasizing that the US activities in space should be directed to peaceful purposes, President Eisenhower in April 1958 proposed the establishment of a new civilian agency which would absorb NACA and would assume responsibility for all space activity other than the directly military. Congress concurred and on October 1, 1958, NASA – the National Aeronautics and Space Administration – came into being. T. Keith Glennan, an experienced administrator, was placed in charge, with Hugh Dryden, the respected former director of NACA, as Deputy Administrator.

Solving Re-entry Problems

Considerable work had already been done by NACA engineers – for example, Maxime Faget working in the Pilotless Aircraft Research Division at the Langley Laboratory in Virginia under the direction of Robert R. Gilruth – on some of the major problems posed by sending a man into space. Research in the early 1950s had shown that the solution to the high temperatures experienced on re-entry into the atmosphere (first encountered during the development of ballistic missiles) perhaps surprisingly lay not in streamlining – which simply distributed heat along the length of the vehicle – but in adopting a blunt forward surface. A "heat sink" surface composed, for example, of beryllium would absorb the intense heat, while a ceramic material such as fiberglass was another possibility, in which the material would "ablate," or vaporize during re-entry, carrying the heat away with it. The concept needed to be worked out, but it helped make the idea of a ballistic capsule containing a man entirely credible to the NACA engineers. What had been dubbed "a man in a can on an ICBM" had the attraction of greater simplicity and less weight than far more sophisticated and complex space vehicles that had been proposed, and it therefore promised the fastest route into space.

By late 1957, Faget and his colleagues were modifying their ballistic satellite or capsule design after extensive study in wind tunnels, and the recognizable shape of the vehicle that was to take the first American astronauts into space had appeared by the late summer of 1958: a blunt and curved face (for re-entry), a truncated cone (the instrumentation and crew compartment) and a cylinder on top of the cone containing the parachute equipment which would return vehicle and astronaut safely to the surface. At about this time plans were drawn up for a solid-fuel tractor escape rocket which would be mounted on top of the capsule to pull it up and away from the launch vehicle should it malfunction during the first few minutes of flight. In addition, Faget proposed a lightweight couch within the capsule which would be contoured to the individual astronaut's body shape, and which experiments showed would enable him to withstand as much as 20Gs.

Thus, even before NASA formally came into being, some solutions to the many problems of putting a man

spacecraft: two forms of protection against the hostile space environment. The spacesuit needed to be designed, and the environment of the capsule chosen: should it be a "shirt-sleeve" atmosphere, with a mixture of oxygen and nitrogen, or one like that in high-performance military aircraft, with pure oxygen breathed and a pressure of 5psi ($0.352kg/cm2$)? The latter was chosen largely because of the system's reliability and weight saving capability. Early on it was decided that, because of the nature of ballistic re-entry, a landing on water offered greater safety and flexibility. The equipment and procedures for such a landing (including the deployment of a large force from the US Navy and other services, which for the last Mercury missions totalled around 18,000 men) had to be developed.

Continual monitoring of the vehicle and its systems

2

3

into orbit and returning him safely to Earth were beginning to appear. In November 1958, Gilruth was placed in charge of a small group of mainly engineers who were to form a separate organization to develop the manned space flight project and who were henceforth to be known as the Space Task Group (STG). Before the end of the year the project had a name, Mercury, for the winged messenger of the Gods.

Redstone and Atlas Launchers

As Gilruth and his team set out on the task of giving Mercury wings, they were faced with evolving an often new and infinitely-detailed technology of great complexity. A major issue was "man-rating" the Redstone launcher and the then largely unproven Atlas missile. The Redstone, which, with a thrust of only 75,000lb (34,020kg), would be used for the early sub-orbital missions, was a direct descendant of the V-2, and as a medium-range missile at least was a proven system. Initially, eight Redstones were ordered early in 1959 at a cost of $1m each. More power was needed, however, to accelerate a capsule to the speeds whereby it could enter orbit. This was to be provided by the USAF's new intercontinental ballistic missile, the Atlas, with a total thrust of 360,000lb (163,300kg). Standing almost 80ft (24.4m) tall and with a diameter of 16ft (4.9m) at its base, an Atlas could lift a payload of 1.5 tons, and nine of the launchers were ordered initially from the Convair company at a cost of $2.5m each. The Atlas was eventually to prove one of the most reliable and long lived of US launchers, but when the STG placed its order there were many trials and tribulations to come before that stage was reached.

There were many other major decisions to be taken and systems to be developed. The astronaut would wear a pressurized space suit and go into space in a pressurized

4

41

in space, and communication with the crewman, were recognized at an early stage to be essential, and this led to the creation for Mercury of a communications and tracking network formed of eighteen ground stations around the world linked by more than 170,000 miles (273,590km) of hard-line circuitry at an estimated cost of over $40m. Computers played a vital role in this network. The detailed means by which a mission was planned, and then every eventuality during the flight allowed for, led to the evolution of a highly-developed flight control system which was perhaps to reach its zenith during the moon missions which were to come. In building the machines that were to take men into space, the industrial contractors – for example, the McDonnell Aircraft Corporation which in January 1959 won the contract to build twenty Mercury capsules – had a totally new reliability and quality control experience forced upon them by the overriding demand to achieve maximum safety for the astronaut in a hostile environment. Although by later standards it was a relatively simple vehicle, there were still more than 40,000 critical parts in the Mercury capsule, and it was at this time that industrial contractors set aside "white rooms" in the drive to achieve surgical-like cleanliness in components.

As this very new world evolved, it was inevitable that

2 *Maxime Faget was a National Advisory Committee for Aeronautics engineer who transferred to NASA on its formation and who subsequently played a prominent role in design and development work, which spanned the years from the Mercury capsule to the space shuttle – and beyond.*

3 *James E. Webb (left) was Administrator of NASA from February 1961 until October 1968 – the eve of the first Apollo flight to the Moon. A lawyer by training, he occupied numerous leading posts in industry and government. Webb is shown here at a presidential mission briefing with Hugh L. Dryden, his deputy, who died a few months after the picture was taken in February 1965. Dr. Dryden subsequently had a NASA space center named for him.*

4 *"Little Joe" was a cluster of solid rockets devised by NASA engineers as an economic means of testing Mercury equipment and procedures. This launch took place on October 4, 1959.*

1 *Launch of "Big Joe" aboard an Atlas vehicle in September 1959. The 2,000 mile ballistic flight was a critical test that led to the ablation shield system being adopted for the Mercury capsule.*

2 *NASA technicians complete preparations for the Little Joe Five launch from NASA's Wallops Station in Virginia.*

1

there would be many setbacks and failures. But from the beginning NASA conducted most of its activities, and particularly its launches, with a complete openness which was admirable and contrasted strongly with the Soviet policy. While the American national fervor to reply to Soviet space initiatives led to sharp criticism from the media and politicians when failures and delays occurred, there was nonetheless an underlying sympathy for the NASA engineers and managers as they fought to catch up with the Soviets, who had achieved their position of superiority as the result of a single-minded drive conducted behind a cloak of almost total secrecy. Nikita Khrushchev secured the maximum propaganda value from the Soviet ability to lift massive satellites into space, but if the STG managers and engineers had any real doubts about their future (early on, manned spaceflight took only about one-eighth of NASA's total budget, and at first no long-term plans had been made for post-Mercury activities), two associated events must have heartened them.

Support from the President

John F. Kennedy won the 1960 presidential election, and from the beginning he regarded space as a new frontier *par excellence*. Despite criticism of the failures and delays being encountered by Project Mercury, he

3

3 *James Chamberlin was chief of the Space Task Group Engineering Division, who played a crucial role in the design and development of the Gemini spacecraft.*

4 *A little-seen informal group portrait of the "Mercury Seven" taken at Ellington AFB (located close to the Manned Spacecraft Center) in 1963. From left to right the astronauts are Alan Shepard, Virgil Grissom, John Glenn, Gordon Cooper, Malcolm Scott Carpenter, Walter Schirra and Donald "Deke" Slayton.*

5 *Walter Schirra enters the gondola of a centrifuge. The effects of acceleration and other forces during a centrifuge ride are well shown in the sequence of images in* **6**.

4

never wavered in his support for it and the hard pressed members of NASA. Following one major disappointment he declared "We are behind … the news will be worse before it is better and it will be some time before we catch up." On another occasion, he affirmed that decisions on whether to go or not to go had to be left to those "making the judgement," whom he would continue to support. This was reassuring for the managers and engineers, but they were in fact to show great steadfastness: as one stated in the face of criticism "We feel no urgency to move the program unsafely."

The second event was Kennedy's appointment

(following the advice of Vice-President elect Lyndon B. Johnson, a long time proponent of American space projects in the Congress) of James E. Webb as NASA Administrator. Trained as a lawyer, he had held high office in governmental and industrial circles and had worked with both engineers and scientists on committees and other organizations. In the challenging times that NASA, and particularly the Mercury team, faced, a better leader, demanding but encouraging, and arguing the NASA case at every level in the land, including at the White House, could not have been chosen.

And what, finally, of the men who would ride Mercury

into space? Late in 1958 an elaborate proposal for a wide-ranging selection process was drawn up, but it was quickly scotched by President Eisenhower who, despite placing manned spaceflight firmly in civilian hands with the formation of NASA, decided that there was a ready-made pool of potential astronauts in the cadres of military test pilots. To be considered, candidates would need to be less than 40 and in excellent physical condition, less than 5ft 11in in height (because of the limited size of the Mercury capsule), possess a bachelor's degree or equivalent, and be graduates of test pilot school with at least 1,500 hours total flying time as well as being qualified to fly jets. Within a few weeks more than 100 qualified pilots had been identified, and a series of demanding physical, psychological, technical and other tests had begun, which led on April 9, 1959, to the NASA Administrator introducing the chosen seven "astronauts" to the American media and public. They were Lt Cmdr Walter Schirra, Lt. Cmdr. Alan Shepard and Lt. Malcolm Scott Carpenter from the Navy; three captains in the USAF: Donald "Deke" Slayton, Virgil Grissom and Leroy Gordon Cooper; and a Lt. Col. from the US Marines: John Glenn.

Long before they ever flew in space an aura of great glamour descended upon the astronauts. The media and public displayed an insatiable interest in their every

The "Seven" entered a cycle of training and of learning which became a pattern for those who were to follow them. There were lectures on such subjects as spaceflight theory and astronomy. But each astronaut was also given responsibility for an intensive scrutiny of a particular aspect of the Mercury systems and procedures on which he represented the astronaut viewpoint, often to constructive effect. They rode cen-

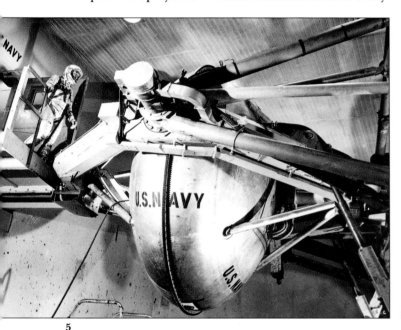

5

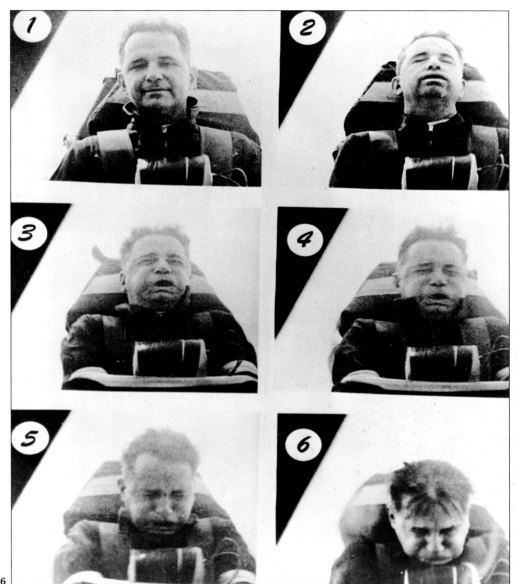

6

7 *Flight Director Christopher C. Kraft in the Mercury Control Center during a manned mission in 1962. He subsequently became director of the Johnson Space Center, Houston.*

thought and act (a $500,000 contract for the personal stories of the astronauts and their wives was negotiated on their behalf with *Time-Life* in August 1959) and there was a tendency to regard them as an ideal of American manhood – possessing "whiter than white" characters as befited heroes who would ride a chariot into the heavens. This they were most certainly not: they were highly-trained, well-motivated and very competitive individuals who were nonetheless far from being paragons of every virtue, whatever the media (and to some degree NASA) might wish them to be.

When they first reported to STG at Langley, the astronauts' precise function was still ill-defined, and there were engineers and others who regarded their future role as that of being passengers in largely automated flights. That was never likely to have been accepted by experienced test pilots, and as Mercury progressed it became obvious that the astronaut had an important role to play in correcting malfunctions in unreliable automated systems. Indeed, at least two of the subsequent Mercury missions ended safely and successfully under the enforced manual control of the astronaut.

trifuges and spent increasing amounts of time in the simulators, where they eventually began to train for their specific missions, so that their responses during the actual spaceflights became almost a matter of reflex actions. For example, before his sub-orbital mission in 1961, Shepard reportedly "flew" 120 simulated Mercury-Redstone flights. When the test came, and whatever the early jibes about being mere passengers or of being on a par with chimpanzees, the value of an active human role in spaceflight was to be powerfully demonstrated.

While many systems could be tested in the laboratory, the ultimate test was their ability to withstand actual launch forces and to function reliably in the space environment; on the assumption, that is, that the launch vehicle itself performed satisfactorily and that it could be mated successfully to the capsule. Thus, from August 1959 onwards there began a series of test launches, some of which were total failures and others which were partial failures. All, however, contributed valuable experience, even if this was no solace to an American public which was eager to begin reestablishing the nation's preeminence in the world of high technology.

"Little Joe," which could be launched from the   7

*Some critical phases of the testing to prove Mercury equipment and procedures were conducted using monkeys and chimpanzees as passengers. In **1** the monkey Miss Sam has been carefully inserted into her life support couch before launch aboard a Little Joe rocket.*

1

NASA station at Wallops Island off the Virginia coast, was a cluster of solid rockets conceived by members of the STG as an economic means (at about one-fifth the cost of a Redstone) of proving the escape and parachute systems; the latter proving surprisingly difficult. There were eight Little Joe launches in all, most of which took place in 1959-60, including the two sub-orbital flights of the monkeys Sam and Miss Sam. "Big Joe" was a single launch in September 1959 of what was in essence an Atlas, and the protection afforded the test capsule by its ablation shield led immediately to the adoption of this system for the orbital missions to come.

But it was the Redstone and then the Atlas launchers that would carry man into space, and it was problems with these and in marrying them successfully to the Mercury capsule that resulted in the many delays which earned NASA increasing criticism from Congress and the media. The launch of MA-1 (Mercury Atlas) on July 29, 1960, was a failure because of structural weakness at the interface between the launcher and the capsule. Worse was to come, however, when less than four months later, and with a dynamic new President, John F. Kennedy, just elected, the first test launch of

Grissom and Shepard, in alphabetical order" had been selected to begin concentrated training for the initial Mercury manned flights. The first of these flights (MR-3) had been set for April, but von Braun's request for one more development flight of the Redstone was approved, even though it was assumed that a Soviet attempt to put a man in space was imminent. The Redstone flight took place successfully on March 24, but less than three weeks afterwards Yuri Gagarin became the first human to venture into space.

Despite the inevitable and intense disappointment, the STG held to its step-by-step plan, and a further Mercury-Atlas as well as a Little Joe flight took place before Alan Shepard, aboard *Freedom 7,* finally became the first American in space on May 5, 1961. His flight lasted 15 minutes 28 seconds, he rose to a height of over 116 miles (187km) and reached a speed of 5,134mph (8,262kph). The sub-orbital flight could not match the Soviet achievement, but America was on its way. (One of the most intriguing if prosaic aspects of this mission received no publicity at the time. The mission, like many before and since, had been subject to numerous delays and Shepard was in the capsule for more than four hours

2   3   4   5

2-5 *As NASA struggled to overcome the challenges presented by manned spaceflight and to respond to Soviet achievements, failures were inevitable. Perhaps the lowest point was reached on November 21, 1960, when – with a new President just elected – the first launch of the combined Mercury-Redstone vehicle failed in a manner which unfortunately invited ridicule. An explosion would have at least been a dramatic failure, but MR-1 rose just a few inches from the pad, settled back and then pathetically (as it seemed) launched its escape tower. The cause was a simple malfunction and success quickly followed.*

a combined Mercury-Redstone (MR-1) failed in the most embarrassing circumstances. The vehicle rose just a few inches, settled back on the pad and then launched its escape tower. The cause, a sequence of events begun by a cable connection that was too short, was quickly located, but the spirits of most members of the Mercury team must have been low. Nonetheless, despite one reference in the press to "Lead Footed Mercury," in less than a month the second attempt (MR-1A) was successful, and the incineration of two dogs in a Soviet space failure was a further reminder of the unforgiving nature of spaceflight, which knew no national boundaries.

The Flight of Freedom 7

The new year began on a better note. The chimpanzee Ham was launched on a sub-orbital flight (MR-2) lasting more than 16 minutes at the end of January, and the success of MA-2 on February 21, 1961, caused a colleague to comment that "Gilruth became a young man again." The STG chief indicated that Mercury was reaching a critical stage when he announced that astronauts "Glenn,

before it was launched. The planning for what should have been a very short mission did not include any provision within his spacesuit for urine collection and before launch, and after consulting Mercury Control, he was forced to relieve himself. The urine pooled toward his back and its passage was recorded by his bio-sensors. Not surprisingly, a urine receptacle was hastily prepared in time for the next mission.)

While Shepard duly received the plaudits of President, politicians and public alike, Virgil Grissom prepared for his sub-orbital flight (MR-4), which took place on July 21, 1961. This was basically similar to the Shepard mission, but Grissom's *Liberty Bell 7* was equipped with a new explosive hatch cover to speed up the pilot's egress. While awaiting the recovery teams as the capsule floated on the Atlantic this hatch cover blew, and in the ensuing drama the capsule sank and Grissom came close to drowning. The reason for the premature firing of the hatch cover was never satisfactorily explained, but it did no harm to Grissom's NASA career.

First American in Orbit

Having accomplished two sub-orbital missions it was time to move on to the orbital flights launched by the Atlas. A 15 minute ride up and down was one thing, but sending a man into orbit demanded more not only of him, flight controllers and others but also of the capsule, which was required to perform critical maneuvers either automatically or under the control of the astronaut and to sustain the astronaut alive and well for a far longer time than during a suborbital flight. Inevitably this meant further development flights, even as the success of the Soviet space program was underlined by the 17-orbit flight of Gherman Titov on August 6, 1961.

In mid-September a Mercury-Atlas, one-orbit flight with a "robot" pilot aboard was successful, and, although a test of the new world-wide tracking system using a Scout launcher was unsuccessful on November 1, the way in which the problems encountered during the MA-5 orbital flight of the chimpanzee Enos on November 29 were overcome meant an American should soon go into orbit, though Robert Gilruth warned early in December that minor problems would postpone the event until 1962.

7

9

6

8

10

Toward the end of January a ritual began at Cape Canaveral that was to take place many times thereafter: the gathering of hundreds of members of the media to cover a NASA launch. On this occasion there was a delay of over two weeks (such delays were also not to be unknown in the years ahead) as first a launcher problem was corrected and then bad weather intervened. But finally, on the morning of February 20, 1962, John Glenn rose into the skies above Florida to orbit the Earth three times in his *Friendship 7* capsule, with an estimated 100 million Americans watching the television coverage. Both astronaut and craft performed well, although drama was provided by an erroneous instrument indication that the spacecraft's heat shield was no longer locked in position. Fixed to the front of the heat shield were the retro-rockets which were fired to slow the spacecraft down and thus commence the return to Earth. The retro-rocket package should have been jettisoned after this action but Glenn was advised not to do so because the straps of the package would help to keep the heat shield in place during re-entry if the signal was accurate. 11

The recovery of US manned space missions at sea was to become almost commonplace as the years passed. In 6, *John Glenn's Friendship 7 is hauled aboard the destroyer USS Noa; in* 7 *two helicopters hover above Walter Schirra's Sigma 7 (at left); and in* 8 *America's first man into (sub-orbital) space, Alan Shepard, is "recovered" on May 5, 1961, by a US Marine helicopter.*

9 *Shepard's flight time was almost 5.5 minutes following lift-off aboard the MR-3 vehicle.*

10 *Many hours of sustained training were required of the astronauts, and Alan Shepard looks pensive in this candid portrait.*

11 *Virgil Grissom is greeted aboard the USS Randolph after a mishap to his Liberty Bell capsule which almost caused his death by drowning.*

1

2
3

It was not, and Glenn returned safely to a hero's welcome – President Kennedy from the White House referring to space as a "new ocean" upon which the US must sail and "be in a position second to none."

Because of doubts about a heart condition, Slayton was removed as pilot for the next three-orbit mission (MA-7), his place being taken by Scott Carpenter. More experimentation and science were planned for this mission, and it was something that the astronaut welcomed with enthusiasm. In fact, during the flight of *Aurora 7* on May 24, 1962, Carpenter concentrated on the experiments, including photography and a study of liquids in micro-gravity, as well as the view from orbit, at the expense of paying close attention to his fast-diminishing supplies of fuel, about which he was cautioned from the ground. The capsule was in an incorrect attitude when the retro-rockets were fired, with the result that Carpenter landed more than 200 miles (322km) off target. The recovery forces knew

The experiments were few, the astronaut delighted in an understated, cool approach to everything ("Same old deal, nothing new, might as well be in an airplane"), he conserved fuel supplies like a miser, and when he splashed down in the Pacific less than five miles (8km) from the aim point few could disagree with his own verdict that it was a "textbook flight."

Mercury was coming to its end, but there was a pause of some months before NASA pushed the existing capsule (which was now being studiedly referred to as a "spacecraft") to its limits in the 22-orbit flight (MA-9) piloted by Gordon Cooper. *Faith 7* was launched on May 15, 1963, in what the astronaut described as the mission of a "flying camera," in which "all I do is take pictures, pictures, pictures!" He conducted other experiments as well, and surprised those back on earth by stating that from orbit he could see individual houses if the lighting and background conditions were optimum. He too, like all the astronauts, saw Glenn's "fireflies"

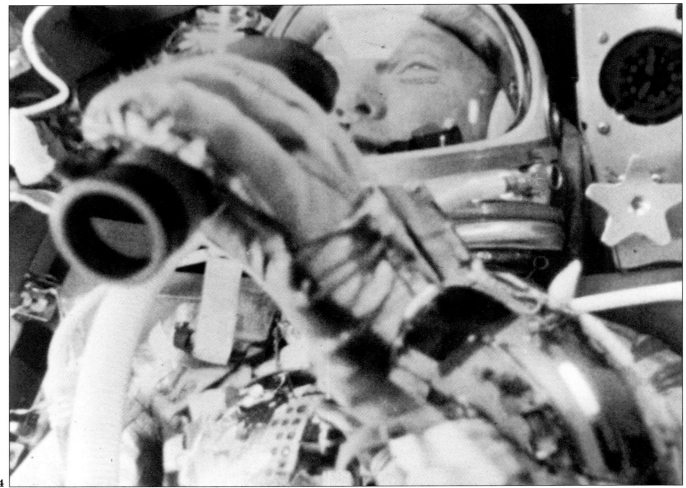

4

almost at once where he had landed but it took several hours for him to be reached and for full radio contact to be established. Carpenter was never in any danger during this period – he was relaxed and enjoying himself – but while the mission had provided valuable additional experience which could be applied in later missions, it seems that some of his astronaut colleagues and members of NASA management regarded it as a sub-standard performance and he never flew in space again. In this, he received much harsher judgement than Grissom, whose performance over the blown hatch was arguably worse.

What was required for the next mission (MA-8 was to be one of six orbits, thereby requiring great attention to ensuring adequate supplies of electrical power and oxygen and to the removal of carbon dioxide from the cabin) was an uneventful flight, and this was provided by Walter Schirra aboard *Sigma 7* on October 3, 1962.

5

6

3 *Following recovery by the destroyer USS* Noa, *Glenn was taken by helicopter to the USS* Randolph.

4 *Before the jubilation there was work to do: here Glenn is recorded by an on-board camera during the mission as he uses a photometer to study the Sun.*

5 *President John F. Kennedy and John Glenn share the delights of success two days after America's first orbital flight. At Cape Canaveral's Launch Complex 14 Glenn here presents his president with a VIP Launch Crew Hard Hat.*

in the spring of 1963 only 500 of the 2,500 NASA staff in temporary accommodation in their new location near Houston in Texas (the STG had by then been redesignated as the Manned Spacecraft Center, led by Gilruth) were working directly on Mercury. The forward planning for Mercury in January 1959 envisaged the final mission taking place in August 1960. It took place almost three years late at a total cost estimated at $450m. It was all part of a learning and development experience which pushed men, management skills and technology to the limits. Soviet missions about the time of Gordon Cooper's flight were achieving four times as many orbits with two craft (one with a woman aboard) in space at the same time. But the gap had closed and the race was on with a vengeance.

GEMINI

In retrospect, the progress of US manned spaceflight from Mercury through Gemini to Apollo and beyond might suggest a highly-structured, long-term plan, but this was not the case. Although it is going too far to describe Gemini (as did an official history) as "something of an afterthought," there is little doubt that it reflected a general desire simply to keep Americans in space in the lengthy period between the ending of Mercury and the commencement of Apollo. More specifically, however, it was the product of two quite separate strands of thinking and planning within NASA that eventually coalesced, with the object of perfecting a number of spaceflight techniques that were essential to a lunar landing.

Within NASA the concept of missions to the Moon had received attention at a very early stage. A ten-year plan published at the end of 1959 proposed a circumlunar flight, with exploration of the Moon and of the nearer planets to follow in the ensuing decade. The assumption was that the mission would be launched direct from Earth, which would require an enormous booster compared with anything then existing. The vehicle specification had been sketched and even given a name: Nova. Direct ascent continued to be favored by top NASA management for much of 1960-61, but it came to be challenged both by Wernher von Braun (who, with his group, had been transferred from the US Army to NASA and was now director of the newly named George C. Marshall Space Flight Center at Huntsville, Alabama) and by engineers and scientists at Langley, where the lead was taken by John C. Houbolt. They argued at length in favor of employing "rendezvous techniques". In Earth orbit, this would employ less powerful boosters than Nova to launch spacecraft sections which would then be joined together and sent on to the Moon – although Houbolt began to argue strongly in favor of rendezvous in lunar orbit, where a dedicated lunar lander would go down to the surface and then come back up to a spacecraft that remained in orbit throughout, prior to acting as the Earth return craft. This received scant support at first, but rendezvous generally was attractive because it promised a much earlier and more flexible method of reaching the Moon than a single, massive development project like Nova.

An Apollo project to orbit the Moon was announced by NASA in July 1960, but at this stage, with the first Mercury sub-orbital lob of a man into space months away, considerations of direct ascent or rendezvous were almost academic. But all this changed in May 1961, when John F. Kennedy threw down the challenge of achieving a lunar landing and return before the end of the decade. A decision about direct ascent or rendezvous was now urgent, because clearly the method of going to the

7

8

6 *Another view of Glenn's launch seconds after the Atlas booster engines had been ignited.*

7 *Walter Schirra enjoys the prospect of his flight as he suits up on October 3, 1962.*

8 *Scott Carpenter is fully suited up aboard his Aurora 7 spacecraft in this picture taken before launch on May 24, 1962.*

and Carpenter's "frostflies" (particles flying off the spacecraft) and had no problem in sleeping soundly at the planned time during his 34 hours 20 minutes in space. The uneventful nature of the flight, however, was not to last, and a crucial electrical failure forced Cooper to control the re-entry manually, which he did to perfection, landing in the Pacific just four miles (6.5km) ahead of the prime recovery ship. It was perhaps a sign of things to come that Cooper gave the first and only "scientific debriefing" following a Mercury mission. In a totally different setting, his ticker-tape welcome along Broadway was witnessed by a crowd of four-and-a-half-million people.

The astronauts, especially Alan Shepard, hoped for one more long-duration Mercury mission. It would have provided useful additional experience, but it was not to be. The Moon and Project Gemini were already beckoning, and as preparations went ahead for MA-9

1 *The most important of the techniques proved during Project Gemini and critical to Apollo was that of rendezvous between spacecraft. A debate centered on whether lunar missions should be launched complete and direct from the Earth; be assembled in Earth orbit following several launches; or indeed involve the rendezvous of a lander craft and a "mother ship" in lunar orbit. The last was the most elegant solution and was strongly advanced – ultimately successfully – by Dr. John C. Houbolt, chief of the Theoretical Mechanics Division at NASA's Langley Research Center.*

1

2 *An early model of the Rogallo wing attached to a Mercury spacecraft. The inflatable and steerable device was scheduled to replace Gemini's ring sail parachute in the later missions, and with its 45ft spread to enable astronauts to guide their spacecraft to a horizontal landing on land. The wing or paraglider project was cancelled in August 1964.*

Moon would affect virtually all of the equipment to be developed and the procedures to be followed. While the Nova concept was theoretically the simplest it depended on a brute force of 12m lb (5,443m kg) of thrust which, even if achievable, must lie many years in the future. Rendezvous, however, despite the problems that would almost certainly be discovered, was a much more manageable prospect. By the end of 1961 it had won the day; and, moreover, the ultimate choice was to fall on lunar orbit rendezvous.

But it was one thing to propose rendezvous and quite another to translate it into practice quickly. The means lay in the second and separate strand of development to which reference was made above. Consideration of post-Mercury manned spaceflight began almost as early as Mercury itself, and in February 1961 Gilruth appointed James A. Chamberlin, a Canadian engineer who was chief of the Space Task Group Engineering Division, to study with the McDonnell company ways of improving Mercury for future manned spaceflight programs. There were many contributions from many quarters – Maxime Faget, for example, suggesting at a very early stage that a two-man version be considered – but Chamberlin tackled the task with great relish and skill and presented his first proposals within six months. A maneuverable "Mark II" two-man spacecraft would be capable of long-duration flight and of rendezvous and docking, with a controlled, precision landing on land. At 430,000lb (195,000kg), the Titan II ICBM offered greater thrust than the Atlas for an inevitably heavier spacecraft, moreover using safer and more manageable propellants. The target vehicle already existed in the Agena B developed by the Lockheed Missiles & Space Company for the USAF. (Chamberlin's thoughts even extended possible use of the Mark II to a lunar landing, and although this was not to be he was the first STG engineer, as distinct from theorist, to declare unhesitatingly for rendezvous, and lunar orbit rendezvous in particular, as the most advantageous method of traveling to the lunar surface.)

The Chamberlin-McDonnell initiative proceeded apace, and much of the groundwork and preliminary design work for what was far more than an "improved"

Mercury spacecraft was complete as 1961 neared its end. The Mark II existed as a development proposal in its own right, Apollo or no Apollo, since controlled flight in a spacecraft, rendezvous and docking, and long-duration missions were obviously the way to the future of spaceflight generally. However, the great pressure on NASA resulting from the Kennedy initiative and the stage which the Mark II plans had reached meant that, perhaps a little by chance, the proposed spacecraft and its capabilities could contribute significantly to Apollo.

Chamberlin's proposals were scrutinized by NASA management and agreement was reached on a project whose primary objective was the development of rendezvous techniques, with long-duration flights, controlled land landing and astronaut training as major secondary objectives. The new project, costing over $500m, was announced in December, 1961, by Robert Gilruth to an enthusiastic audience in Houston, where the MSC's new home was to be built, but it was to be a few weeks before its official title of *Gemini* (appropriate for a project to orbit a two-man space vehicle) was announced.

Gemini's Sophistication

The Gemini spacecraft was far more than a redesigned Mark II Mercury. Its external shape showed its parentage, but it was essentially a new vehicle. Chamberlin was determined that it should be operational: Mercury had been a crash program, where essential equipment was crammed into the minimum space, and it was a maze of stacked equipment, which made changing or repairing components difficult and time consuming. In Gemini, systems were modular and placed in accessible packages; a rear adapter section, for example, contained most of the crew's oxygen and coolant supplies as well as batteries, fuel cells and thruster propellants. The spacecraft was meant to be flown with maximum crew control – thrusters providing the means of maneuvering in orbit to achieve rendezvous and docking, with offset gravity in the spacecraft yielding a degree of aerodynamic lift for greater control when landing. Each astronaut had a large hatch for ease of entry and egress (particularly valuable in an emergency), which also provided the means for conducting an EVA – extravehicular activity or spacewalk – as well as being essential for the ejection seat system which was incorporated in place of Mercury's tiresome escape tower.

Although the development and production of the spacecraft itself did not suffer the major problems encountered in man-rating the Titan and qualifying the Agena target vehicle, its progress was not smooth, particularly where new systems were involved. The weight penalty of batteries was prohibitive for long missions and the answer lay in fuel cells, in which oxygen and hydrogen reacted together to produce power and water (supposedly for drinking). Development problems resulted in the early, shorter missions carrying batteries, and the fuel cell did not qualify for flight until May 1965, in time to power the Gemini V mission. The more sophisticated thrusters in both the orbital maneuvering and re-entry systems persistently failed to achieve the required cycle of firings, with the ablation linings of the thrust chambers being particularly prone to burning through. It took until the spring of 1964 for this issue to be resolved.

Ejection seats were favored by the ex-test pilot astronauts, but the Gemini requirement, for a seat located at the top of a launcher 148ft (45m) in the air to be fired in a trajectory stable enough to clear a booster explosion and high enough for the parachutes to open, could not be met by existing seats. The development of a satisfactory system was difficult and lengthy (on several occasions it was proposed that an escape tower should replace the ejection seats) but flight qualification was finally achieved just days before the first Gemini mission was launched in March 1965.

No such success awaited one of the most intriguing Gemini concepts. An STG engineer at Langley, Francis Rogallo, had worked for some years on the design of a flexible kite, or wing, developing lift, which could be deployed from a returning spacecraft in the final stages of a controlled, precision landing. The idea was developed for Gemini in the form of an inflatable wing and the North American company received a development contract in 1961 for what by that time was being called a *paraglider*. A landing on land obviously had many attractions, not least that of dispensing with the need for the deployment of a large recovery force at sea. Unfortunately, in the time allowed no solution was found to the fundamental difficulty of how to eject and inflate the wing from a spacecraft weighing more than 8,000lb (3,628kg) that was plunging down through the atmosphere at high speed. The paraglider was deleted from Gemini plans in August 1964.

The Problem of POGO

The paraglider was not essential to Gemini but the Titan II was. It was being developed for the USAF as an ICBM and was powerful enough to launch the heavier Gemini spacecraft into orbit. Moreover, it used *hypergolic* propellants (a fuel and an oxidizer which burned spontaneously on contact), which were regarded as both safer and more manageable than the fuel system using liquid oxygen that powered Atlas and similar rockets. Titan made good progress in its role as a missile, but the early optimism of the Gemini engineers was confounded by a phenomenon encountered shortly after lift off: a rapid, lengthwise vibration which would subject astronauts, already experiencing 2.5Gs at launch, to bursts of a further 2.5Gs eleven times a second for half a minute. This was plainly unacceptable, especially in the case of any emergency demanding a fast response from the crewmen, and NASA demanded that the oscillation (nicknamed POGO) must be reduced to a maximum of 0.25G. Solving the problem was made more difficult by NASA having to deal with the contractor, the Martin Company, through the USAF as its agent, at a time when the air force was well satisfied with Titan's progress as a missile. Launch failures for other reasons and a lack of thrust in the second stage compounded the problem, but eventually resonance in fuel lines was identified as the cause of the oscillation, and this was corrected. But this took until early 1964, with the first successful Titan launches of unmanned Gemini spacecraft not taking place until April 1964 and January 1965.

Agena, the target vehicle already being produced for the USAF and which NASA engineers originally thought could be used almost as it came off the assembly line, was not as critical to the success of Gemini as Titan, which was fortunate, as was the fact that docking with it was not scheduled until later in the program. Many changes in fact were required to be made to the vehicle. NASA's demand that the main Agena engine be capable of five firings in orbit proved difficult for Lockheed to achieve, and the development of a complex new command and communications system created major problems. Agena was not qualified for flight until March 1966, and by then it had a rival in the shape of a cheap target built at short notice by McDonnell and sometimes known as the "glob," although more properly as the Augmented Target Docking Adapter (ATDA).

2

Thus Gemini followed Mercury in presenting major engineering and organizational problems. Costs constantly rose, and early in 1963 estimates of the total cost of the project were predicted at over $1,000m, twice the original, admittedly highly-optimistic estimate. Personnel changes were made and these included the replacement of James Chamberlin (who as an engineer had done so much to create Gemini) as project manager. The program published at the end of 1961 had allowed for the first Gemini manned flight to take place in May of 1963: it was to be almost two years late.

Additional Astronauts Selected

But if the machines had been difficult the men were ready and waiting. The planned schedule of ten flights at approximately two-monthly intervals clearly demanded an increased pool of astronauts. A group of nine was introduced to the public in September 1962 and a third group of 14 just over one year later. With one exception, all the 1962 astronauts flew on Gemini missions (four of them twice), together with three of the Mercury astronauts and five from the 1963 group. The backgrounds of the new astronauts were broadly similar to the Mercury Seven (though the test pilot qualification was dropped from group three onwards) and, like the first group, they were given individual system specialisms to study as well as entering a busy schedule of general training – lectures, centrifuge rides, water egress and survival techniques, in addition to maintaining their flying skills – before beginning intensive training specific to a mission once they had been selected. Virgil Grissom and John Young trained for over nine months as "prime" crew for the first manned Gemini mission, with Grissom, for example, spending more

3

3 The Gemini spacecraft started its life as a Mercury Mark II spacecraft masterminded by James A. Chamberlin (a Canadian born engineer) and his colleagues in NASA's Space Task Group Engineering Division working together with engineers of the McDonnell company. It was, however, far more than an improved craft: it was designed for maximum operational efficiency and to give the crew of two astronauts extensive control during flight.

1 *Controllers and technicians at consoles were to become the subject of countless photographs. This one was taken during a Gemini mission in 1965.*
2 *Family fun in the shape of Michael McDivitt – son of Gemini IV astronaut James McDivitt – playing with a model of a Mercury capsule.*

than 77 hours flying the mission in ground simulators and Young even more, with 85 hours. Before they lifted off, the two astronauts had taken part in almost 400 simulations of launch abort procedures.

Understandably, Gemini 3 (it followed the two, earlier unmanned proving missions) was set limited objectives. Basically, these were to evaluate the performance of the spacecraft and its systems, its maneuvering capabilities, the performance of the worldwide tracking network, and of re-entry and recovery procedures. Grissom and Young were launched on March 23, 1965, on a three-orbit mission. There was little drama, although history was made when the crew maneuvered the spacecraft to alter its orbital height and, even more significantly, the plane of its orbit. It was perhaps an indication of the low-key nature of this flight that it is chiefly remembered for two matters that were in no way concerned with its efficient demonstration of Gemini in action.

3 *The lighting of cigars in Mission Control after a mission or a marked success became a tradition early on. Here (left to right) flight director Christopher C. Kraft, astronaut Gordon Cooper and director of the Manned Spacecraft Center Robert Gilruth celebrate the successful rendezvous between Gemini VI and VII.*

4 *A powerful picture of some leading members of NASA manned spaceflight management taken during a Gemini press conference. Second from right is Dr. Gilruth, and at extreme right is George E. Mueller, Associate Administrator for Manned Spaceflight for six years until December 1969.*

With memories of what happened to his Liberty Bell 7 capsule, Grissom decided to name Gemini 3 *Molly Brown* after the "unsinkable" heroine of a Broadway stage success. NASA management was not amused and, while the name assumed a quasi-official status, Grissom's was the last spacecraft to be formally allowed a name until the Apollo command and lunar modules flew separately in space for the first time in 1969 – two craft obviously demanding separate call signs. (For good measure, all Gemini missions after the first were referred to by Roman numerals.) The second event was the smuggling aboard the spacecraft by Walter Schirra (the back-up commander) of a corned beef sandwich for Grissom. Conventional forms of food can be eaten only with great difficulty in zero-gravity conditions, and tests of specially prepared space food were planned for the Gemini missions. The joke caused something of a storm and led to the introduction of stringent new rules about what personal items could be taken on space flights.

Success of Gemini IV

James McDivitt and Edward White (both second group astronauts) spent almost a year training for their Gemini IV mission, which was certainly not low-key. This was NASA's first long-duration mission, and there was great interest in how the astronauts would withstand four days in space and the effects of return to normal gravity. In addition, White was to conduct a brief EVA at the end of an umbilical which, besides safely tethering him to the spacecraft, supplied him with oxygen and suit coolants. An experimental gas-gun was to be used to help the astronaut control his movements. Cosmonaut Alexei Leonov had already achieved the world's first walk in space on March 18, 1965, but the NASA plan for an EVA had been announced well before that date. Over 1000 media representatives requested accreditation to cover the mission from the MSC at Houston, a number which was only to be rivaled in 1969 on the occasion of the Apollo 11 moon landing mission.

Gemini IV was launched on June 3, 1965, and returned after almost 98 hours on June 7. White made his EVA safely and was outside the spacecraft for over twenty minutes, during which time his actions were recorded on 16mm movie film and (with considerable skill) by McDivitt using a Hasselblad 70mm still camera. Having to end the spacewalk was "the saddest moment of [White's] life," but there was a significant portent of problems to come when, after only a short period on EVA with no work to perform, White returned to the spacecraft physically exhausted, with sweat pouring into his eyes and with his helmet faceplate fogged.

Another event was of equal significance. The flight plan (for this, the first of the long-duration Gemini missions, the new Mission Control Center in Houston assumed responsibility, with flight control teams divided into three shifts around the clock) called for McDivitt and White to attempt to station-keep with the expended second stage of their Titan launcher. They attempted to close with it by firing thrusters, much as they might try to catch up with a car by pressing the accelerator. They failed, and thereby they and most of those on the ground learned that orbital rendezvous needed far greater sophistication. Speeding up a spacecraft moves it into a higher orbit than the target, and paradoxically the faster-moving spacecraft actually *slows* relative to the target because its orbital period (a direct function of its distance from the center of gravity) has also increased. What must be done is for the spacecraft to *reduce* its speed, which drops it to a lower and thus shorter period orbit, which allows it to gain on the target. A burst of acceleration at the correct moment then lifts the spacecraft

to the target's orbit, and if the maneuver is skillfully executed this brings the two vehicles close enough to one another for minor adjustments to effect rendezvous or docking. (This problem was extensively studied by the third group astronaut Edwin "Buzz" Aldrin, who took the pilot's role in rendezvous for his doctoral thesis at the Massachusetts Institute of Technology.)

Worries about McDivitt's and White's health after four days in space largely dissipated when rigorous medical examinations after their return revealed no major problems. But there were small indications of what would give rise to concern in the future when a loss of bone mass and also loss in the volume of blood plasma was detected. In addition, the astronauts agreed that a systematic exercise program would be required to maintain crew fitness during very long missions.

Such a mission followed from August 21-29, when Mercury astronaut Gordon Cooper was joined by Charles "Pete" Conrad on Gemini V; their flight patch showing a covered wagon with the motto "Eight days or bust." They achieved that target despite major fuel cell and thruster problems. The former meant that evaluation of the rendezvous, guidance and navigation system using a separate radar pod had to be abandoned, but a simulated rendezvous with a "Phantom Agena" was successful. As a result of the thruster problem, for much of the flight the spacecraft had to be allowed to

5

6

drift, with the attitude control system only being used to correct excessive tumbling when it developed. Overcoming such problems was valuable experience for both astronauts and the flight control teams in Mission Control. Cooper and Conrad worked hard at experiments and, despite the length of time they were confined within the spacecraft, it was a basically-healthy crew that returned to Earth. However, the trend of blood plasma volume and bone calcium losses in the astronauts, which was increasing on the longer missions, continued to worry the doctors, and the need for more exercising in orbit was again stressed.

Gemini VI was scheduled to launch on October 25, 1965, shortly after an Atlas-Agena was launched. This was to be the first rendezvous mission, and the last spacecraft using batteries, which would limit it to a duration of two days. Walter Schirra and Tom Stafford were already in their couches when the Agena was launched from a nearby pad at the Kennedy Space Center (renamed in honor of the late president in November 1963), but shortly after lift-off the vehicle

5 March 23, 1965 – and astronauts Virgil Grissom and John Young lift off for the first Gemini mission.

6 Broad smiles on the faces of Mrs. Pat White (left) and Mrs. Pat McDivitt as they talk to their husbands orbiting the Earth during the Gemini IV mission.

which the two crewmen had carefully prepared. Such mundane things as stowage of equipment to be used during the 14 days, and of refuse became extremely important, as did hygiene. Special lightweight suits incorporating a soft hood were to be worn, but over-heating in them was to cause the plan for one crew member to be in a suit at all times as a safety measure to be abandoned after a few days into the flight. The spacecraft was launched on December 4, 1965, and serious attention was given to medical experiments. The days passed and the crew progressed well; working, sleeping, eating, exercising and resting. In the words of the official history "good humor and good spirits prevailed," and the orbit was adjusted to await the visit of Gemini VI-A.

Value of Test-Pilot Experience

Schirra and Stafford were ready to go on December 12 when equipment failure caused an engine shutdown on the Titan II after 1.2 seconds. Instrumentation in the spacecraft showed that the vehicle had lifted (which would have required an immediate use of the ejection seats), but the two astronauts considered this to be an error and remained in the spacecraft, a first class example of an instant decision based on lengthy test-pilot experience. There was more success on December 15, and in less than six hours Schirra and Stafford had executed the world's first space rendezvous – the final approach being conducted appropriately with the stars Castor and Pollux (in the constellation Gemini) aligned with the target spacecraft. For three revolutions the two craft stayed together at distances ranging from one foot (30cm) to 295ft (90m). Schirra and Stafford found maneuvering so precise that they concluded docking would not present a problem. The two crews settled down to sleep some ten miles (16km) apart, after which, as Stafford humorously reported seeing a satellite which appeared to be transmitting *Jingle Bells* over the communications link, Gemini VI-A headed home at the end of a mission lasting a little over one day. The spacecraft landed in the western Atlantic only seven miles (11km) from the planned landing point and in full view of TV cameras that transmitted their pictures back home via satellite.

For Borman and Lovell the ensuing, final three days dragged. Between working and reading, fuel cell and thruster problems provided some interest, as did Lovell's musing that the Gemini spacecraft was perfect for a man without legs because (if no EVA were being conducted) the legs served no useful purpose. On their return, one week before Christmas, 1965, the onset of G-forces "felt like a ton," but the crew landed even closer to the aim point than Gemini VI-A. Although very weary, Borman and Lovell managed the helicopter retrieval with little difficulty and walked across the recovery ship's deck without stumbling. To the pleasure of the doctors, it also appeared that the critical bone and blood changes were no worse than during the earlier, four-day mission, and that a degree of adaptation had taken place. MSC director Robert Gilruth called 1965 "a fabulous year for manned space flight," but there were to be some dramas in 1966.

They started with Gemini VIII on March 16. Commanded by Neil Armstrong, who had flown the X-15 rocket research aircraft, and with David Scott scheduled to conduct NASA's second EVA, the spacecraft was launched in pursuit of an Agena target vehicle which at last had achieved orbit successfully. In less than seven hours the astronauts achieved the world's first docking of two vehicles in space. Within minutes, however, the craft began to roll and, without any indication of the

exploded. Within a matter of three days it was decided that the Schirra/Stafford mission would be postponed until December, when the Gemini VII spacecraft, with Frank Borman and James Lovell (both second group astronauts) aboard on a 14-day mission, would act as the target. The two vehicles could not dock, but docking in any case was a much simpler exercise than rendezvous. The critical issue had been whether pad 19 at the launch site could be refurbished after the launch of Gemini VII in time for Gemini VI-A (as it was now designated) to be prepared and launched, and it was decided that it could. In the weeks that followed Borman and Lovell took precedence in the simulators, with Schirra and Stafford, who had been fully prepared for their mission, taking occasional sessions.

Gemini VII was primarily intended to gain experience of the length of time in space that would be required for a lunar landing mission. It was the last long-duration mission in the Gemini series and one for

3

3 *The world's first rendezvous in space took place on December 15, 1965, when Gemini VII – with astronauts Frank Borman and James Lovell aboard – acted as the passive target for Gemini VI-A, which had been launched less than six hours before. This splendid photo was obtained by Tom Stafford in the right hand pilot's seat, with Walter Schirra occupying the commander's seat. After keeping station at varying distances for three revolutions of the Earth, the two craft separated and Schirra and Stafford headed for home after little more than a day in orbit.*

4 *Happy landings! With the flotation collar and inflated raft securely fastened, "Pete" Conrad egresses (NASA's favored jargon) the Gemini XI spacecraft at the end of a record-breaking mission in September 1966. Pilot Richard Gordon had still to open his hatch when the photograph was taken from a hovering helicopter.*

source of the malfunction, the crewmen separated from the Agena. The roll of the Gemini VIII spacecraft then speeded up until Armstrong and Scott were being subjected to one revolution per second. They became dizzy and their vision blurred. Having tried everything else to correct the malfunction they cut out the thrusters of the orbit attitude and maneuvering system (OAMS) and activated those of the re-entry control system. The spacecraft responded and the motion ceased.

An Early Return

For safety reasons, mission rules stated that once the re-entry control thrusters had been used the mission must end as soon as possible, and Armstrong and Scott landed in a secondary recovery area in the western Pacific less than 11 hours after taking off. Subsequent tests showed that an electrical fault had caused a thruster to stick in the "open" position. A modification was introduced to prevent this happening in the future, but the frustrating thing was that the fault developed when Gemini VIII was out of radio contact with the ground controllers and engineers. Telemetry would have shown those on the ground where the malfunction was and they could have advised the crew what corrective action to take. Nonetheless, the docking had been flawless and the Agena vehicle was boosted to a higher orbit to await a later mission.

A tragedy preceded the next mission. Two astronauts from the second and third groups respectively, Elliot See and Charles Bassett, were selected for Gemini 4

1 *The Gemini IX-A mission lifted off on top of its Titan booster on June 3, 1966.*

2 *Astronauts Stafford and Eugene Cernan successfully rendezvoused with their target docking vehicle but found that a protective shroud had only opened partially. This prevented any attempt at docking and the target vehicle went down in history as the "angry alligator."*

1

3

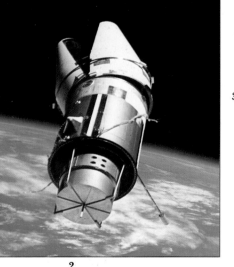

2

3 *A second, even more powerful image of the "angry alligator," with flare from the Sun out of frame seeming to create appropriate special effects.*

4 *The last two Gemini missions conducted an experiment in orbital mechanics in which the spacecraft was connected by a 100ft tether to the target docking vehicle – an Agena in this image taken during the Gemini XII mission in November 1966. The surface below includes parts of southeastern Arizona and southwestern New Mexico.*

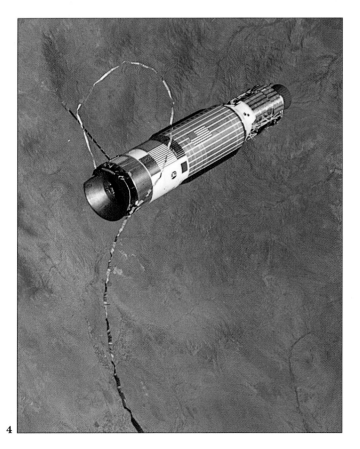

IX but they were killed when their T38 jet aircraft crashed as they were flying to the McDonnell plant at St. Louis. Tom Stafford (making his second Gemini flight) and group three astronaut Eugene Cernan were chosen to replace them. Stafford seemed to attract problems at launch. His new mission's target Atlas-Agena lifted off on May 17, 1966, but a short circuit in the Atlas caused the combination to plunge into the Atlantic. NASA promptly prepared to launch the McDonnell stand-by target vehicle, which had been developed in the face of constant Agena problems, and this was launched on June 1. Stafford and Cernan prepared to follow but the rendezvous mission had only a very short launch "window" of 40 seconds (for the spacecraft to achieve the correct position in space relative to the target) and equipment problems caused a postponement until what was now termed Gemini IX-A was at last launched on June 3.

The three day mission was to concentrate on rendezvous, docking and extravehicular activities. Various rendezvous maneuvers and station keeping were performed successfully, but docking proved impossible when a shroud protecting the docking port on the target vehicle only partially opened and could not be jettisoned. To Stafford it looked like an "angry alligator," a description well borne out by the mission photographs.

4

Although the warning signs were there during Ed White's space walk, the inherent problems of EVA had tended to be underestimated. They were underlined in full during Gene Cernan's shortened stay of just over two hours outside the spacecraft. Although footbars, handbars, stirrups and velcro patches had been added to the outside of the craft along which he was scheduled to proceed to the aft, adapter area, where he was to don a USAF-developed Astronaut Maneuvering Unit (AMU), they proved inadequate. Everything took much longer than in simulations and he discovered the uncomfortable truth of one of Newton's laws, for every movement of arm or leg in free space triggered a reaction from his body. Minute movements upset his equilibrium and set his body in motion. He could not maintain his body position and he kept floating out of control. He had to fight the inevitable "stiffness" of his pressurized spacesuit, the umbilical was a "snake" that was extremely difficult to control and he had constant problems with fogging on the inside of his face plate. Having got to the adapter, Cernan decided that, although it would be

5 Command pilot Tom Stafford (right) and pilot Gene Cernan seem perfectly relaxed as they await a ride in a helicopter at the conclusion of their Gemini IX-A mission in June 1966. Cernan conducted the first American EVA (extra-vehicular activity in the jargon) since Ed White's spacewalk from the Gemini IV spacecraft in June 1965, and the problems encountered underlined that there was still much to learn.

easy to connect himself to the AMU when it was fixed to the spacecraft, he would have the greatest difficulty in releasing himself from it when he was in free space. Mission control agreed and the EVA was abandoned forty minutes early.

When the crew returned on June 6 there were mixed feelings: the advanced rendezvous maneuvers had been impressive and the docking failure was the direct result of faulty preparation of the adapter. But EVA was proving very difficult.

John Young went back into space as command pilot of Gemini X when it was launched on July 18 on a mission lasting just under three days. He was joined by Michael Collins, a third group astronaut, who would conduct two EVAs. The target Agena behaved immaculately and docking took place without any difficulty. An hour later the Agena's main engine was

commanded to fire and the two astronauts were taken to a height of over 468 miles (753km), the highest point to which human beings had traveled, although the magnificent view tended to be blocked by the Agena in front of their hatch windows. Subsequently, while still docked to the Agena, Collins conducted a stand-up EVA in his hatchway. Much of this was spent in astronomical photography and it was a gentle prelude to the second EVA which was to follow.

After 39 hours docked to their own Agena target vehicle, Young and Collins undocked and prepared to maneuver toward the Gemini VIII Agena, from which Collins was to remove a micrometeorite package as the major task of his second EVA. The rendezvous was successfully achieved and Young kept station a few metres away while Collins opened the hatch. He, too, faced the problems that Cernan had encountered, but

6 Gemini IV pilot Edward White floats in space at the end of a tether which not only secured him to the spacecraft but supplied oxygen for him to breathe and coolants to control the temperature inside his spacesuit. White found his spacewalk, lasting a little over 20 minutes – the first of America's space program – exhilarating but exhausting.

1 *"Buzz" Aldrin demonstrated during the Gemini XII mission – this image was taken by James Lovell – that handrails, a waist tether and boot restraints in the work area all contributed to a successful solution of previous EVA problems.*

2 *The problems of spacewalking were encountered by pilot Dick Gordon during the Gemini XI mission – despite Pete Conrad's exhortations to "Ride 'em Cowboy!" The difficulties were not basically resolved until "Buzz" Aldrin's two EVA's during the final Gemini mission in November 1966.*

by dint of operating a nitrogen powered "zip" gun, and pulling on his umbilical, the astronaut secured the package from the outside of the Agena, although his movements when in contact with spacecraft or target vehicle upset the spacecraft's equilibrium, which had to be corrected by Young firing the thrusters, much to Collins' concern lest his spacesuit should be damaged.

The mission improved docking and rendezvous techniques still further and demonstrated that another vehicle's engine could be used to boost a spacecraft into a different orbit. Michael Collins had performed his two EVAs with a cool determination but they had been limited; he joined Gene Cernan in advocating much better and more numerous restraints and handholds. If spacewalking was delightful, it was also "dangerous, difficult and deceptive.".

taut – with the intention of studying the value of tethered flight for lengthy, unattended station-keeping between two spacecraft, and also whether imparting a spin to the configuration would create a small amount of artificial gravity.

For the EVA, handholds at the docking end of the Agena had been modified, better foot restraints fixed to the adapter section of Gemini XI, and the umbilical was shortened to 30ft (9m). But things went no better. Even before starting the EVA, Gordon had to spend time and energy wrestling to clip a sun-visor to his helmet faceplate. Once outside, he pushed off toward the Agena but moved in an arc which missed the target and took him to the aft-end of the Gemini, with Conrad having to pull him back to the hatchway by the umbilical. Gordon started again and reached the Agena, which he

1

2

3 *A classic image of the early space era: the best of the excellent sequence of photographs that command pilot James McDivitt took of Ed White during his spacewalk, using a Hasselblad 70mm still camera.*

The effervescent "Pete" Conrad went back into space as command pilot of Gemini XI, which was launched on a three-day mission on September 12, 1966. He was joined by another US Navy pilot, Richard Gordon, in what was a highly-ambitious flight.

Apollo lunar orbit rendezvous would take place during the first lunar lander revolution after launch, so this task was set for Gemini XI. (Other missions had typically conducted rendezvous during their fourth revolution). The launch window was just two seconds and both Atlas-Agena and Titan-Gemini performed admirably. The docking was accomplished as planned, with on-board computations only (as distinct from ground control assistance) being used. So well did things go that, again for the first time in Gemini, both command pilot and pilot were able to practise docking and undocking with the target. The Agena was involved in another new experiment. The spacecraft carried a 100ft (30m) dacron tether that Gordon was to attach to the target during an EVA. The plan was that the two vehicles would then be separated, with Conrad endeavoring to keep the line

sat astride but found great difficulty in fixing the tether, despite cries of exhortation from Conrad to "Ride 'em cowboy!" He eventually fixed the tether, but by then was sweating profusely inside the spacesuit and the sweat began burning his eyes. Gordon could scarcely see and Conrad, realizing how exhausted his pilot was becoming, ordered him back inside. The EVA lasted for less than one-third of its scheduled duration.

EVA Problems Resolved

Nevertheless, Gordon recovered quickly, and next day the Agena's main engine was fired to take the astronauts to a new world record height of over 850 miles (1,369km), which provided the opportunity for over 300 photographs. Concern about possible radiation hazards at the much higher altitude (which was very important in the context of the Apollo missions) proved unfounded. After two revolutions Gemini XI's orbit was lowered to a more usual height and Gordon conducted a stand-up EVA in the hatchway. This was much more enjoyable: he carried out astronomical photography experiments and on one occasion Conrad reported that they had both dozed off, himself inside the spacecraft and Gordon hanging outside the hatch on his tether. After undocking from the Agena considerable time was devoted to the tethered flight experiment: while Conrad achieved several different rotation rates the results were inconclusive, being described officially as "interesting and puzzling." Gemini XI was the first to use completely automatic re-entry procedures, much as Apollo would do, and the crew landed less than three miles (5km) from the recovery ship.

There was only one mission left and, looking toward Apollo as well as for its own satisfaction, Gemini XII had to solve the EVA problem. James Lovell returned to space in the four-day mission which began on

3

November 11, 1966, accompanied by group three astronaut "Buzz" Aldrin. A number of the by-then-normal events took place, with the rendezvous and docking exercise benefiting from Aldrin's expertize when the rendezvous radar failed. There was also a period of tethered flight. But everything was subordinated to EVA and to demonstrating that man could work outside a spacecraft. One of the casualties of this was the USAF Astronaut Maneuvering Unit, which was never tested during Gemini and was sacrificed on the last mission

because of the need to concentrate on performing simple, basic tasks.

In preparing Gemini XII, there was no dramatic difference from the early missions. But whereas IX-A had carried nine restraints to assist in body positioning, Gemini XII had 44. One new feature was a waist tether, which enabled the pilot to use tools (such as a wrench) and retrieve packages or experiments without having to use one hand to hold on to the vehicle. Both spacecraft and target had handrails and handholds, together with

1 *John Glenn used a basically standard Ansco Autoset (Minolta) 35mm camera – shown here – during his Friendship 7 mission. Photography in space was to progress quickly and many of the images taken during the Gemini missions of 1965-66 are still unsurpassed in quality, particularly views of the Earth's surface.*

2 *The Indian sub-continent looking northeast; an image taken during the Gemini XI mission from an altitude of over 400 miles.*

rings on which Aldrin could hook the waist tether. Perhaps most important of all, "golden slippers" (overboot restraints) were fixed to the adapter area at the rear of Gemini XII, where Aldrin was to work. The experience gained on the earlier missions also helped shape Aldrin's approach. The first EVA (like that of Michael Collins on Gemini X) was stand-up and therefore constituted a somewhat gentle introduction. During the second and most important EVA, Aldrin worked slowly and deliberately, taking frequent rests. Pre-mission training (particularly in the new underwater zero-gravity simulator), mechanical aids, careful planning and astronaut attitude paid off, and EVA problems seemed to disappear magically in a performance justly described as flawless. Before they made an uneventful return both Lovell and Aldrin established new world space records; the command pilot over his two Gemini flights having spent the longest time in space (425 hours) and Aldrin's three EVAs totalling over five-and-a-half hours.

Interspersed with the major Gemini requirements outlined here, the missions conducted a total of 52 experiments in orbit. Of these, eight were medical in nature; 13 (largely originating from the Manned Spacecraft Center itself) could be described as a mixture of operational, technological and scientific in nature; 14 originated from the Department of Defense and 17 were scientific. There had been (and continued to be) an uneasy relationship between engineers charged with achieving basic spaceflight objectives and scientists whose experiments tended to introduce complications into an already complex situation. Nonetheless, a Manned Space Flight Experiments Board was established in January

During the 1960s, the US also attempted to develop two manned military space projects. The first was the *Dyna-Soar:* a contraction of Dynamic Soaring. This was a small space plane with a single pilot. The program had its roots in two different lines of development. The first stemmed from various studies of rocket-powered bombers and reconnaissance aircraft made during the 1950s. The second was the need for a follow-on research aircraft to the X-15, whose mach-six speed was well short of that required for orbit.

The Dyna-Soar (X-20) program underwent a number of changes. Originally it was to be launched on a sub-orbital flight by a Titan I ICBM, later changed to a Titan II. In 1962, the booster was again changed, with a Titan IIIC scheduled to put Dyna-Soar directly into orbit without any sub-orbital launches. It was also planned that an operational system including orbital bomber, reconnaissance or space station resupply would be developed. However, in December 1963, with no operational system in sight and the cost too high to justify a research aircraft, the Dyna-Soar was cancelled.

Its replacement was the Manned Orbiting Laboratory (MOL). This was a small military space station composed of a modified Gemini craft attached to a pressurized module that would be launched by a Titan IIIC (later changed to an uprated Titan IIIM when the spacecraft's weight grew). The two-man crew would transfer through a hatch cut in the Gemini's heat shield into the module, from which during a 30-day flight they would conduct photo and ELINT reconnaissance.

Like the Dyna-Soar, MOL had problems. During the 1960s it was not known if humans could survive 30 days in orbit, or whether they could undertake militarily useful activities there. The biggest problem, however, was the Vietnam war. It delayed the program and raised the cost, which by 1969 had doubled to $3,000m, at which point it was cancelled by the incoming Nixon administration. During the shuttle era some military tests have been performed, but these appear to have been only a shadow of those planned for the MOL.

CURTIS PEEBLES

3 *From an altitude of almost 400 miles Gemini XI looks down on the southern Red Sea and the Gulf of Aden. The view at the top is eastwards, and the Bab el-Mandeb strait is at lower center. The image is an excellent demonstration of plate tectonics in action: if the land mass at left is in imagination drawn towards the right it will be seen that there is rough fit between the two coastlines – i.e. that the land masses were once joined.*

4 *The Bahamas region from Gemini IV reveals typically delicate shades of blue. Most of the lighter blue area at left is the shallow Great Bahama Bank. At right is the narrow Great Exuma Island, beyond which the sea floor falls away sharply into the deep Exuma Sound.*

1964, the failure of the paraglider released more space aboard the Gemini spacecraft for experiments, and by the final Gemini mission the experimenters' representative was allowed to occupy a console in the holy of holies: the main control room of Mission Control. Inevitably, individual astronaut attitudes varied (Schirra commenting perhaps predictably that on his mission "we couldn't afford to *play* [sic] with experiments"), but by the end of 1966 much had been achieved. Despite the best efforts, some of the more demanding experiments had limited success, but others achieved results which, even at the beginning of the 1990s, stand comparison with anything subsequently achieved. A large proportion of the experiments involved some form of photography, and two of them, S-5: Synoptic Terrain Photography and S-6: Synoptic Weather Photography, were each performed on seven missions and entered into with enthusiasm and no little competence by most of the crews. It is a fit comment on these particular results that in their continuing value today they represent Gemini's most durable legacy.

The Achievements of Gemini

If it had stuttered at first, and even if it were regarded as an "afterthought," Gemini contributed much to Apollo. As Gemini developed equipment and procedures, and as the flight controllers and astronauts wrestled with difficulties and sometimes emergencies, minor and major, elements were added to the learning curve. Gemini's total cost ($1,147m) was double the original forecast, but was not as high as seemed likely at one time and was surely worthwhile in terms of forwarding the Kennedy goal. The space environment is too dangerous for spaceflight ever to be regarded as "routine," but with the coming of Project Gemini it could certainly be described as "operational," and the flights took place with such frequency that, by the close, media representatives attending launches had dwindled five-fold. No matter, if the race with the other super-power was not over yet, the United States was undoubtedly winning. During the period of the ten Gemini flights between March 1965 and November 1966, when the US was setting one space record after another, not one Soviet cosmonaut orbited the Earth.

4

5

APOLLO – MAN ON THE MOON

1

1 President Kennedy addressing Congress in May 1961, when he set the nation the task of going to the Moon. NASA had been working for years on the demands such a task would make, but henceforth it would be time for action rather than words.In the same speech, Kennedy proposed a $500m increase in NASA's 1962 budget.

2 Without the Saturn rocket there could have been no Apollo successes. As the S-IC first stage of a Saturn V is moved from a building at the Marshall Space Flight Center, it provides a formidable demonstration of its size compared with attendant technicians.

3 Flight tests of the Saturns began in 1961. This is the very first launch – of a Saturn I on October 27 of that year. The objectives were achieved and nine more Saturn I test launches took place over the next four years, paving the way for subsequent tests and manned launches of the Saturn IB and the Saturn V launcher that would take men to the Moon.

On May 25, 1961, President John F. Kennedy, in an address to the Congress, uttered 31 words which were eventually to lead to one of the greatest events in the history of mankind: "I believe that this nation should commit itself to achieving the goal, before this decade is out, of landing a man on the moon and returning him safely to the earth." There had already been much discussion at the highest level about possible initiatives to counter the Soviet Union's apparent superiority in spaceflight. The Moon was an obvious target; NASA's Administrator James Webb had already estimated the possible cost of reaching it at between $20,000m and $40,000m; and at a slightly more superficial level a name had been awaiting the new project since the middle of the previous year: "Apollo," because "the image of the God ... riding his chariot ... gave the best representation of the grand scale of the proposed program."

But now the challenge had been made. NASA and the US, with just one sub-orbital lob into space achieved, had to stop the talk and act. Robert Gilruth, who led what by the end of 1961 would be known as the Manned Spacecraft Center at Houston, later professed himself to have been "simply aghast." The complex problem could be stated simply: to take off from a point on the Earth traveling at 1,000mph (1,610kph) as the Earth rotated, go into orbit at some 18,000mph (29,000kph), accelerate to over 25,000 mph (40,230kph) at the appropriate time to break out of Earth orbit on a journey of around a quarter of a million miles to another body in space which was itself traveling at 2,000mph (3,200kph) relative to the Earth, to go into orbit around it, to send a machine with men inside down to the surface to explore and to leave scientific equipment behind when they returned to orbit, and return to Earth, repeating the journey not once but as often as NASA was directed.

By the end of 1962 the major decisions about modes of flight and the machines to pursue them had been taken. There were three major elements. The first was an immensely-powerful, three-stage launch vehicle: the *Saturn V*.There was a *command module* in which three astronauts would travel to the Moon and return to Earth: this would be a blunt body with an ablative heat shield, shaped and ballasted to achieve a controlled re-entry into the Earth's atmosphere at the high terminal speeds of the return from the Moon and landing at sea by parachute. For all but the final part of the journey, the command module would have attached to it a

service module, with a propulsion engine for altering the trajectory, attitude control thrusters, fuel cells for generating electric power, and tanks of fuel: hydrogen and oxygen, (water for the crew being a by-product of combining oxygen and hydrogen in the fuel cells). The command and service modules were normally regarded as a single unit often referred to as the *CSM*. Thirdly, there was the two-part lunar vehicle which would transport two astronauts to the Moon's surface and then return them to the command module, which would remain in orbit throughout. The lunar vehicle was originally called the *lunar excursion module* (LEM), but at some stage it seems this was thought to create too

2

frivolous an impression of activities on the Moon and the name was changed to *lunar module* (LM).

The diagram reproduced here shows the major stages of the journey to and from the Moon. The two lunar discs represent the movement of the Earth's satellite in its orbit during a lunar landing mission and the dotted lines indicate a loss of communication with Earth as the astronauts passed around the far side of the Moon. Lunar orbit insertion took place when the service module's engine was fired to allow the Moon's gravity to capture the combined Apollo vehicles, the

(as well as to save fuel), but the first mission to the Moon, Apollo 8, did not make a midcourse correction, since, as a safety measure in case the CSM engine failed to fire behind the Moon, its initial translunar injection placed it on what was termed a "free return" trajectory, whereby it would swing around the Moon and return to Earth. (Astronauts on occasion referred to such trajectories, and also coasting flight between the Earth and Moon, as "letting Sir Isaac Newton do the driving.") Mid-course corrections on the way to the Moon resulted in a departure from a "free return" and meant that an engine firing *had* to take place at the Moon otherwise the crew was in grave peril. Because of the greater gravitational effect of the Earth on the Apollo vehicle, the journey to the Moon took about 20 percent longer in time than the return to Earth.

Structure of a Mission

The departure of an Apollo mission from Earth was timed according to lighting conditions at the landing site at the time of spacecraft arrival in lunar orbit. For the final landing, the lunar module crew required some shadow to enable them to interpret the detailed characteristics of the lunar surface as they approached it, but the shadows must not be too extensive; that is, the Sun must not be too low. Similarly, the Sun must

4 The last three Saturn I launches in 1965 placed Pegasus meteoroid satellites in orbit. Here Wernher von Braun, by this time director of the Marshall SFC, points at a TV monitor in launch control at the Kennedy Space Center. On his right is Kurt Debus, director of KSC.

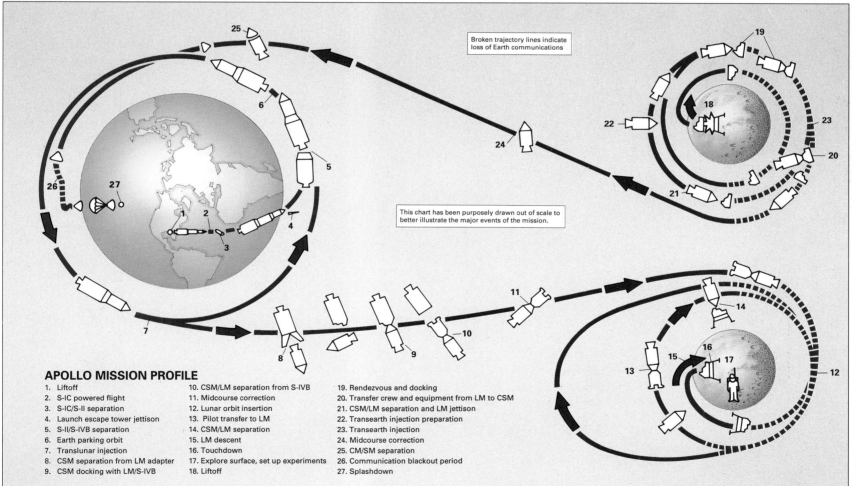

APOLLO MISSION PROFILE

1. Liftoff
2. S-IC powered flight
3. S-IC/S-II separation
4. Launch escape tower jettison
5. S-II/S-IVB separation
6. Earth parking orbit
7. Translunar injection
8. CSM separation from LM adapter
9. CSM docking with LM/S-IVB
10. CSM/LM separation from S-IVB
11. Midcourse correction
12. Lunar orbit insertion
13. Pilot transfer to LM
14. CSM/LM separation
15. LM descent
16. Touchdown
17. Explore surface, set up experiments
18. Liftoff
19. Rendezvous and docking
20. Transfer crew and equipment from LM to CSM
21. CSM/LM separation and LM jettison
22. Transearth injection preparation
23. Transearth injection
24. Midcourse correction
25. CM/SM separation
26. Communication blackout period
27. Splashdown

same engine being fired later to break the vehicles free from lunar orbit for the return journey. After separation from the command module, the lunar module's descent engine was fired briefly to lower its orbit closer to the Moon, and on approach to the landing site it was started again and fired continuously until the moment of touch down. Lift-off of the LM ascent stage was timed precisely relative to the passage of the CSM above to facilitate rendezvous and docking.

Midcourse trajectory corrections were executed to achieve greater accuracy on arrival at the Moon or Earth

not be too high, otherwise surface detail as revealed by shadows disappeared from view. The temperature on the surface was another consideration. Mission rules generally held that a solar elevation above the horizon of between 15 and 45 degrees was acceptable, and therefore the whole mission sequence was keyed to that requirement. For operational and navigational reasons, the landing sites of the earlier Apollo missions were restricted to a zone approximately five degrees either side of the lunar equator and between 45 degrees W and 45 degrees E in longitude. As experience was gained

This diagram is explained in the text. It shows in schematic form the concept of Earth parking orbit and of lunar orbit rendezvous, which was decided on by NASA after much debate. Placing the combined Apollo craft in Earth orbit for a few revolutions enabled checks to be made, and lunar rendezvous meant that only the lander descended to the lunar surface.

1 *President Kennedy points out a feature of interest to Robert Seamans (left) and von Braun during a visit to the launch center at Cape Canaveral on November 13, 1963. Just nine days later the president was assassinated.*

2 *Rollout of a Saturn V from the Vehicle Assembly Building at the Kennedy Space Center. The enormous size of the building is legendary, but the claim that clouds form in it was (and is) a myth!*

3 *It took the "crawler" about six hours to transport the Saturn V vehicle and mobile launcher the 3.5 miles from the VAB to the launch complex.*

4 *An Apollo mission lifts off for the Moon. The 7.7m lbs of thrust generated by the five Saturn V first stage engines assaulted the eardrums of the nearest viewers over three miles away. No recording ever did the indescribable sound justice.*

1

2

5, 6, 7, 8 *A protected automatic camera positioned at the 360ft level on the mobile launcher records the lift-off of Apollo 15 in July 1971. Above the silver-colored service module and conical-shaped command module is the launch escape tower.*

the restriction on latitude was eased.

NASA had been pondering the launch vehicles required to get men to the Moon long before President Kennedy issued his May 1961 challenge to the United States. As early as the time of Sputnik 1, Wernher von Braun and his colleagues at the Army Ballistic Missile Agency proposed a "Super Jupiter" vehicle developing a thrust of 1.5mlb (680,390kg). The concept progressed under the name Saturn, the next planet out from Jupiter, with both the project and von Braun's group moving across to NASA in 1959-60. What came to be known as Saturn I clustered eight engines together to develop the required thrust – inevitably inviting the

3

witticism when it first appeared on the test stand of "Cluster's Last Stand." Perhaps more perceptively, in 1959 NASA contracted with the Rocketdyne Division of North American Aviation for the development of an engine first proposed by the USAF (and known as the F-1) which, *as a single unit,* would develop 1.5m lbs of thrust. At the beginning of the next year, President Eisenhower granted Saturn and the "super booster" concept the status of highest national priority.

Out of the intense thinking and debate eventually came the shape of the Saturn family of launch vehicles which would take man to the Moon. The Saturn I was an early, two-stage proving vehicle and there were ten launches between 1961 and 1965. The Saturn IB was a more advanced two-stage vehicle, clustering eight engines each of around 200,000lb (90,720kg) thrust in the first stage, which was to conduct Earth orbital tests of Apollo hardware and in due course take astronauts into Earth orbit. Saturn V was a giant, three-stage vehicle. Its first stage (designated S-IC) would be powered by five F-1 engines developing a total 7.5mlb (3.4mkg) of thrust. The second stage (S-II) would also have five engines, but these would be of a different type, designated J-2, which would each develop over 200,000lb of thrust. The third stage (S-IVB) would use a single J-2 engine, with a restart capability, to place

well over 220,460lb (100,000kg) in low Earth orbit, or to send a payload of over 88,185lb (40,000kg) to the Moon. It would also act as the second stage of the Saturn IB launcher.

The first stage (S-IC) was intended to be conventional in that the F-1 engine used the well established propellants liquid oxygen and RP-1 (a form of kerosene), with the overall design being "state of the art." But, like the rest of the S-IC structure, the very size of the engine and the demands that it made on advanced metallurgy, for example, (with three metric tons of fuel and oxidizer being consumed every second) necessitated considerable innovation. The stage, for which Boeing received the contract at the end of 1961, was 33ft (10m) in diameter and over 150ft (46m) in height. The fuel tank contained 200,000 gallons (almost 757,100 litres) of RP-1, and the LOX tank over 330,000 gallons (almost 1,249,200 litres) of oxidizer, resulting in tank sizes which made unprecedented demands on welding engineers and technicians. Each stage took 14 months to complete and a number of F-1 engines exploded during tests, but by 1966 the engine had been qualified for manned missions. The S-IC was never a threat to the Apollo timetable.

Problems with the S-II

The same could not be said for the S-II stage which was being built by North American Aviation. Many of these problems resulted from the development of the J-2 engine, which did not use a conventional fuel but liquid hydrogen. The attraction of this was that it gave a 40 percent performance advantage over an RP-1 engine of a similar size. (Liquid hydrogen was not used in the first stage of Saturn, however, because, since it was much less dense than kerosene, a far greater tank volume would have been required with disadvantageous effects on vehicle weight and aerodynamic design.) But it had the drawback of a very low boiling point of -423°F (-253°C), which necessitated devising an insulation system capable of preventing the transfer of heat from the outside and the comparatively warm -297°F (-183°C) liquid oxygen. This was a major problem during S-II development, as was welding – the stage's 3,280ft (1000 metres) of welded joints having to be surgically clean and flawless, with many accurate to less than 0.02in (0.5mm). Rocketdyne's J-2 suffered the inevitable development problems of a new engine and NASA became seriously worried about the lack of progress in the production and testing of the S-II. However, management changes at North American were made and the tragic fire at the Kennedy Space Center in January 1967 provided more breathing space to achieve an eventually-successful outcome.

The S-IVB was built by the Douglas Aircraft Company and it, too, faced the insulation problems presented by a tank holding more than 60,000 gallons (227,125 litres) of liquid hydrogen, as well as J-2 engine failures and explosions. But this stage, which was scheduled to play a vital part in the lunar missions, was never a major concern to NASA and Douglas delivered the first flight stage in 1965.

Overall management of Saturn, including overseeing the work of contractors, was in the hands of Wernher von Braun and his staff at the Marshall Space Flight Center in Alabama. It was a powerful and very experienced team which tended toward conservatism in its engineering concepts and procedures. In developing a major new vehicle such as Saturn V, the time-honored practice would have been to test fly each individual stage thoroughly before taking the next step of adding another stage. But in the autumn of 1963, because of budgetary constraints and time pressures if President Kennedy's goal was to be achieved, George Mueller, the new Director of the Office of Manned Space Flight at NASA Headquarters, required the centers to pursue "all-up" testing of the Saturn V launch vehicle. The move was initially greeted with "shock and incredulity" at the Marshall SFC, but von Braun and his colleagues accepted the decision, and four years later Saturn V made its first all-up flight test with complete success.

There is a tendency to describe the Saturn V system in terms of enormous size and power; taller than the Statue of Liberty, developing as much power as 85 hydro-electric dams, akin to launching a naval destroyer by weight and with one F-1 engine "bell" being big

The Saturn was reputed to have three million different parts, and the IU computer played a major part in the automated check out of the vehicle, which took three months or so before a launch as some 30,000 control documents were worked through. This was the responsibility of the technicians and specialists in the Launch Control Center at the Kennedy Space Center in Florida. Back in 1961 it was realized that the US Air Force launch site at Cape Canaveral was not large enough to accommodate the Saturn V, so 140,000 acres (56,656 hectares) of sand, scrub and sunken land were purchased across the Banana River on Merritt Island. It was to this site that barges and the custom-

enough to swallow the entire Mercury capsule. However, Saturn was at the same time a creation of great precision and detail, and nothing represented this better than the Instrument Unit: a 24-panel cylinder, 3ft 3in (1m) high and just under 26ft (8m) across, positioned between· the S-IVB stage and the lunar, command and service modules above. The IBM-built unit was the brain of Saturn.

An inertial guidance system sensed the distance traveled by the vehicle and compared its path with that in the IU computer's memory, any deviations being corrected by computer instructions to the engines which gimbaled (moved) to effect the changes. The IU was in full charge of the Saturn from a few seconds before launch. Besides guidance and control, it was responsible for the command and sequencing of all vehicle functions, including engine cut-off and the separation of stages, and for the insertion of the vehicle into Earth orbit. On lunar missions, the IU also controlled all functions which sent the Apollo vehicle on its trajectory to the Moon, as well as maneuvering itself and the S-IVB stage from that flight path. During the six-hour flight of the Saturn V, it relayed information on vehicle position and functions back to Earth at the rate of 300 pages of data per second. A reflection of Saturn's success was that the IU was never called upon to carry out an automatic abort sequence.

modified Pregnant Guppy and Super Guppy aircraft ferried launcher stages and other equipment, which were brought together in the appropriately-named Vehicle Assembly Building – an enormous structure with a volume of just under 130m cubic feet (almost 3.7m cubic metres), where folklore, for it is no more, had it that clouds formed and rain fell. Once assembled, the unfueled Saturn and its mobile launcher were moved down to one of the newly constructed launch pads (39A and 39B) three-and-a-half miles (5.6km) away by the transporter, or "crawler," a 3000-ton vehicle bearing a load of 6000 tons and moving on a specially-built trackway at a maximum one mile per hour (1.6kph). The Saturn would eventually depart from the launch Center at a somewhat higher speed.

Command Module Development

As early as 1960, engineers of the then Space Task Group had been working on a command module configuration which in essence changed little in the subsequent years. It was the vehicle in which three astronauts would travel to and from the Moon. It was a slightly flattened cone in shape: 13 feet (4m) in diameter, 11 feet (3.4m) long at its widest and weighing six tons. For the astronauts who flew on Gemini it was positively roomy, particularly in zero gravity, with four times the volume. Like Mercury

1, 2 *The Apollo crew escape system reverted to the Mercury concept of a rocket-propelled escape tower. The Apollo command module was much larger than the earlier manned spacecraft, so thoroughgoing tests of the new system were required. These two pictures show the commencement of one test at the White Sands Missile Range, and the moment when the tower separated from the "boiler plate" command module, but before the recovery parachute had opened.*

it was provided with a rocket-powered launch escape system and its ablative heat-shield was wrapped around the entire module, so as to withstand the 5,000°F (2,760°C) heat build-up as the vehicle plunged into the Earth's atmosphere at 25,000mph (40,230kph) on its return from the Moon. An additional value of the wrap around shield was that it would protect the crew from radiation beyond the Earth's atmosphere. Compared with Mercury and Gemini, Apollo was to have highly-sophisticated navigation and other equipment, complete with a powerful computer. Twelve small thrusters were used to control its attitude in space.

Fuel cells with the necessary tanks of liquid hydrogen and oxygen for generating electrical power as well as drinking water for the crew; oxygen for the

Chaffee (a group three astronaut) – were in the command module on top of the launch stack at the Kennedy Space Center to conduct a test that would verify that the combined launcher/spacecraft could operate on independent power. They were fully space suited and sealed in the command module as for an actual launch. There had been many delays and it was early evening before the test proper was ready to begin. Suddenly a shout was heard over the radio circuit from inside the module: "There is a fire in here." Shortly afterwards a sheet of flame flashed from the spacecraft. The hatches could not be removed until more than five minutes after the shout was heard. Doctors found the three crewmen dead, and subsequent post mortems showed that they had died from asphyxiation caused by the inhalation of toxic gases.

Of course, it had always been realized that spaceflight was dangerous – but the shock generated by the tragedy was perhaps compounded by the fact that it took place on the ground. NASA immediately launched a full-scale investigation, which reported in a little over two months in a document of almost 3000 pages in length. There was detailed condemnation of carelessness – a wrench had been found imbedded in the spacecraft's wiring, for example – but there was much more emphasis on an overwhelming failure to realize the dangers from an accumulation of developments which had made a tragedy almost inevitable. Although the precise cause of the fire could not be found, wiring was vulnerable, combustible material existed throughout the spacecraft, plumbing carried a combustible and corrosive coolant, provision for fast crew escape was totally inadequate and, most important of all, on the launch pad the spacecraft atmosphere was of pure oxygen at a pressure of 16 psi (1.1kg/cm²), whereas in space it was 5 psi (0.352kg/cm²): under the former conditions even steel could be set on fire.

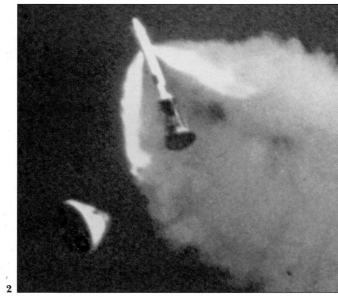

1

3 *Flight tests concentrated on different objectives. The command module here being mated to its service module at the Kennedy Space Center was placed on top of the first Saturn V to be launched, on November 9, 1967. The mission was designated Apollo 4 and combined a test of the launcher with the command module's performance when subjected to entry into the Earth's atmosphere at the velocity of a returning lunar mission.*

environmental system (exhaled carbon dioxide was not discarded but "scrubbed," using lithium hydroxide and recycled into the system as oxygen); and a 20,500lb (9,300kg) thrust engine, which fired to put the crew into orbit around the Moon and again to send them back to Earth, were located in a service module which was attached to the blunt end of the command module. The engine was powered by *hypergolic* propellants, which burned spontaneously when mixed – the absence of an ignition device being a safety factor, as was the use of pressure from helium gas rather than temperamental pumps to inject the propellants into the thrust chamber. The service module was a cylinder with a diameter (like the command module) of 13ft (4m), a length of 24ft (7.3m) and fully loaded it weighed 26 tons. It had 16 low-power thrusters for attitude control, and on the journey to and from the Moon it typically rolled about its long axis in the so-called "barbecue mode" to distribute the heat from the Sun.

The command and service modules went out to tender in 1961 and North American Aviation was successful. Two versions of the command module were planned: a Block I, which would be used in Earth orbit and had no provision for joint flight with the lunar module, and a Block II, which would have the necessary docking equipment for such flight. It took many months for the design details to be defined, but production of the Block I modules began in September 1964. Almost two-and-a-half years later disaster struck.

Tragedy at Kennedy Space Center

The program had progressed to the stage where the first manned launch was scheduled for February 21, 1967. On January 27, the crew of Apollo 1 – Virgil Grissom, Edward White (both veterans) and Roger

Major Equipment Modification

The NASA reaction was thorough. Almost 1,700 detailed changes were considered and over 1,300 were implemented – one of the most important being the change to an atmosphere in the command module while on the launch pad of 40 percent nitrogen and 60 percent oxygen, which would be slowly changed to pure oxygen after orbit was reached. A new hatch was developed which could be opened from outside or inside within five seconds. The safety changes were quite logically extended to the lunar module as well as further afield; for example, the nylon, outer layer of the pressure or space suits was replaced by an inorganic substance that would neither catch fire nor produce toxic fumes. In addition, there were major changes in both NASA and North American management after the tragedy, and the recovery took 14 months. As a memorial to the crew the designation *Apollo 1* remained unused by later missions: sadly, a further reminder of the dangers of the new frontier occurred within less than three months when Soviet cosmonaut Vladimir Komarov was killed flying a new spacecraft, *Soyuz 1*.

The lunar module could be regarded as the first, and so far only, true space ship used in manned spaceflight. It was designed for lunar operation only, and the absence of any appreciable atmosphere on the Moon made the typical aerodynamic shape and ruggedness demanded of a spacecraft facing re-entry into the Earth's atmosphere totally unnecessary. In the elegant engineering concept adopted, a combined, two-part craft composed of a descent stage and an ascent stage went down to the lunar surface. At the conclusion of the mission the ascent stage blasted off from the descent

4

5

7

4 The scene of a tragedy. The Apollo 1 command module 012, showing the effects of the flash fire which killed three astronauts on January 27, 1967, during a test of the CM/launcher power systems.

5 The crew of Apollo 1: (left to right) Virgil Grissom, Edward White and Roger Chaffee photographed shortly before the accident. Chaffee had not flown in space before.

6

6 These four photographs were included in the final report of the Review Board and show different parts of the fire-damaged interior of the 012 command module. Photographs were taken as each of 1,000 components was removed – approximately 5,000 images in all being taken during the ten week investigation. All four areas recorded here were in the general location of commander Grissom's couch.

7 Seated at the witness table before the Senate Committee investigating the Apollo 1 accident are: (left to right) Robert Seamans, NASA Deputy Administrator; James E. Webb, Administrator; George Mueller, Associate Administrator for Manned Space Flight; and Maj. Gen. Samuel Phillips, Apollo Program Director. Well over a thousand changes were implemented by NASA, two of the most important being a total redesign of the CM hatch and a change in the atmosphere of the module during the conduct of tests on the ground.

stage, which was used as a launch platform and left on the surface. Once in orbit, the ascent stage, using its 16 attitude thrusters, docked with the command module following which it, too, was jettisoned. Thus, the LM could be purely functional, and in essence comprised one box on top of a larger box which had four landing legs attached. It was somewhat flimsy, with paper-thin walls and spindly legs, but large, weighing 16 tons and standing 23 ft (7m) tall. In the words of one senior NASA manager, it looked like a "gargantuan, other-world insect."

While work on the Saturn launcher and the command module could proceed very early on, the lunar orbit rendezvous mode of visiting the Moon was not fully

approved until well into 1962, and it was January 1963 before the Grumman Aircraft Engineering Company of Bethpage, New York, was awarded the contract. Even then, while the general concept and some detailed aspects of the project were agreed and could proceed, it was some time before certain crucial elements could go forward. For example, with no information until a relatively late stage on the nature of the lunar surface, the design of legs and landing pads was held up. (Eventually NASA ordered that the pads were to be increased to over three feet, almost one metre, in diameter, and in order for the LM during launch to fit into a protective adapter between the service module and the Saturn S-IVB stage the legs had to be retracted.) These uncertainties obviously affected the design of the overall spacecraft, and considerations such as weight, which in turn affected the allowance for propellants and so on. In fact, it was not until well into 1965 that most details were settled. Large windows for good visibility during landing, for instance, had to be abandoned for structural reasons, and seats for the astronauts were omitted because if they stood during the landing their view of both instruments and the lunar surface through the redesigned, small triangular windows was improved, besides achieving a saving in both weight and space in the module.

Thus, through no fault of its own, Grumman started late on a project which steadily came to be regarded as the most critical in the entire Apollo venture. While there was much complex equipment to be developed for the LM, especially that for rendezvous, guidance

1 *Apollo can be regarded as the ultimate in technological achievement. It seems somewhat bizarre, therefore, that for some time it was proposed that the astronauts should descend to the lunar surface from the LM by means of a knotted rope! This picture shows a simulation being conducted at the Grumman plant in Bethpage, New York, in 1964.*

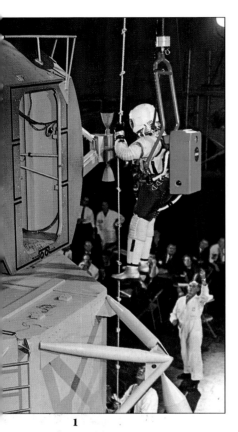

1

and navigation, with the inevitable difficulties involved, the major concern centered on the two main engines. The descent stage was powered by a throttleable engine, the thrust of which, for control during landing, could be as high as almost 10,000lb (4,536kg) and as low as 1,000lb (almost 454kg). This was a new concept in manned spaceflight and the development of the engine and its reliable operation can be regarded as the biggest challenge and the most outstanding technical development of Apollo, but along the way the difficulties were many. Although the ascent stage engine, developing 3,500lb (1,588kg) of thrust, was less complicated it, too, suffered from problems, such as unstable combustion, and the achievement of maximum reliability was essential, since this engine *had* to work if astronauts were not to be stranded on the Moon. Risks could be limited by sensible engineering, the selection of batteries rather than fuel cells for electric power and of hypergolic fuels for both main engines, but the problems were numerous. The weight grew inexorably despite weight-reduction programs, and at a very late stage stress corrosion cracks in the lander's aluminum structure were discovered.

Nonetheless the challenges were met and, although delayed, the lunar module became available in time to play a brilliant part in Apollo's success: how odd that at one early stage of this high-technology project it was being seriously proposed that astronauts move between the crew compartment and the lunar surface by means of a knotted rope!

During the years of Apollo, "Mission Control" had high visibility, both for those in the program and the general public. The scenes of what might be termed "purposeful and efficient calm" in the main control room were frequently shown on television and somehow seemed to suggest both the infinite attention to possible life-saving detail and at the same time the awe-inspiring technological achievement that was Apollo.

when the command module was orbiting behind the far side of the Moon. The mass of data coming back to Earth and being routed to Houston indicated the status of the spacecraft and the numerous sub-systems, the location of the spacecraft in space, and, by means of biosensors, the condition of the astronauts. The data were routed to five IBM 7094 model II computers, which processed the information and displayed it almost instantaneously on consoles in front of the specialist flight controllers in the control room, who operated in shifts under a flight director. Each of the controllers was able to call upon the advice and expertize of more specialists in support rooms adjacent to the main control room and elsewhere in the USA.

Much of the work was routine, in that the controllers monitored the progress of a mission and advised the astronauts when it was time for various procedures to be started or ended, watched over numerous matters of detail and sought further information from the crew if they had doubts about anything. But interspersed with long hours of routine were occasions when major decisions had to be taken in split seconds in a potentially life-threatening situation. One of the most dramatic of these was when computer program alarms began sounding in the Apollo 11 lunar module as it was close to landing. This was caused by two radar data streams, one of which was not in fact required for the landing, overloading the onboard computer, but recognition of this fact by the responsible flight controller, 27-year-old Steven Bales, was required instantly, otherwise the landing would have to be aborted in a matter of seconds. His "Go" was emphatic, and in due course earned him an award.

Dramas in Space

When, in November 1969, Apollo 12 lifted off from the launch pad through heavy cloud there was a light-ning strike which seriously affected the on-board electrical systems. Swift but systematic checking of the problem enabled the controllers to advise the crew of the corrective action to take and the mission proceeded to the Moon. The fact that the Saturn V and the spacecraft, at the insistence of Marshall SFC engineers, operated under separate control and guidance systems could also have been a major contributing factor to this successful outcome. Docking problems between the command module and lunar module on Apollo 14 had to be simulated on Earth before they could be resolved, and a lunar module computer problem during the same mission necessitated the development, verification and loading of a new computer program over the course of several hours before the landing could proceed.

However, the resourcefulness of Mission Control and the effectiveness of flight control/astronaut cooperation was demonstrated to its fullest extent during the Apollo 13 mission, which lifted off on April 11, 1970. After two uneventful days in space, and not long after the "capsule communicator" (CapCom) Joe Kerwin in Mission Control had told the crew, "The spacecraft is in real good shape as far as we are concerned. We're bored to tears down here," there was an explosion and command module pilot Jack Swigert uttered the words that were to become famous: "Houston, we've had a problem here." A later investigation was to show that an oxygen tank in the service module had suffered damage during various procedures prior to launch, and that an electrical failure caused it to explode, which in turn damaged the second tank, from which all the oxygen escaped.

The mission now faced crisis. The command module was rapidly losing both its supplies of power and oxygen,

2

2 *Apollo 12 faced a disturbing situation at launch in November 1969 when lightning affected onboard instrumentation. This image shows a lightning bolt to one side of the mobile launch tower some 36 seconds after the crew lifted off.*

Mission Control was the last element in a complex chain extending from the Moon. As an Apollo mission progressed a stream of data and telemetry was transmitted back to the Manned Space Flight Network (MSFN) on Earth. This was a development of the system that started to be established in Mercury days, and which for Apollo included the large NASA deep-space "dishes" or antennae located in Spain, Australia and California. The addition of these stations strategically located around the Earth meant that Apollo lunar-mission crews were never out of contact with Mission Control, save

3

4

6

7

5

8

9

Apollo 13 may be described as a heroic, brilliant failure, based on teamwork between the astronaut crew and mission control directing a full scale support effort on the ground.

3, 4 *Flight directors, astronauts and NASA management confer in these two images. Standing in 3 are astronaut James McDivitt and, at right, astronaut chief Deke Slayton. In 4 the latest weather forecast for the proposed landing area in the South Pacific is being studied.*

5 *A picture taken by the Apollo 13 crew shortly before re-entry shows the extent of the damage to the service module.*

6 *Mary Haise, wife of the Apollo 13 lunar module pilot Fred Haise, has the situation explained to her by astronaut Gerald Carr in the viewing room at mission control.*

7 *Thomas Paine, NASA Administrator, reports on the telephone to President Richard Nixon.*

and as the result of a midcourse correction firing of the service module engine, which was now inoperative, the command module *Odyssey* and the lunar module *Aquarius* were not on a free-return trajectory to Earth. There was almost immediate agreement between crew and Mission Control to regard the lunar module as a "lifeboat." Systems were shut down in the command module and all three astronauts moved into *Aquarius*,

primarily to use the oxygen which had been due to last two men for two days: now it had to last three for almost four days.

With as much power as possible switched off in the LM, it became cold and damp, and water as well as food was limited. More important, however, was the fact that the module's environmental control system could not cope with the amounts of carbon dioxide being

Scenes in mission control as the Apollo 13 crew return home safely.

8 *The display screen shows the command module descending to the water beneath three deployed parachutes.*

9 *The cigars have an extra meaning on this occasion.*

1 *As early as 1964, studies were being undertaken to determine the problems astronauts might face in working in the Moon's one-sixth gravity. Shown here at NASA's Langley Research Center is a system of slings supporting most of the test subject's weight and thus simulating lunar conditions.*

2 *Simulation of a different kind: Apollo 13 astronauts James Lovell (left) and Fred Haise practice procedures for a lunar traverse at Kilauea in Hawaii. Various sites were selected to give the astronauts experience in recognizing, selecting and documenting geological specimens.*

generated by the three crew members. Back at Mission Control a system was worked out to modify the lithium hydroxide canisters in the command module so that they could be used in the lunar module. Instructions were transmitted to the crew and they made the modifications using tape, plastic bags and cardboard, "just like building a model airplane." So the crew was temporarily safe if very uncomfortable.

But there was still the problem of the trajectory. Again *Aquarius* came to the rescue, since its descent engine, designed to take mission commander James Lovell and lunar module pilot Fred Haise down to the lunar surface, could be fired in place of the service module engine to alter the trajectory for return to Earth. Mission Control worked out the navigational details and a 35 second "burn" of the LM descent engine took place five hours after the explosion. As the Moon

1

approached, Mission Control decided on another engine firing to reduce the time taken for the return journey. It was essential for the crew to get a navigational "fix" on a star to ensure that the spacecraft attitude was correct and that they were accurately aligned for this, but there was so much debris from the ruptured oxygen tank accompanying them that a star sighting proved impossible. Once again, it was another suggestion from Mission Control that solved the problem: to use the Sun (the strong light from which would not be affected by the debris) for the alignment, and this was sufficiently accurate for the five minute burn to take place successfully. As the crew neared Earth they went back into the command module, which would safeguard them during re-entry, separated from the damaged service module which they photographed, and then said goodbye to *Aquarius*, which had saved their lives.

On April 17, 1970, at the end of a mission that might be described as a brilliant failure, the astronauts touched down in the Pacific to find that many millions of people the world over had been following their exploits. The sobering thought was that if this had been Apollo 8, which flew to the Moon without a lunar module, the crew would have died.

The reference to navigation raises an issue which was of vital importance to the success of the Apollo missions generally. The complex problem of navigating to the Moon was solved by cooperation between the

2

spacecraft and those back on Earth. Both command and lunar modules were equipped with an inertial guidance system which effectively gave the precise *direction* of the spacecraft at any time. Information as to its *position* was obtained by accelerometers linked to the onboard computer, with updating being supplied either via telemetry from Earth or by an astronaut taking a sighting on a target star and entering the information into the computer. The guidance and navigation system was developed for NASA by MIT (the Massachusetts Institute of Technology). For the most part Earth-based radar tracking was able to compute a spacecraft's position and the parameters of its orbit or trajectory with great accuracy, and the data was then transferred to the onboard computers, although the systems on the spacecraft would have been of vital importance if communications with Earth had broken down. However, while the radars and computers back on Earth did the bulk of the work, for critical events such as engine burns and lift-off from the lunar surface, information was required from the astronauts on the alignment of guidance platforms and on the position of target stars before subsequent computations of trajectories and timings could be calculated.

The importance of this work was revealed in subsequent comments, such as that by Buzz Aldrin, the Apollo 11 lunar module pilot, who stated that if a second firing of the command module engine in lunar orbit during the mission had been as little as two seconds too long the crew would have been on an impact course to the other side of the Moon. It is seen, too, in the accuracy of landings on the Moon and on the return to Earth. Thus, Pete Conrad on the Apollo 12 mission to Oceanus Procellarum put the lunar module *Intrepid* down just 535ft (163m) from his target, the unmanned, soft lander Surveyor III, and Alan Shepard during Apollo 14 landed the LM *Antares* only 175ft (53m) from the targeted site.

At the beginning of the program, it was stressed that the command module returning to Earth from the Moon, after traveling for almost a quarter of a million miles, had to hit a 27-mile (43.5km) corridor "where the air was thick enough to capture the spacecraft and yet thin enough so as not to burn it up." When the Apollo 14 command module *Kitty Hawk* landed in the Pacific on February 9, 1971, it was less than 3,300ft (1,000m) from its landing site. In fact, there was a suggestion that recovery ships should keep a little further away from landing sites for fear that one would be hit by the returning command module!

The Astronaut Challenge

Whatever the dependence upon computers and technology, the essential objective of Apollo was to land men on the Moon, and they too had to meet the challenge. With the selection of six scientist astronauts in June 1965, and a further 19 pilot astronauts in April 1966, a pool of some 50 or so men was available for assignment. All the missions that flew were commanded either by Mercury veterans (Schirra and Shepard) or those with Gemini experience. Understandably, crews for the earlier missions contained astronauts with more seniority, but in due course men from the fifth (1966) group, too, were selected for a trip to the Moon. In 1963 Deke Slayton was placed in charge of flight crew operations, which meant in essence that he selected crews. In general, his selections appeared to be based on general competence and reliability, with added emphasis attached to experience and seniority. While assignment to a particular Apollo crew and mission was on merit as interpreted by Slayton, luck could play an important part in the actual outcome. Thus Michael Collins was originally scheduled to fly on Apollo 8 but surgery caused him to be shifted to Apollo 11. Again, in time a system emerged whereby the crew backing up a "prime" crew could expect to become a prime crew itself three missions later. Pete Conrad commanded the back-up crew for the mission which eventually flew as Apollo 9, while Neil Armstrong commanded the back-up crew for the Apollo 8 mission. However, if all had gone according to the original plan, the Earth-orbiting Apollo 9 mission would have flown *before* the Apollo 8 mission that orbited the Moon at Christmas 1968, in which event Conrad and not Armstrong would have been the first man to walk on the Moon.

But luck had little to do with the long hours of training faced by the crews. There is no doubt that Gemini flights were a valuable experience for the

dangerous than the LM – Neil Armstrong for one having to eject from it on one occasion to save his life.

However, the single most important element in the training was that spent in the command module and lunar module simulators, which were built at both the Manned Spacecraft Center in Houston and at the Kennedy Space Center in Florida. These were computerised and simulated the events of a mission, in sound, vision and movement, with the equipment controllers being able to present the crews with virtually any form of emergency. (When the emergency was not dealt with successfully, there was usually an exchange of black humor between crew and trainers about the demise of the former. The external appearance of the simulators was also the subject of humor, John Young describing the command module simulator as looking like "the great train wreck.") But usually the business

3 The lunar landing training vehicle ("flying bedstead") was developed to familiarize Apollo crews with flying the lunar module. In this photograph the LLTV is being flown by Apollo 12 commander Charles "Pete" Conrad.

4 The training vehicle appeared to be more dangerous than the lunar module! On May 6, 1968, astronaut Neil Armstrong had to use the rocket-propelled ejection seat, which is shown here lifting him away to safety.

5 Later, as the LLTV burned on the ground, a 16mm documentary motion picture camera recorded Armstrong floating to the ground by parachute.

3

4

astronauts but, as in Project Gemini, dedicated training for a particular Apollo mission to come was even more valuable. The training included experience of zero-G in KC 135 aircraft flying the so-called "Keplerian trajectories," and longer periods of time in neutral buoyancy water tanks. In preparation for EVAs on the lunar surface, crewmen experimented with a device that partially supported them, so that an impression of walking and working in the Moon's one-sixth gravity was obtained. They also worked spacesuited in the vacuum of an environmental chamber, and underwent background geological training (including field trips), as well as working in spacesuits in a specially-prepared area where they rehearsed the activities to be conducted on the lunar surface. Lunar landing commanders gained some practical experience of what it would be like to fly the craft by piloting the lunar landing training vehicle (LLTV), which was affectionately known as the "flying bedstead" and which appeared to be considerably more

in the simulators was serious, and many hours were spent there. And those hours increased with the complexity of the mission. Thus, in preparation for Apollo 7, the first Earth-orbiting mission, the three-man crew spent 600 hours in the command module simulator operating the 725 manual controls. James McDivitt, commander of Apollo 9, estimated that he and his crew spent seven hours in training for each of the 241 hours spent in space, while the Apollo 11 crew reportedly spent a total of 2,000 hours in simulators.

Crew experience of living in space, and their reactions to it, gave insights into them as persons as well as affecting changes in flight conditions where appropriate. Of considerable importance on the Apollo 8 and 9 missions, for example, was the occurrence of nausea, possibly caused in both cases by the amount of movement permitted to the crewmen by the greater space available in the Apollo vehicles compared with the Gemini spacecraft. More prosaic issues affected crew

5

APOLLO – MAN ON THE MOON

attitudes. As time went on firm attempts were made to improve the forms of food that could be eaten in zero-gravity, but crews continued to grumble about items tasting like, for example, "granulated rubber." While urination aboard the spacecraft presented few problems, defecation into bags was awkward, unpleasant and time consuming. Water condensed on pipes, windows were obscured and there tended to be noise in the command module from clattering fans, which over a period of time became wearing, with sleeping during a mission presenting problems. For the earlier lunar landing crews, the quarantine isolation on return to Earth (lasting a nominal 21 days in the case of Apollo 11) was understandably and strongly disliked.

Other events gave the public some appreciation of astronaut personalities and attitudes. The Apollo 9 crew chose the call signs *Gumdrop* and *Spider* for their command module and lunar module respectively, while Apollo 10 opted for *Charlie Brown* and *Snoopy*. A senior public affairs officer at NASA HQ expressed a wish that later call signs would be somewhat more "dignified," and Apollo 11's *Columbia* and *Eagle* well represented the more appropriate tone of those that followed.

When the Apollo 10 lunar module crew separated the ascent stage from the descent stage in lunar orbit, the former lurched wildly (because of a switch being in the wrong position) and caused lunar module pilot Gene Cernan to shout "Son of a bitch!" This ejaculation caused something of a stir at the time but seems quite restrained compared with what might have been said in such a situation. Less restrained were the comments of Apollo

It was stated earlier that astronaut seniority was one element in Deke Slayton's choice of crews. From Apollo 13, which was launched in April 1970, onwards, members of the fifth group of astronauts selected in April 1966 began to be appointed as prime crew members, ahead of any of the scientist astronauts who joined NASA before them. Eventually, one of that latter group, Harrison "Jack" Schmitt, a geologist, landed on the Moon during Apollo 17, the last of the lunar missions. But a previously-chosen group five pilot astronaut had to make way for him, and the Schmitt selection was a recognition by Slayton that there would have been uproar if a scientist had not gone to the Moon before Apollo ended. This would have resulted from a situation which had existed for years.

Scientists Against Apollo

Since the beginning of the space age, a steadily-growing number of scientists had seized the opportunity presented by the new technology greatly to expand study of the near-Earth environment and deeper space in a manner impossible from the Earth's surface. From its birth NASA had sought to promote and fund this space science, which was carried out mainly by scientists in the outside, academic world. Virtually all of the work was conducted using unmanned probes, and very early on there appeared a strongly-held opinion on the part of many of these scientists that instruments sent into space aboard unmanned probes were far more productive scientifically than man, and usually cost less. They were appalled at the implications of the Apollo program, and its potential

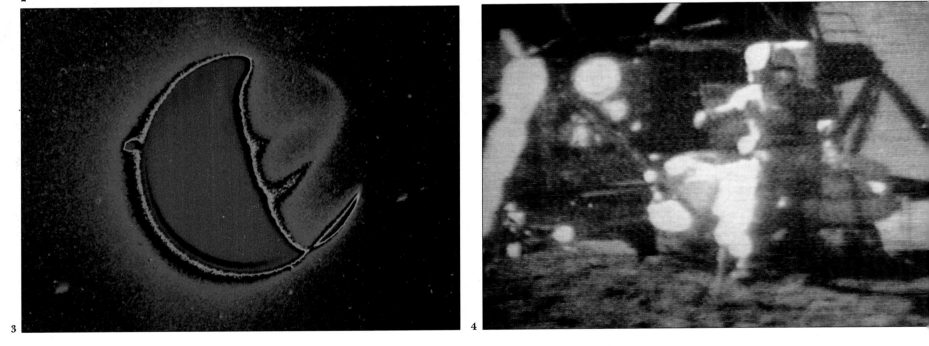

1 Homer Newell, head of NASA's Office of Space Sciences, who faced the difficulty of reconciling the differences between engineers and scientists during Project Apollo.
2 A photomicrograph taken by polarized light of a thin section of lunar rock.
3 A false-color version of a far ultraviolet image of the Earth taken by the Apollo 16 astronauts in 1972.
4 A hammer and a feather dropped at the same time by Apollo 15 commander David Scott hit the lunar surface simultaneously, demonstrating Galileo's assertion that in the absence of air resistance objects fall at the same velocity.

16 commander John Young when walking on the lunar surface. He complained in four-letter-word terms about the effect of drinking too much of the potassium-rich orange drink which the doctors had recommended, not realizing that he was transmitting the message to the world. Apollo was in deadly earnest, but it did not stop the inimitable, 5ft 6in (168cm)-tall Pete Conrad from proclaiming as he jumped down from the last rung of the Apollo 12 LM ladder onto the surface of the Moon: "Man, that may have been a small one for Neil [Armstrong] but that's a long one for me." A very different aspect of astronaut behavior was that of the Apollo 15 crew, which initiated an arrangement for the exploitation of some postal covers they had taken to the Moon, and for which action they received an official reprimand from NASA.

impact on NASA funding of space science, and many were never reconciled to the program, despite the heroic efforts of Homer Newell, head of NASA's Office of Space Sciences (later Office of Space Science and Applications) to bridge the gap. The situation was not helped by the decision that the science due to be conducted by the Ranger and Surveyor probes to the Moon must be subordinated to Apollo's desperate need for information about the lunar surface. Against this background of an extensive opposition to Apollo was a more limited viewpoint, which was that, if it must proceed, then science must be at least a major element in the planning, with scientists going to the Moon. And this is where dissension arose.

From its outset, the development of manned spaceflight had been shaped essentially by operational requirements, which dictated that the safety of the crew was paramount, followed closely by the successful accomplishment of mission objectives. In the early stages, the latter requirement was largely devoted to evolving the complex equipment, techniques and procedures of spaceflight itself. The target of landing men on the Moon had greatly increased the extent as well as the complexity of this requirement. To some at least of the engineers, specialists and managers wrestling with the problems, "science" was at best a distraction and at worst a menace which greatly compounded their problems. Hence arose a view among some scientists that the Manned Spacecraft Center was unsympathetic or downright obstructive to the cause of space science.

When the first group of scientist astronauts was selected, in June, 1965, it was made plain to them that they had to train as pilots, if they were not already, and that they must undertake all the normal training chores besides endeavoring to keep up with their chosen scientific specialisms. It was Slayton's belief that in carrying out a mission a highly experienced astronaut with a piloting background would always react more constructively to an emergency than a scientist, even if the latter had some flight training. As regards conducting science, the view was that pilot astronauts should be trained to perform scientific experiments. The assigning of a non-scientist astronaut, who was regarded as the best geology student in his crew, as command module pilot during the mission, which meant he would not go down to the lunar surface, was regarded by at least some of the scientist astronauts as typical of "operational" attitudes, but even more important was more junior pilot astronauts beginning to be assigned to Apollo missions. This led in 1971 to unrest, which caused Homer Newell to visit Houston for discussions with the scientist astronauts; his report was considered by NASA senior management and it was this that led eventually to scientist-astronaut Harrison Schmitt being assigned to the Apollo 17 mission.

Science to the Fore
There was undoubtedly considerable reason for this scientific concern, but there was another side to the story. Homer Newell, caught between the contending worlds, spoke of space scientists' "presumption of special privilege," which sometimes verged on arrogance. And there certainly was support among NASA managers at all levels for conducting science When the Kennedy challenge was thrown down, Administrator James Webb stated that he would not take responsibility for a program that subordinated all else to the lunar landing, and once the initial Apollo flights had been successfully accomplished, George Mueller, in charge of manned spaceflight at NASA's Washington HQ, instructed the centers to concentrate henceforth on lunar exploration. In 1967, a Lunar Receiving Laboratory was built at Houston. This was partly intended to house lunar samples under optimum conditions and partly to act as a quarantine for astronauts and specimens, to prevent any possible contamination of the Earth's environment by alien microorganisms if they existed, but some NASA senior managers, including James Webb, hoped that it would become a Lunar Science Institute. At the MSC, Max Faget, who had risen to be director of engineering and development, expressed the somewhat more prosaic thought that "it wouldn't look very good if we went to the Moon and didn't have something to do when we got there." The appointment in 1967 of a distinguished scientist, Wilmot Hess, as chief of a newly formed Science and Applications Directorate at the MSC was a step in the right direction, even if he only remained for a

relatively short time. One battle he fought successfully was to change a decision made during 1968 that, for reasons of operational safety, the Apollo 11 lunar landing mission should carry no scientific instruments at all.

Whatever the scientists might say, there had always been an implicit assumption in Apollo planning that, once it had been demonstrated that man could land safely on the lunar surface and return, operational limits and conservatism would be relaxed to yield greater geological and scientific returns. This policy was demonstrated most dramatically during the last three lunar landing missions. The command modules were modified to incorporate a Scientific Instrument Module (SIM) which conducted high-resolution photography and other remote sensing of the lunar surface, while, below, besides the deployment of a battery of instruments (ALSEP: the Apollo Lunar Surface Experiments Package), modifications to the lunar module enabled the two crewmen to stay on the surface for much longer, to bring back a greater weight of rocks and to go further afield aboard a four-wheeled, electric-powered *lunar rover*. The relaxation of landing constraints, which had limited the LM to a few degrees either side of the lunar equator and to a safe, relatively-flat region, meant that the last three missions could go further afield and to areas of greater geological interest: Apollo 15, for example, landing at more than 20°N, close to the lower slopes of the lunar Apennine Mountains. Apollo 15 may be regarded as the coming of age of Apollo lunar exploration, and a scientist from the prestigious California Institute of Technology paid tribute to it as "one of the most brilliant missions in space science ever flown."

There had been many disagreements and problems along the way but evaluations such as that should be placed in the scales of any final balancing of Apollo achievements.

Manned Missions Begin
As all the strands came together, then, unmanned test flights were begun. Between May 1964 and January 1966 four flights of Little Joe II vehicles demonstrated the reliability of the rocket-powered launch escape system. In February and August 1966, two Saturn IB launches put both the launch vehicle and the command and service modules through their paces in sub-orbital tests. Then, on November 9, 1967, towards the end of

5 Many of the photographs taken in orbit around the Earth and during the lunar missions reflected the engineering and scientific demands of Apollo. But, not infrequently, crew members secured images which appealed to the heart and senses more than the brain. Such is the case with this Apollo 7 picture of the Sun shining from space across the seas surrounding Florida. The Earth seems a warm and welcoming place.

6 The Saturn V is tested unmanned for the first time as Apollo 4 on November 9, 1967, reached an apogee of 11,400 miles before the command module returned in a 25,000 mph re-entry.

7 Apollo 6 on April 4, 1968, provided further opportunities for research and development.

1 *Apollo 7 tested the command and service modules in Earth orbit and Apollo 8 tested them in lunar orbit. By the time of Apollo 9, in March 1969, the lunar module was ready for its first tests in the relative safety of Earth orbit. Here the CM "Gumdrop" is photographed in separate flight from the LM "Spider." The docking mechanism is visible in the nose and a high-gain antenna just below. A NASA official was reported to have wished for more appropriate code names for the vehicles on future missions!*

2 *Another of the splendid views obtained during the Apollo 9 mission. Here CM pilot David Scott stands in the hatchway of Gumdrop and is photographed by Russell Schweickart, who was in the lunar module along with mission commander James McDivitt. The LM ascent stage fills the foreground of the image.*

a year which had begun so tragically, came the major success of Apollo 4, with the first launch of a Saturn V and a re-entry of the CSM from a height of 10,000 miles (16,000km), which simulated return from the Moon. Despite the very high temperature outside, inside the cabin it remained at a very comfortable 70°F (21°C). An unmanned LM was flown in January 1968. The performance was not perfect, nor was that of a second Saturn V launch during the Apollo 6 mission in April 1968, but the problems shown up were understood and worked on. The equipment was now deemed ready for men.

Launched aboard a Saturn IB on October 11, 1968, Apollo 7 was a mission of just under eleven days in Earth orbit to demonstrate the functioning of the new Block II command module. As commander, Mercury astronaut Walter Schirra flew his last space mission, accompanied by group three astronauts Donn Eisele and Walter Cunningham. The fuel cells and electrical system developed some faults, but in general the mission was highly successful, particularly the performance of the CSM main engine, which was fired eight times as a demonstration of its readiness to put men into and take them out of lunar orbit. The decision to put a small black and white television camera aboard was only taken at the last moment, but the broadcasts proved very popular, as did the flotation bags which righted the CM after it had turned upside down as it hit the sea south of Bermuda on October 22.

Some time before Apollo 7 flew, George Low, who took over as spacecraft program manager after the Apollo fire, proposed that if the mission were successful, then Apollo 8 should be sent to orbit the Moon. This was daring and had not been planned: in the typical NASA step-by-step progression, the next mission should have been a test of the lunar module in Earth orbit, but the LM was still not ready. It appeared to Low, and to an increasing number of senior managers who began to warm to the idea, that any Earth orbit mission in the circumstances then existing would be of limited value. So the idea won the day, and the CSM bearing Gemini veterans Frank Borman and James Lovell, together with group three astronaut William Anders, lifted off from Pad 39 on top of a Saturn V on December 21, 1968. Robert Gilruth later summarised its importance: "Technically, it gave us information on our communication and tracking equipment for later missions, a close view of our landing sites and experience in cislunar space with a simplified mission. Politically, it may have assured us of being first to the Moon, since the stepped-up schedule precluded the Russians from flying a man around the Moon with their Zond [spacecraft] before we reached the Moon ...."

The stages of the mission progressed and the crew shared them with the millions on Earth by means of their black and white TV camera. On the afternoon of December 23 an historic event in human history took place when for the first time a body other than the Earth began to exert a gravitational pull on a manned spacecraft. Up to that point the Apollo 8 CSM had been slowing, but under the influence of the Moon it began to accelerate. At almost 70 hours into the mission came another historic first: the engine fired to put the spacecraft into orbit round the Moon and the crew began describing, tele-vising and photographing the lunar surface 70 miles (112km) below. They stayed for ten two-hour orbits, during which time they read from the Book of Genesis to "all of you on the good Earth." On the last orbit the critical engine firing to bring them back took place on the far side of the Moon, and when Apollo 8 once again came into view Jim Lovell told Houston "Please be informed there is a Santa Claus." Two days later the burned and scarred command module splashed down in the Pacific south of Hawaii.

Apollo 9 was to spend ten days in orbit, and at long last mission commander James McDivitt would have the chance of flying the lunar module in space for the first time. With him on the mission were David Scott, who had flown during Gemini, and another group three astronaut, Russell Schweickart. Shortly after launch into orbit on March 3, 1969, the CSM undocked from the S-IVB stage, turned in space, and docked with the lunar module which had been riding below, protected within an adapter attached to the top of the S-IVB. This manoeuvre – known in the jargon as *transposition and docking* – was to be repeated on every mission thereafter once the cluster had been boosted out of orbit toward the Moon by the third stage of the Saturn V. One hour later, the docked spacecraft were ejected from the S-IVB and the LM *Spider* and CM *Gumdrop* could start work.

As the mission progressed McDivitt and Schweickart tested the lunar module in separate flight for the first time. They extended its legs, activated the attitude thrusters and fired the descent engine. In due course they jettisoned the descent stage and fired the ascent engine also for the first time in space. From a position 10 miles (16km) below and 80 miles (almost 130km) behind Dave Scott in the command module, they then began a rendezvous and docking maneuver in much the same way as it would be performed in lunar orbit.

Earlier in the mission Schweickart had suffered some nausea, but he recovered and proceeded to test the Apollo spacesuit, complete with its lunar surface back-pack or PLSS (Portable Life Support System) containing coolant water and oxygen, during a stand up EVA in the porch of the LM. With Scott emerging from the command module the astronauts took a number of dramatic pictures of each other. The numerous images

3 Apollo 8 brought back to Earth the first high resolution color photographs of the Moon. The crater Langrenus is at lower center.

3

exposed during the last few days of the mission, however, before the CM splashed down in the Atlantic on March 13, emanated from an important multispectral remote sensing experiment.

There was a suggestion at one stage that Apollo 10 might land on the Moon, but the lunar module was of an early design that was too heavy to make a successful return into orbit if it had touched down. So the all-veteran crew of Tom Stafford, John Young and Gene Cernan that lifted off on an eight day mission on May 18, 1969, was essentially to build on what the Apollo 9 crew had done in Earth orbit: test the LM in separate flight but this time under lunar conditions, to take it down close to the lunar surface and then rendezvous and re-dock with the command module. This was also an excellent rehearsal for the specialists and controllers back on Earth, but without the ultimate, additional complexity of a landing. For the general public there was the added attraction of a color TV camera being

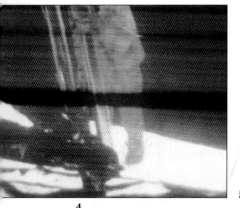

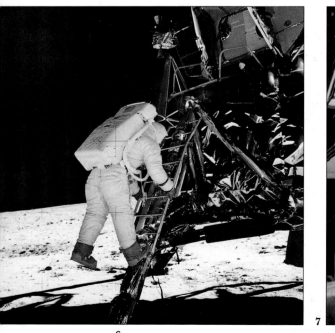

aboard the command module. The rehearsal flight of *Snoopy* and *Charlie Brown* went well, much time was spent on studying and photographing potential landing sites, and the way for Apollo 11 was prepared.

A television audience estimated at 1,000 million reputedly watched the saga of Apollo 11 unfold in July 1969, and since then it has been chronicled in many ways, many times over. Michael Collins subsequently expressed the view that he regarded the chances of a successful landing and return as "no better than evens." But Neil Armstrong, Buzz Aldrin and Collins succeeded in the daunting task, backed by the tens of thousands of workers, specialists and managers who had labored in the cause of Apollo. No matter that the stay time of the astronauts out on the lunar surface (the hatch was open for just over two-and-a-half hours) was short compared with those who were to follow. This was the

first time, and it was primarily "operational" within the NASA meaning: to test and demonstrate equipment and procedures carefully, as quickly as possible and safely. The elaboration would come later.

Apollo 11 can be remembered for some felicitous phrases. As command and lunar modules separated in lunar orbit there were Armstrong's words "The *Eagle* has wings!" There was his comment seconds after landing: "Houston, Tranquility Base here. The *Eagle* has landed," followed by the CapCom's inelegant but graphic reaction: "Roger, Tranquility. We copy you on the ground. You've got a bunch of guys about to turn blue. We're breathing again. Thanks a lot ...." And, although it started a lengthy debate on the absence or otherwise of a spoken indefinite article, there was Armstrong's comment on stepping from the LM's footpad: "That's one small step for [a] man, one giant leap for mankind." On the lunar surface, Aldrin spoke of its "magnificent desolation," while still remaining on one of the legs of the lunar module descent stage left behind is a plaque showing two hemispheres of the

1 *The beginning: 9:32 AM (EDT) July 16, 1969, and the Apollo 11 crew of Neil Armstrong, Edwin "Buzz" Aldrin and Michael Collins lifts of for the Moon.*

2 *Mission accomplished – July 21. Eagle approaches Columbia.*

9

3 *Some hours before launch Neil Armstrong is helped by a technician to "suit up."*

4 *Four days later Armstrong takes the epoch-making first step down onto the lunar surface, the event recorded for posterity by a black and white TV camera. The black bar was a television system anomaly.*

5 *All of the photos of an Apollo 11 astronaut on the lunar surface released by NASA at the time were of Buzz Aldrin, photographed by Armstrong. Research conducted by Englishman H.J.P. Arnold almost twenty years later identified this picture as being of Armstrong working in the shadow of the LM. It is the only original, still photo taken of the Apollo 11 commander on the lunar surface.*

6 *Aldrin carefully backs down the LM ladder to join Armstrong.*

7 *Flight controllers display justified elation in Mission Control as the Apollo 11 mission is successfully completed.*

8 *The general quality of Armstrong's photography was excellent. In this classic image of Aldrin, his convex visor shows the lunar module, the US flag, the staff-mounted TV camera – and Armstrong himself taking the picture.*

9 *Astronaut boot and footprint.*

8

1 *Nine months separated the aborted Apollo 13 mission and Apollo 14, which landed in the Fra Mauro region in January 1971. Alan Shepard returned to flight status and is shown here working with geological tools. In front is a small transporter which helped lighten the astronauts' workload.*

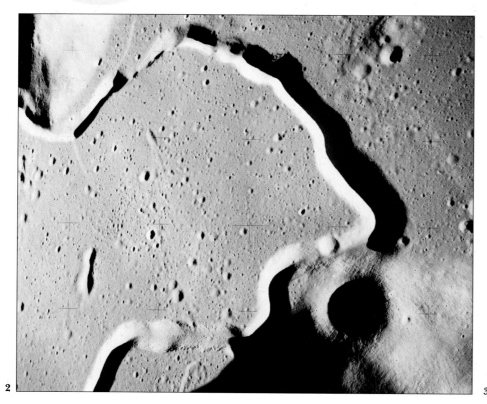

2 *Apollo 15 landed close to Hadley Rille, a valley that winds sinuously along the Apennine Front of the Moon. This high-resolution image shows the actual landing site, which was a little above and to the right of the "nose" feature.*

3 *Apollo 12 landed with precision close to the Surveyor III unmanned lander. The LM Intrepid can be seen on the horizon.*

Earth with the words "Here Man from the planet Earth first set foot upon the Moon, July 1969 AD. We came in peace for all mankind." When *Columbia* touched down on July 24, 1969, at the end of the eight day mission it contained over 46lb (21kg) of priceless lunar rocks. The three crew members faced quarantine and then a lengthy period of inevitable and exhausting publicity. What else could there be for history-making heroes? None of them flew in space again.

The late President's challenge having been met successfully, lunar exploration could begin. In a ten day mission in November, 1969, Apollo 12 went west, to a mare region of the Oceanus Procellarum (Ocean of Storms), whereas their predecessors had gone east. Aboard the lunar module *Intrepid* Pete Conrad and lunar module pilot Alan Bean made a pinpoint landing close to the Surveyor III unmanned probe. Working carefully and unhurriedly, they conducted two periods of EVA in which they collected a total of over 75lb (34kg) of rocks and deployed a full scale ALSEP, while above in *Yankee Clipper* command module pilot Dick Gordon was carrying out a multispectral photographic experiment which would contribute to site selection procedures for future missions as well as providing general knowledge of the lunar surface. The crew splashed down on November 24 1969, and because of the near tragedy of Apollo 13 it was to be over 14 months before men walked on the Moon again.

Then, as commander of Apollo 14, Mercury veteran Alan Shepard returned to space, together with group five astronauts Edgar Mitchell and Stuart Roosa. The target of their nine-day mission beginning on January 31, 1971, was the same as that of Apollo 13: the Fra Mauro region, which was assumed to consist largely of material ejected when the massive Mare Imbrium basin was formed. Once again two EVAs were conducted; as the learning cycle progressed, they were longer than those conducted on the previous mission and the astronauts also had the use of a two-wheeled transporter,

Apollo 8. The addition of an electric-powered lunar rover enabled the astronauts to travel far more extensively, and a color television camera aboard the LRV, which could be commanded from Mission Control, enabled both scientists and public on Earth to join the crews on those journeys, and to watch their departure as the LM ascent stage blasted off from the lunar surface. The three missions demonstrated quite graphically the increased yield from each succeeding flight.

Apollo 15, crewed by David Scott, James Irwin and Alfred Worden, was launched on July 26, 1971, and landed in the Hadley-Apennine region. After the lunar module *Falcon* touched down, Scott and Irwin conducted

4 *The lunar module was no sleek aerodynamic shape: just two highly sophisticated boxes set on a quartet of landing struts. Here the Apollo 14 LM Antares moves away from the CM Kitty Hawk. Plainly visible at center is the docking tunnel entrance and at each corner the highly reflective landing strut pads.*

5 *The bright Sun shines from behind Antares and creates what the crew called a "jewel-like" effect.*

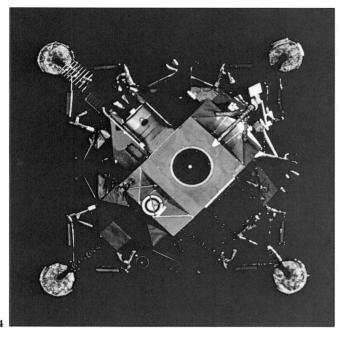

6 *On the way back from the Moon, the Earth comes between the Apollo 12 CM and the Sun, creating a solar eclipse for the crew's eyes only.*

7 *Apollo 12 commander Pete Conrad photographs lunar module pilot Alan Bean, and his own image is reflected in the visor. Bean is holding a lunar soil container.*

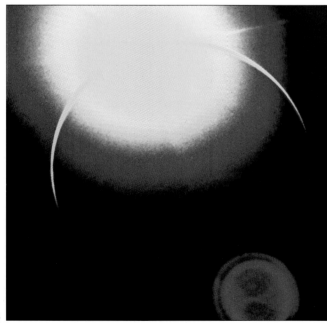

which they pushed or pulled and dubbed the "rickshaw." Some 95lb (almost 42kg) of rocks were loaded aboard *Antares* and another ALSEP deployed. Roosa's duties in the command module *Kitty Hawk* included high-resolution photography of the lunar surface with a special mapping camera.

Apollo 14 was a bridge between the early Apollo missions and the last three, in which both the command module and lunar module were modified and increased in weight to yield more data and a longer stay time on the lunar surface. The Saturn V was uprated to boost a payload of 116,000lb (over 52,600kg) to the Moon for Apollo 17, compared with 80,000lb (over 36,280kg) for

1 *On the lunar rover was a color TV camera operated by remote control from Mission Control in Houston. Here for the first time viewers see the moment of lift-off of an Apollo lunar module ascent stage – that of Falcon at the end of Apollo 15 surface activities. The unretouched, raw image has the appearance of an impressionist painting, and debris from the force of the engine firing fills the lunar sky with myriad rainbows.*

2 *Apollo crews recorded numerous Earth rises over the lunar horizon. This was one of the best, significantly from Apollo 17, the last lunar landing mission.*

3 *Apollo 15 commander David Scott takes the first ride aboard the rover.*

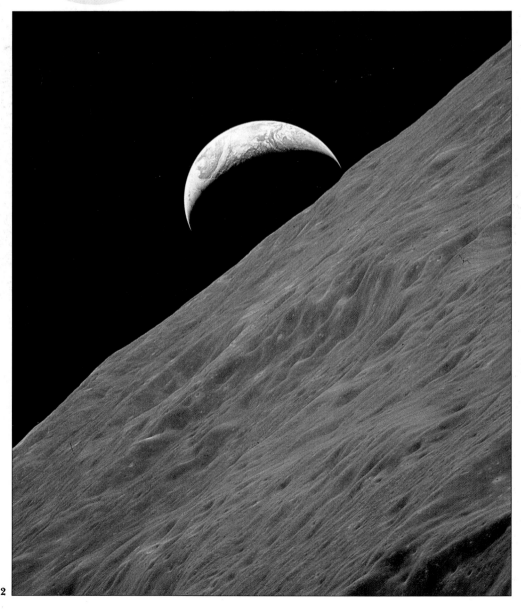

three major EVAs totaling over 19 hours in length, deployed an ALSEP and collected 169lb (almost 77kg) of rocks. Above, in *Endeavour*, Worden operated the new SIM Bay, which necessitated a spacewalk by him on the way back to Earth to retrieve film magazines from specialist cameras in the bay. The mission lasted for just under 260 hours.

Apollo 16, which landed in the Descartes Highlands region of the southern hemisphere, lasted for almost 266 hours, having been launched on April 16, 1972. The three EVAs conducted by John Young and Charles Duke from the lunar module *Orion* totaled over 20 hours in length and 208lb (almost 95kg) of rocks were brought back to Earth. Command module pilot Thomas Mattingly in *Casper* operated the SIM bay instruments.

The Apollo 17 mission to the Taurus-Littrow area, in the Moon's eastern hemisphere well to the north of the lunar equator, continued and completed the progression. Veteran Gene Cernan commanded the crew, which included geologist Harrison Schmitt, who joined him aboard the lunar module *Challenger*, and Ronald Evans, who remained aboard *America* operating the battery of instruments. The mission lifted off on December 7, 1972, and was the only night-time launch during the whole Apollo program. The three EVAs lasted for over 22 hours. The fifth ALSEP was deployed and 243lb (more than 110kg) of rock were collected. When the appropriately-named *America* splashed down in the Pacific on December 19, 1972, to bring the Apollo lunar landing program to a close it also established a record for the length of an Apollo mission; almost 302 hours, or twelve-and-a-half days.

Remaining in silence on the Moon, well to the north and a little to the east of that on the descent stage of *Eagle*, Apollo 17 left another plaque: it, too, showed the two hemispheres of Earth, but in addition there was a map of the lunar near side, with the Apollo landing sites marked and the words "Here man completed his first exploration of the Moon, December 1972 A.D." Twenty years have passed and we cannot predict yet when humans will begin the second exploration.

Legacy of Apollo

The NASA Administrator James Webb was accurate in his forecast. Apollo cost an estimated $25,000m and at its height 400,000 Americans, spread over NASA, the universities and 20,000 companies, were devoting themselves to it. A total of 838lb (almost 380kg) of lunar rocks were brought back to Earth, together with more than 33,000 photographs and 20,000 reels of magnetic tape, while four of the geophysical stations left behind continued operating until switched off by NASA in 1977. While the fundamental question of the Moon's origin is still unanswered, Apollo enabled geologists and scientists to reconstruct its geological history in great detail and to learn much about its structure, composition and internal temperature. The work continues and much of the large quantity of rocks returned is still available, awaiting the development of new laboratory tests and procedures. The largely-unsung Apollo Earth orbital missions carried out a number of scientific experiments and took around 800 terrain images – an

4 *Apollo 15 astronaut Alfred Worden spacewalks to retrieve films from specialist cameras in the service module.*
5 *The cameras are seen here.*
6 *Astronaut and rover at Hadley Rille.*

1 *This chart of Apollo lunar landing sites shows how the later missions, because of the project's growing experience, moved further away from the lunar equator.*

2 *Harrison "Jack" Schmitt, the only professional scientist to go to the Moon, is photographed by Apollo 17 commander Eugene Cernan during one of their EVAs in the Taurus-Littrow area. The US flag points to a gibbous Earth in the sky.*

3 *One of the science experiments included on some Apollo missions was the placing of a specially built reflector on the lunar surface. Laser beams directed from Earth and reflected back enabled researchers to calculate the Earth-lunar distance with great accuracy.*

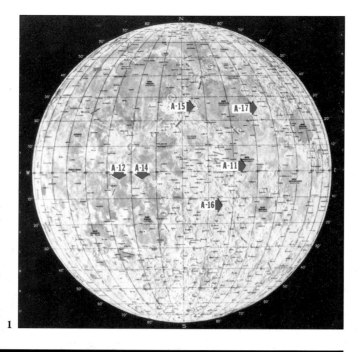

4 *In the Lunar Receiving Laboratory, Houston, a laboratory technician examines an Apollo 14 lunar rock specimen, which was one of the largest returned. "Big Bertha" weighed almost 20 lbs.*

acting as a surrogate for war in the rivalry between the two super powers – and America, it seemed, had other things to worry about.

President Johnson hoped to build a "Great Society," but the years of Apollo on the way to the Moon were marked by riots and violent demonstrations in American cities as the civil rights movement fought for what it considered to be its just dues, and "the establishment" was challenged by various sections of society on numerous fronts. Before the end of 1968, man had flown to the Moon, but it was also the year in which Martin Luther King and Robert Kennedy were slain, together with 15,000 Americans in Vietnam. NASA and others could point out that now was the time to reap the rewards of the new technology; that in the years since 1959, in sending man to walk on the Moon and probes to the planets, far less was spent on space than went in one year to the Department of Health, Education and Welfare; and that the view from deep space had shown the apparent fragility of Earth, thereby aiding the cause of the rapidly-emerging environmental movement that was so critical of the "waste" of money devoted to the space program. It was pointless, too, to point out the "spin-off" value from the space effort or the employment provided; pointless to argue that there was no guarantee that funds not devoted to space would be directed to other causes considered more worthy, or to argue that the success of Apollo, albeit in achieving a clear-cut target, demonstrated what could be achieved by a nation when given the political will, the spirit and the resources to do it. No, Apollo was yesterday's story; NASA's budgets were cut, three planned missions to the Moon were cancelled, and it had to strive to find a meaningful role for the future.

In the context of spaceflight and space science, Apollo – to some degree as it was taking place and strongly afterwards – was criticized on at least two grounds. At its height it consumed well over 60 percent of NASA's total budget, and it was argued that this bias took funds from unmanned space missions and applications. But there was no firm evidence to support this contention. With Apollo, NASA had effectively been given a world *realpolitik* task to perform, and the funds devoted to that may well never have been devoted to other projects in space. Secondly, it was argued that Apollo took the American space program off in a direction that was a dead end. What could be done, the argument ran, with very expensive Saturn Vs and Apollo equipment suitable only for going to the Moon when you were not going there any more? This approach implied that there was a much more flexible and logical way of going into space which was far more open ended. With hindsight it is possible to argue this, but, again, in the absence of the Kennedy challenge it was by no

Apollo 9 experiment pointing the way to a multispectral remote sensing application which reached its culmination in the unmanned Landsat satellites launched from 1972 onwards.

But President Kennedy did not throw down the Moon challenge for science: it was set as a counter to current Soviet superiority in space and to re-establish US pride and self-belief. It was, as he said, "time for a great new American enterprise." While there was a strong motivational purpose and it was a major initiative on the new frontier, in a hard-headed way he was fully aware of the risks, since no space project would "be so difficult or expensive to accomplish." It was to be a long haul and many Americans began to tire of it well before the goal was met. Moreover, once man had landed on the Moon then the precise nature of the challenge worked against the exploitation of the superb achievement. The race was run and won – perhaps

5

means certain that the funds would be provided for a different approach. Moreover, while the Apollo technology did not make use of reusable elements, given the political will and sound technical guidance it was a way to the future, for it could have led to a decade or more of major Earth-orbiting space station activity, as well as the possible creation of a semi-permanent base on the Moon with no significant new technology needed. While this was going on a new generation technology could have been evolved. What was missing was the national resolve to do it.

Apollo can be regarded in many ways. On a personal level, for those outside NASA and following it at the time there will be individual memories: of the forward leaning lope of astronauts in the one-sixth gravity of the Moon; of the exchanges full of jargon between astronauts and Mission Control, with the typical pause of a second or so resulting from the distance between

Earth and its satellite; and of individual traits, such as Pete Conrad chuckling and humming to himself as he worked with Alan Bean in a corner of Oceanus Procellarum. But, of course, Apollo was infinitely more, too. Even if its origins lay in world politics, it assumed an enormous philosophical significance for mankind as the first step was taken from the cradle described by Konstantin Tsiolkovskii. And, just "for one priceless moment," men and women almost the whole world over *were* as one.

There is a story that on the night of July 20, 1969, a small bouquet of flowers was placed on John F. Kennedy's grave in Arlington with a card on which were written the words: "Mr. President, the Eagle has landed." It would be nice to think that it actually happened.

5 *The last Apollo mission, in December 1972, obtained a magnificent color image of a full disc Earth. Since it was the northern hemisphere's winter, the planet tilts away from the spacecraft (and Sun line) at the top. The rapidly-growing environmentalist lobby criticized the space program for its heavy expenditure of funds, but from the beginning the views of Earth from deep space (emphasized by the astronauts' words) underlined that the planet was a very fragile world – a "lifeboat" in space – which needed to be cherished and safeguarded by its inhabitants before irreparable damage was done.*

6

UNMANNED EXPLORATION OF THE MOON AND PLANETS

US: FROM MARINER 2 TO THE VIKINGS

Scientists' knowledge of Mars and Venus in the early 1960s was severely limited. All that could be seen on Mars were vague markings and variable polar caps, while Venus hid its secrets below a brilliant white cloud deck. If life was going to be found anywhere else in the solar system, then Mars seemed to be the most likely host. But even Venus was not ruled out, though the surface conditions were a complete mystery. Despite its failures, NASA's Ranger was a capable spacecraft. In fact, it was very much over-designed for its lunar role because NASA used it as a test bed for more distant ventures. For the Moon, it could have flown without the large antenna, the complex three-axis attitude control system and the solar panels. Heat sterilization of its components, which had proved to be a major factor in Ranger's failures, was unnecessary to prevent contamination of the hostile Moon by terrestrial organisms, but it was required for what might prove to be the more benign planets.

1

2

So Ranger provided the basis for the first US planetary mission. The Soviets had already made two unsuccessful attempts on Venus in 1961 and were expected to launch follow-ups in the summer of 1962. The relative movements of the planets around the Sun create the best conditions for looping probes between them at fixed intervals. "Launch windows" (as they are termed) to Venus repeat about every 19 months, with those to Mars opening at 25-month intervals. Venus was a more tempting target for NASA in 1962: it is slightly easier to reach and scientists preferred to wait until the 1964 Mars window for a more intensive scrutiny of the Red Planet.

The planetary Mariner family was born. The spacecraft looked remarkably similar to the first two test Rangers, with the tower, rising from the hexagonal bus, carrying seven science instruments. The 450lb (204kg) probe was limited to about 20lb (9kg) of experiments and, as the probing of surface conditions at Venus was considered to be of paramount importance, there was no TV camera. Instead, two radiometers would detect microwave and infra-red radiation to provide temperature and pressure estimates of the surface and in the opaque clouds. Other detectors registered magnetic fields, particle radiation and cosmic dust. Thermal protection was increased to cope with the almost-doubled solar heating at Venus' distance from the Sun, and the main liquid thruster was enlarged for the mid-course correction. No upper stage to this day is accurate enough to hit a planet, and the Agena of the early 1960s could not even guarantee reaching the Moon without en-route corrections. Mariner's thruster could be fired only once, but it was the beginning of deep space navigation that in subsequent years has allowed Voyager to journey with astonishing accuracy among the outer planets.

It was a brave decision to go for a planetary mission in 1962. Ranger at that time could not be coaxed to survive the 64 hours to the Moon, yet JPL hoped to stay in contact with the similar Mariner for almost four months. Confidence was dented on July 22, 1962, when Mariner 1, through no fault of its own, could not even reach beyond Earth's atmosphere. Its Atlas booster strayed off course as it headed out over the Atlantic and the Cape's flight rules dictated detonation of the range safety package. It was an embarrassing failure, caused by a hyphen being omitted from the booster's guidance program.

3 *Although crude by later standards, this image of the Martian surface was described by a leading scientist as "one of the most remarkable scientific photographs of this age." It was the eleventh of 21 images transmitted back to Earth by Mariner 4 after its flyby of the "Red Planet" on July 15, 1965. It shows an area known as Atlantis, and appeared to indicate that Mars had a lunar-like, cratered surface. Impacts creating the smaller craters obviously occurred after the impact that originally created the much larger crater, which shows evidence of weathering. The images were recorded on tape and transmitted back to Earth later – transmission of each image taking about eight hours with a power about ten times weaker than a household light bulb.*

4 *An Atlas-Agena launcher sends Mariner 4 on its way to Mars on November 28, 1964. The flight distance was about 325m miles (over 520m km) and took a little under eight months. While there would be disappointments in the years to come, Mariner 4 set a precedent of longevity when it continued to send back data to Earth for three years.*

Mariner 2 was a different story, though. Two days after a Soviet Venera was stranded in Earth parking orbit, the second Mariner was boosted away from its home planet on August 27 by the Agena's second burn. Without correction, Mariner would have missed Venus by more than 230,000 miles (370,000km), but the 28-second course-correction burn of September 4 swung the trajectory to within 25,460 miles (41,000km) of the planet. Missing it with the unsterilized spacecraft was a deliberate decision.

Mariner remained healthy throughout its voyage, breaking records as it went. It glided past Venus' dark side on December 14, 1962, and in that half-an-hour changed all preconceptions of the planet. The radiometers' scans showed that the surface temperature was not around 212°F (100°C) as expected, but at least four times higher. Also, the atmosphere was almost bare of oxygen or water. There appeared little chance of any Earth-like lifeforms on Venus. Nor did the magnetometer detect any evidence of any fields akin to Earth's complex magnetosphere.

Mariner 2 had proved a magnificent success, and it contributed to Ranger's survival in the face of continuing failures. JPL controllers remained in contact with it until the New Year, when the more than 53m-mile (86m km)

distance and the 129 days of continuous operations set new standards.

The month before Mariner 2 arrived at Venus, NASA had approved a dual assault on Mars for the late 1964 window. The first foray had demonstrated the wisdom of launching identical craft, establishing the dual launch pattern usually followed almost into the 1990s. (Missions such as Mars Observer and Galileo have become so complex, and therefore costly, that solo flights are now the norm.)

Mariner was already evolving away from Ranger's configuration, but there was no mistaking the heritage of the Mars probe. The bus had become octagonal, the longer flight required the course correction thruster to be re-startable and the weaker sunlight needed four solar panels. Pressurized nitrogen jets, gyros and Sun and Canopus star sensors were to hold Mariner with the panels towards the Sun and its instruments directed at Mars during the flyby. Reliability had to be extended to more than seven months and communications maintained over almost 150m miles (240m km). In addition to the now familiar collection of radiation, dust and magnetic fields detectors, Mariner-Mars carried a single TV camera to transmit the first close-up views of the planet.

1 *Engineers at the Hughes Aircraft Culver City, California, plant conduct tests on one of the Surveyor spacecraft built for NASA. Surveyor was a soft lander, its descent being slowed by three vernier engines, with the impact of landing being further reduced by the crushable aluminum footpads of the tripod legs.*

2 *An artist's impression of Mariner 4 at Mars. Mariner 3 failed, but while Mariner 4 was still operating in space its backup spacecraft was sent in 1967 as Mariner 5 to Venus, which confirmed the inhospitable nature of the planet, with temperatures of almost 1,000°F and pressures about 75 times greater than on Earth.*

1

2

3 *A powerful Atlas-Centaur launches the first Surveyor to the Moon in May 1966. Besides taking numerous TV images of the lunar surface and conducting various soil tests, the lunar landers showed convincingly that fears of Apollo spacecraft disappearing beneath a sea of dust were groundless. Seven Surveyors were sent to the Moon and five were successful. Surveyor 2 crashed somewhere near the crater Copernicus and contact was lost with Surveyor 4.*

As with Mariner-Venus, however, the first twin suffered from unreliable booster hardware. Its Agena successfully injected it into interplanetary space on November 5, 1964, but the protective fairing refused to separate. Mariner 3 never got the chance to spread its wings. Mariner 4 followed on November 28 under a modified shroud and was released into a long arc around the Sun, heading for an estimated Mars miss distance of a little over 149,100 miles (240,000km). The thruster fired briefly seven days later and modified the flight-path to within 6,085 miles (9,800km) of the planet. Forty minutes before that closest approach on July 15, 1965, the scan platform began tracking the camera across Mars' face to allow 21 pictures to be recorded on tape. About half-an-hour later, as viewed from Earth, Mariner dipped behind the planet. Its radio signals briefly passed through the planet's atmosphere, their fluctuations providing evidence on density.

The following day, Mariner 4 began transmitting the stored images to Earth. Each comprised 40,000 dots and required more than eight hours to relay through the low-power transmitter. The 85ft (26m) diameter radio telescopes of NASA's Deep Space Network had to receive the data from a source ten times weaker than a household lightbulb at the distance of Mars. By contrast, Voyager transmitted its pictures from Jupiter almost 14,000 times faster only 14 years later.

Mariner's images were both startling and disappointing. They revealed a crater-pitted surface more like the barren Moon's than the harbor of life hoped for by many scientists. Not only that, but atmospheric pressure was put at only 0.005 of Earth's; 15 times lower than some of the pre-mission estimates. Daytime surface temperature was believed to be -148°F (-100°C). Mars was clearly an inhospitable place for any organism.

If Mariner 4 was disappointing for the exobiologists, it was a resounding success for NASA's unmanned space program. It followed hard on the heels of Ranger's successful conclusion at a time when the Soviets were still struggling with the Moon, Mars and Venus. Mariner 4 lived on until December 1967, by which time it had been joined in space by its backup spacecraft. The solo, cut-price Mariner 5 was modified to fly within 2,485 miles (4,000km) of Venus, ten times closer than Mariner 2. As it had to venture towards the Sun instead of the colder, outer solar system, thermal protection was increased and the stubbier solar panels were flipped over. The science instruments were similar, but the TV camera was replaced by a device to look for hydrogen and oxygen in the high atmosphere.

Launched by Atlas-Agena on June 14, 1967, it flew past Venus on October 19. As with Mariner 4 at Mars, the passage of its radio signals through the atmosphere provided evidence of surface conditions: 980°F (527°C) and a pressure of at least 75 Earth atmospheres. These figures contrasted sharply with the Soviets' Venera 4 results. Only a day before Mariner 5's arrival, the first Soviet capsule had penetrated the atmosphere and transmitted direct measurements apparently showing conditions of 536°F (280°C) and 15-22 Earth atmospheres at the surface. It later emerged that Venera's capsule had not survived down to the surface but had collapsed under the increasing pressure while still at a height of 17 miles (27 km).

Probes to the Moon

Apart from those Venus missions of 1967, the mid-1960s belonged to unmanned lunar exploration. The year 1966 began with the Soviets' Luna 9 transmitting the first pictures from the Moon's surface and Luna 10 becoming the first lunar satellite, but NASA took the

3

initiative with a new generation of spacecraft and never again relinquished dominance. In the 19 months from May 1966, no less than seven Surveyors and five Lunar Orbiters flew a remarkable set of missions. In conjunction with Mariner, they taught controllers how to handle problems as they arose and to plan activities on lengthy flights.

Surveyor was able to benefit at last from the powerful Atlas-Centaur combination, still used today to launch geostationary satellites. Science took a back seat once the program became dedicated to the manned Apollo effort, demonstrating landing techniques and investigating the lunar surface at sites selected for their interest to Apollo. Luna 9 had made the first instrumented landing, but Surveyor was the first true soft lander, using thrusters to guide it to a gentle touchdown on three legs.

Surveyor 1 departed from the Cape on May 30, 1966, intended as an engineering demonstration and with no great expectation of the outstanding success it was to prove. The one-metric-ton lander was separated from its Centaur and locked on to the Sun and the bright star Canopus using small nitrogen thrusters. Sixteen hours out, three liquid thrusters throttleable over 30-103lb (13.5-47kg) thrust fired briefly to compensate for the launch errors; Surveyor being far more flexible than its Ranger predecessor. Development difficulties of these thrusters had contributed to the tortuous Surveyor development by JPL and prime contractor Hughes, which saw the overall cost rise to more than $2,000m in current values. Approaching the Moon at 5,960mph (9,600kph) the altitude radar triggered the 1,430lb (650kg) solid-propellant retro-motor nestling at Surveyor's core. The thrust of almost 9,000lb (4,080kg) cut the descent rate to almost 250mph (400kph) over the 40 second burn before the spent motor was ejected at an altitude of about 25 miles (40km) to clear the legs for landing.

The liquid thrusters then took over, controlled by the radar system, while JPL controllers looked on unable to intervene. As programmed, the thrusters switched off still 13ft (4m) up to avoid disturbing the surface, and Surveyor 1 touched down on June 2 near the crater Flamsteed and only 9.3 miles (15km) off target. The first TV image was transmitted within the hour, showing one of the footpads; an event repeated by Viking almost

exactly ten years later on Mars. Afterwards, more detailed 600-line pictures were relayed through the high-gain panel antenna sitting on top of the 10ft (3m) high lander. Scanning the camera around under ground control revealed the craft to be sitting in a shallow, ancient crater. Imaging the footpads showed that Apollo had no worries about sinking helplessly beneath a thick layer of dust.

Over the next two weeks, Surveyor transmitted almost 11,000 TV pictures of what we now know to be a typical lunar landscape. As lunar night fell on June 14, Surveyor was placed in hibernation in order to survive the two weeks of temperatures falling well below -148°F (-100°C). Although reluctant to revive, it eventually responded and returned another 600 pictures before its imaging system succumbed during the next period of darkness.

Surveyor's good fortune ran out with the second spacecraft in September 1966. One of the three thrusters refused to ignite for the course correction, leaving the craft spinning wildly until it impacted somewhere near the crater Copernicus. But Surveyor 3 more than made up for the failure, although only after a few heart-

stopping moments. The control system somehow lost sight of the surface and Surveyor touched down with the thrusters still burning. It was half a minute before ground controllers realized what was happening and commanded shutdown, while Surveyor hopped around. Fortuitously, the multiple bounces gave the camera several footprints to photograph, providing valuable information for the engineers designing Apollo's lander. This was also the first craft to carry a robot arm to test surface conditions. During the next day, directed by a JPL engineer, it dug trenches and tested the soil.

Surveyor 3 failed to respond after its hibernation during the lunar night, but its career did not end there. More than two years later, Apollo 12 astronauts Pete Conrad and Alan Bean landed 525ft (160m) away and trekked over to see how the lander had fared. They took samples and even brought back the whole camera.

4 Surveyor 3 was the first spacecraft to carry a robot arm. This was used to scoop up lunar soil and dump it on a footpad for examination.

5 Surveyor 3 was visited by the Apollo 12 astronauts in November 1969. Their examination was valuable in establishing the effects of a two-year exposure to space on the Surveyor, and some parts were removed for transporting back to Earth.

6 Engineers knew that a landing error had caused Surveyor to bounce around its landing site until its engine was switched off. Apollo 12 photographs provided graphic evidence of the fact.

Some excitement was generated when an Earthly micro-organism was at first believed to have survived on the returned samples, but it was later concluded that they might have been contaminated on the return journey.

Surveyor 4 headed for the same target as number 2 – Sinus Medii in the center of the Moon's face – but there was another failure. Contact was lost during the landing sequence and never regained; it is still not known if touchdown was achieved. Surveyor 5 also was almost lost, but quick thinking saved the day. Pressurizing gas leaking from the liquid thruster system would have run out during a normal landing sequence, but engineers re-programmed Surveyor to brake late and hard. There was little leeway for error, but they achieved the series' third successful landing, this time within 18.5 miles (30km) of where Apollo 11 would land within two years. In place of the mechanical arm, an alpha-particle

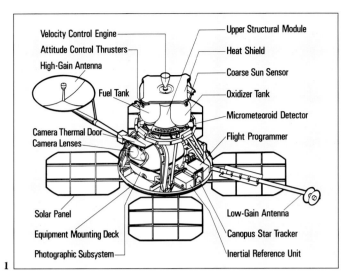

1

2

scatterometer was carried to make the first basic analysis of the soil, revealing a basalt-type material.

In November 1967, Surveyor 6 at last reached Sinus Medii, completing Apollo's interest in the equatorial band. Surveyor 7 was therefore directed at the more scientifically-interesting bright crater Tycho in the far south. It landed on the debris blanket some 19 miles (30km) north of the crater rim and showed a rough area scattered with boulders. There were suggestions for landing one of the last Apollo missions near Tycho but it was deemed too hazardous.

Orbiter Missions

Contemporaneous with Surveyor was the Lunar Orbiter program, a series of five spacecraft designed to map a 10-degree band around the lunar equator at 3.3ft (1m) resolution for pinpointing Apollo and Surveyor landing sites. LO was essentially a flying camera and, at around 860lb (390kg), was still small enough to fly on the Atlas-Agena instead of the more expensive Atlas-Centaur. Its structure was built around a pressurized shell housing a photographic system of wide-angle and telephoto lenses. Up to 212 image pairs could be recorded on high-

definition black and white aerial film, with the film being moved slightly during exposures to avoid image "smear" resulting from the high speed of the vehicle relative to the surface below during the low pass of each orbit. Development was by a transfer system and the images were then scanned by a fine beam of light for transmission to Earth. It was a complex approach (reputedly based on military reconnaissance satellite technology), but the resolution of TV cameras as carried by Surveyor was then too coarse for the job.

The first US Lunar Orbiter (in fact, the first American orbiter of any celestial body) arrived on August 24, 1966, eventually lowering its path to come within 31 miles (50km) of the surface for the most detailed imaging. Although the anti-smear system failed on the telephoto shots, the wide angle views were excellent, covering a number of Apollo candidate sites and returning the first good views of the far side. A major and unexpected discovery took some time to explain; the spacecraft's orbit was affected by the Moon's lumpy gravitational field, caused by the lava-filled seas facing Earth. Without Lunar Orbiter's discovery of these "mascons," Apollo would have landed far off course.

3

Its film supply exhausted, the first Orbiter was deliberately crashed onto the Moon in October 1967 to avoid radio interference on future missions. Orbiters 2 and 3 completed mapping for Apollo, covering almost 6m square miles (15m sq km) and pinpointing Surveyor 1 on the surface. The remaining two spacecraft were used for more scientific purposes, both being placed in high polar orbits to map the whole lunar disc. The fourth Orbiter covered 99 percent of the lunar near side, while the fifth concentrated on the far side.

The way to the Moon was now open for manned expeditions. Since Surveyor 7 in January 1968, the only US unmanned lunar craft have been two small satellites released by Apollo 15 and 16 in 1971-72 and the Explorer 49 radio astronomy satellite that used the Moon's bulk in 1973 to block terrestrial radio interference. Lunar science has suffered while NASA turned its attention to other celestial bodies and as a result of the anti-Apollo backlash. The next stage is a polar orbiter to map the Moon's resources, but funding for this Lunar Observer was denied throughout the 1980s.

Return to Mars

Mariner 4's pioneering voyage to Mars was a remarkable success but it had provided only a tantalising glimpse of the Red Planet's secrets. Its simple imaging system had covered only one percent of the surface and, while atmospheric density had been estimated, quite accurately as it turned out, the constituents still had to be identified. Mariner 4's size had been constrained by the lack of rocket power in 1964, but by the time of the Mars window in the spring of 1969 NASA had the Atlas-Centaur at its disposal. The two new Mariners could be 265lb (120kg) heavier, allowing them to carry not only a more versatile imaging system, but also remote-sensing instruments that would yield data on atmospheric composition and surface temperature. Another innovation presaged a new era of flexibility: the onboard computer could be reprogrammed in flight to meet unexpected requirements. Responding to changing needs has since become a major feature of NASA's planetary investigations. Without this capability, for example, Voyager 2's mission would have ended long before reaching Uranus or Neptune. The Soviets have always operated with more modest computer power, which has consequently led to more limited results.

After a flight of 157 days, Mariner 6 flew past Mars on July 31, 1969, at a height of 2,110 miles (3,400km) and returned 24 close-up pictures. They showed intriguing features around the south pole, so Mariner 7 was re-programmed to slew its camera platform further south than planned during its near-encounter on August 5. In fact, the second craft was fortunate to be still operational. A few days before, as engineers later reconstructed, its single battery had cracked and sprayed its contents out like a rocket thruster. The trajectory shifted slightly and electronics were knocked out by short circuits. However, although fearing the worst, controllers were delighted to discover that all the science instruments worked when they were turned on shortly before encounter.

The two Mariners returned 57 close images from the total of 202, confirming the lunar-like starkness discovered by Mariner 4 and adding desert-like plains and chaotic terrain featuring ridges and valleys. The remote sensing instruments revealed the thin atmosphere to be almost totally composed of carbon dioxide and that the surface temperature swung wildly over the day. The equator's 61°F (16°C) noon-time temperature plummeted to -100°F (-73°C) at night, while that of the south polar cap fell to -193°F (-125°C), indicating a covering of frozen carbon dioxide rather than water ice.

On the whole, Mars still appeared to present a rather barren face. But the Mariner twins had studied only a tenth of the surface. What was needed now was an orbiter in the manner of Lunar Orbiter to provide comprehensive coverage over weeks, if not months. Mariner 9 in 1971 did just that, and at last revealed the planet's rich diversity of features. It also laid the groundwork for later landing missions.

Mariners 8 and 9 appeared much like their predecessors, but the 292lb (132.5kg) thruster and its propellant tanks for braking into Mars orbit more than doubled their launch weights to almost a metric ton. Atlas-Centaur could still be used because the Earth-Mars alignment in the 15-year cycle was at its most favorable in 1971. The remote-sensing instruments were much the same as before except for the more sophisticated imaging system. This time, there were two 700-line wide-angle and telephoto TV cameras.

But the first of a Mariner pair was lost again. The launcher's Centaur autopilot failed and the planetary craft ended its voyage in the Atlantic. Mariner 9 departed successfully on May 30, 1971, to slot into an 870x11,200-mile (1,400x18,000km) polar orbit around Mars on November 14 with a 15-minute burn of its main thruster. It was the first man-made satellite of another planet, arriving only 13 days before the Soviets' Mars 2 orbiter and lander.

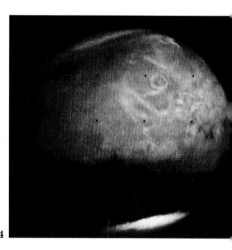

4

4 *Mariners 6 and 7 flew past Mars in the summer of 1969. The more than 200 images transmitted back to Earth still broadly showed a lunar-like surface, although some additional features were revealed.*

5 *The lunar comparisons ended as soon as Mariner 9 started to return images from orbit around Mars in 1971. This mosaic shows what came to be called Mons Olympus – a volcano that is far larger than any on Earth.*

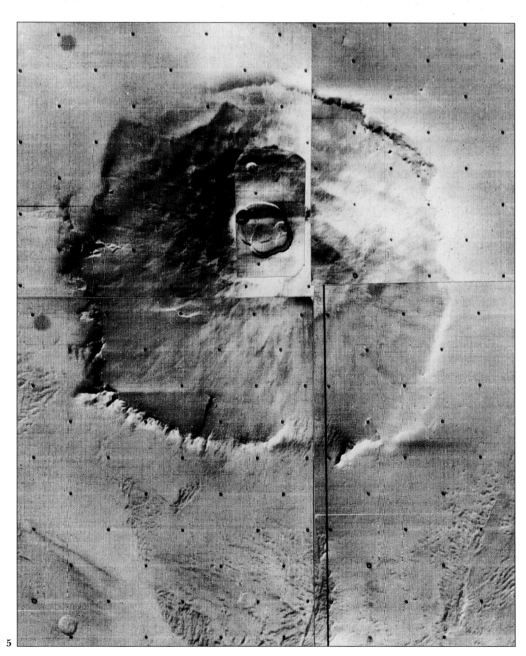

5

1 *Most of the Pioneer 10 and 11 experiments were not of the kind to interest the general public, but an imaging photopolarimeter devised by principal investigator Dr. Tom Gehrels of the University of Arizona (shown here at a mission briefing) and his colleagues produced images of Jupiter and Saturn which had great impact. The instrument was primarily meant to study dimlight phenomena such as the zodiacal light during the Pioneers' cruise to the planets, as well as the constitution of the planets' atmosphere. But by using the spin of the spacecraft and a small telescope to scan strips of the planets, images in blue and red light could be built up on Earth using computer processing.*

2

2 *The second Pioneer spacecraft was launched on April 6, 1973, thirteen months after Pioneer 10. It is shown here at the Cape Kennedy launch center during tests involving a mock-up of the Atlas-Centaur booster's third stage.*

3 *The Pioneers were the first spacecraft to penetrate the Asteroid Belt between Mars and Jupiter and to explore the outer gas planets. The distances involved resulted in the most prominent feature of the spacecraft being a 9ft (2.7m) communications dish.*

The achievement was muted by a global dust storm transforming the planet into an almost featureless sphere. While controllers waited over the next two months for the dust to subside, Mariner's orbit took it within range of Mars' tiny moons, Phobos and Deimos. For the first time, astronomers could see that they were battered lumps appearing suspiciously like captured asteroids. Systematic mapping of Mars began in January 1972, almost immediately proving how previous missions had given the wrong impression of the planet's character. Four dark spots showing through the dust storm were revealed to be huge volcanoes, the highest of them rising 17 miles (27km) above its 372-mile (600km) diameter base. Olympus Mons, as it was later named, easily surpasses any such terrestrial feature. More dramatic evidence for at least the beginnings of tectonic activity is the Valles Marineris, a canyon-like feature some 2,485 miles (4,000km) long, 62 miles (100km) wide and up to 3.7 miles (6km) deep as it winds along a fifth of Mars' circumference. Also in the south are the 1,242-mile (2,000km) diameter Hellas and 560-mile (900km) diameter Argyre impact basins.

Some features on Mars strongly resemble dried-up river beds, suggesting that the planet has experienced running surface water at some stage in its history. Mariner 6/7 indicated that the south polar cap is frozen carbon dioxide, but the theory was modified in the light of Mariner 9's findings. The carbon dioxide layer is now believed to vaporize in summer, leaving behind a smaller cap of water ice.

JPL worked with Mariner throughout most of 1972, systematically mapping what was now known to be a highly-diverse body. Scientists were more optimistic: life was possible there, albeit in a primitive form. On October 27, after returning more than 7,300 pictures covering most of the surface at better than 1.3 mile (2km) resolution, and a small proportion at down to 325ft (100m), Mariner 9's attitude-control thrusters ran out of nitrogen. It had done its job magnificently. The $120m mission allowed planning to proceed to the next phase: the arrival of twin orbiters and landers in 1976. This mission was too complex to be ready for the 1973 window, which created a hiatus in exploration of the inner solar system that was exploited by a truly pioneering project: penetration of the outer system beyond the Asteroid Belt.

To the Outer Planets

In the 1960s, NASA conceived a Grand Tour of the outer planets using the planetary alignments of the late 1970s to swing several sophisticated spacecraft from target to target; an opportunity that would not recur until the year 2155. But the agency's budgets were diminishing as enthusiasm for the space program dwindled, and the expensive Grand Tour proposal was finally axed in 1972. It was deftly replaced by the cheaper Voyager, which ended up achieving most of Grand Tour's targets anyway. However, before Voyager could be committed to the unknown hazards beyond Mars, NASA decided to dispatch a small, cheaper precursor. Whereas the Mariners and Voyagers were complex three-axis stabilized craft carrying a large array of instruments on scan platforms, the new Pioneers were simpler, spin-stabilised vehicles with rigidly-mounted scientific instruments. The mass was held to around 573lb (260kg) to allow the Atlas-Centaur, augmented by a solid third stage, to be used instead of the more powerful, and more costly, Titan-Centaur.

Beyond Mars, the Sun's light is too weak to power solar arrays, so Pioneer carried two small nuclear generators working on plutonium. The distance involved also meant that Pioneer was dominated by the 9ft (2.7m)

dish used to transmit the eight watt radio signals back to Earth. All the electronics, computer and control systems were housed in the hexagonal bus, where the 11 scientific instruments were also mounted. As usual, most were concerned with particle radiation, cosmic dust and magnetic fields. Pioneer was too small to include a relatively heavy TV camera, but the imaging photopolarimeter was designed to scan a planet's face in strips as the craft rotated at five rpm. The images are relatively crude but Pioneer was not designed to undertake planetary surveys: it was intended to show that Voyager could survive the harsh radiation trapped around Jupiter, the rock and dust hazards of the 174m-mile-wide (280m km) Asteroid Belt, and the long flight times.

Pioneer 10 began its 620m-mile (1,000m km) journey to Jupiter on March 3, 1972. The third stage accelerated it to almost 32,100mph (51,650kph), making it the fastest man-made object ever and giving it sufficient energy to fly straight out of the solar system. It crossed the Moon's orbit in a mere 11 hours, six times faster than typical lunar craft. Pioneer began its nine-month traverse of the Asteroid Belt the following July, but fears of crippling celestial debris striking at 15 times the speed of a bullet proved grossly over-estimated – fewer than 60 small dust impacts were recorded.

Heading in towards Jupiter, Pioneer crossed the orbit of the outermost moon Sinope on November 8

3

1973, and penetrated the planet's magnetic field 18 days later. The closest approach, on December 5, had been set at a conservative 80,730 miles (130,000km) above the swirling cloud tops but, even so, the detectors were almost swamped by the radiation trapped in Jupiter's dense magnetic field – 19,000 times stronger than Earth's. The polarimeter had been returning pictures with greater clarity than Earth-based telescopes since December 2, but the radiation affected the electronics and some images were lost around closest encounter. Nonetheless, more than 300 images from the swingby were returned, including views of the centuries-old Great Red Spot: an anti-cyclone almost 14,300 miles (23,000km) across and rising five miles (8km) above the surrounding clouds.

4 *A Pioneer 11 view of Saturn and its rings obtained from a distance of 1.5m miles (2.4m kms) on August 29, 1979. The detail is already better than that seen by Earth-based telescopes: the moon visible at bottom is Rhea.*

5 *Pioneer 11 begins its twenty-month journey to Jupiter. It took just under another five years to reach Saturn.*

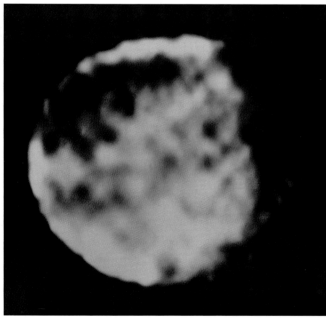

Pioneer 10's success allowed its near-twin Pioneer 11, launched on April 6, 1973, to approach Jupiter to within 26,720 miles (43,000km). The radiation environment was much harsher here, but because Jupiter's strong gravity accelerated the spacecraft to almost 107,000mph (172,000kph) it suffered a lower total dose. The new craft, though, had a more intricate flight path: for the second time a swingby of one planet was used to bend the trajectory to reach a second target. On December 2, 1974, Pioneer 11 sped over Jupiter's south pole to travel far above the plane of the solar system, before being pulled back down to meet Saturn on September 1, 1979, after a journey of 1,988m miles (3,200m km), less than 20 years after the first failures to reach the Moon. Pioneer came within 13,050 miles

(21,000km) of Saturn's cloud tops and discovered its magnetic field, as well as two rings and a moon, which it missed by only a few thousand kilometres.

Both Pioneers remain in good health and continue to return data on interplanetary space to their controllers at NASA's Ames Research Center. Pioneer 10 became the first man-made object to leave the known solar system by crossing the orbit of Neptune, the outermost

6 *Jupiter's largest satellite, Ganymede. Although not of high resolution, the Pioneer 10 image enabled scientists to speculate about craters and other likely surface features.*

7 *December 1, 1973, and Pioneer 10 reveals details of Jupiter's clouds together with the famous Red Spot and the shadow cast by the moon Io.*

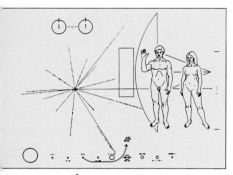

1 *The idea that plaques combining graphic illustrations and binary data about the spacecraft, humans and the Earth's position in the cosmos be placed on the Pioneers was enthusiastically supported by Dr. Carl Sagan of Cornell University. The message was intended for any beings that might encounter the spacecraft as they journeyed beyond the solar system.*

planet until 1999, in June 1983. The nuclear generators aboard the spacecraft and continuing refinements in the sensitivity of Earth receivers should allow scientists to stay in contact with Pioneer 11 until at least 1994, and with Pioneer 10 until the year 2000. By then, they might even have broken out from the Sun's huge magnetosphere and crossed into true interstellar space. The pair and their Voyager successors are being very precisely tracked as they depart the solar system in different directions. If there is a large tenth planet hidden beyond Pluto, or a dark star companion to the Sun deep in space (as some theorists have proposed), then the paths will be affected to different degrees. Should the inert probes be recovered in the far distant future by alien beings, each carries a small, gold-plated plaque with line drawings of naked humans, the solar system and its position relative to 14 special types of stars known as pulsars.

Gravity Assist Technique

NASA's next planetary mission, launched after the Pioneers but completed while Pioneer 11 was still on its way to Saturn, was the first to employ the gravity-assist technique fully. Scientists pressed for a flight to Mercury because, being so close to the Sun, very little was known about it. But its position means that flinging even a small flyby probe at it by normal means would require the costly Titan-Centaur. However, JPL analysts calculated that if they could hit a 250-mile-wide (400km) window swinging past Venus in 1974, then an Atlas-Centaur could be used to deliver a Mariner-type space-craft to Mercury. True, it would be too small to house a retro-motor to enter into orbit for extended observations, but further analysis revealed that, by controlling the flyby very carefully, Mariner 10 could loop around the Sun and meet Mercury again, and then again. By flying only one spacecraft, the cost of the project was held down to $300m at 1984 rates, significantly less than Mariner 4's single Mars flyby.

The 1,100lb (500kg) spacecraft was typical of the Mariner class. The main design problem was countering the intense heat from the Sun at Mercury's orbit: a "parasol" was released after launch to shade the bus, and the two solar panels could turn away from direct sunlight. The scan platform's wide-angle and telephoto TV cameras were accompanied by several spectro-meters and radiometers to measure temperatures and

identify atmospheric gases. Following launch on November 3, 1973, the cameras were directed towards the Moon for testing, coincidentally providing some of the best views ever of northern lunar regions. Several course corrections by the main 50lb (22.4kg) thruster delivered Mariner to within 15 miles (27km) of the perfect aim point at Venus on February 5, 1974. For the first time, a probe transmitted TV pictures of the planet. Instead of the dense, bland veil seen at normal wavelengths, by using an ultraviolet filter Mariner was able to reveal banding in the clouds. Little was known about the planet's atmospheric circulation but, astonishingly, the upper atmosphere was shown to be rotating once every four Earth days, whereas radar observations had already shown that the planet itself was rather lazy, turning once every 243 Earth days. Windspeeds of several hundred miles per hour were not uncommon, although they reduced to almost nothing in the syrupy lower layers.

Mariner moved rapidly on to its ultimate target for a 437 mile (703km) flyby on March 29, 1974. Mercury proved to be almost unbelievably Moon-like, with the images looking as though they could have been taken by Lunar Orbiter. The planet was covered with craters, faults and lava-flooded features, but dominated by the 808-mile-diameter (1,300km) Caloris basin. Trajectory deviation indicated a density similar to Earth's, unusually suggesting a metallic core accounting for some four-fifths of Mercury's mass. While no magnetic field had yet been detected at either Mars or Venus, this small planet was found to have one one-sixtieth the strength of Earth's. Its origin remains to be explained.

Orbital mechanics, with a little help from three course adjustments, returned Mariner to the planet on September 21, 1974, although then it approached to within only 29,825 miles (48,000km). However, the third and final pass, on March 16, 1975, was planned to dive close in to slice deeply through the unexpected magnetic field. The 200-mile-swingby (327km) also produced the highest-resolution pictures yet, showing features as small as 460ft (140m) in some cases. Further encounters were theoretically possible, but Mariner's attitude control and course correction propellants were limited. The spacecraft was shut down at the end of the month, having completed the most complex planetary mission of the time. Its observations remain unique and there are no firm plans to launch a follow-up mission. The same celestial arrangement that allowed Mariner's repeat visits also kept one hemisphere hidden. Bearing in mind the false impression created by the early Mars flybys, Mercury might still be holding surprises in store for a future orbiter.

3

Challenge of Viking

It was also originally planned that 1973 would see the most complex and expensive missions depart for Mars. When NASA approved the enterprise in 1968, it was expected that Apollo would be followed by manned flights to Mars in the 1980s, using information gathered by Mariner-1971 and the paired orbiters and landers. However, President Nixon's administration rejected the ambitious post-Apollo proposals submitted in 1969 (only the Shuttle survived), and Viking became the sole means for the foreseeable future of exploring Mars in detail. Postponed to 1975 because of NASA's falling budget, it grew into the most expensive planetary mission ever undertaken. Its price tag of well over $3,000m at today's rates greatly exceeds even Voyager's, which visited four planets over 12 years at a third of the cost.

Viking's sheer scope remains unequalled. Two enlarged versions of the Mariner 9 craft would each carry a 1,455lb (660kg) lander into Mars orbit packed with instruments to probe the surface. At sites selected from Mariner 9 photos, they would chemically analyze the atmosphere and soil, search for signs of life using samples scooped up by a robot arm, return routine weather reports from meteorological instruments, listen for seismic activity and transmit stereo pictures from cameras able to swivel through 360 degrees. Discovery of any lifeform, however simple, was a priority, almost guaranteeing future missions. While the orbiters could rely on 32ft-span (9.75m) solar panels for power

4

2 *A mosaic of about 300 Mariner 10 images in 1973 is centered over latitude 55°S and longitude 100°E on Mercury. The closest planet to the Sun, Mercury was revealed as a lunar-like, heavily cratered world with a magnetic field.*

3 *September 8, 1976: NASA Administrator James Fletcher (right foreground) briefs President Gerald Ford (extreme left) and White House staff members on the two Viking missions to Mars. Viking 1 had landed on July 20 and Viking 2 on September 3.*

4 *A Viking lander is tested in a US desert to simulate conditions on Mars as far as possible. Martin Marietta was the prime contractor for the NASA mission, which was by far the most complex conducted to that date.*

The concept behind the use of one planet's gravity to bend a spacecraft's trajectory to reach another, first demonstrated by Mariner 10, was considered as long ago as the 1920s, but it was not until the 1960s that the theory was worked out in detail. Even then it took time to be accepted because at first sight it appeared to break the law of the conservation of energy.

Basically, the intermediate planet acts as a free-accelerating or braking rocket stage, drawing on its momentum about the Sun as propellant. The key lies in that movement about the Sun. Catching up from behind, a probe will be tugged along by the planet's own motion and thus speeded up. But there is never something for nothing; the acceleration must be paid for by the planet itself slowing down.

The disparity in sizes, however, makes the planet's change undetectable. Conversely, a probe passing ahead will lose energy by giving it to the planet. This is what Mariner 10 did, flying ahead of Venus to lose energy so that it would fall in towards Mercury.

This "celestial billiards" requires precise targeting, as any error will be magnified by journey's end. The benefits are enormous, however, opening up the solar system to exploration by small, chemically-propelled craft. The Jupiter Galileo mission could not have been launched by the shuttle in 1989 towards the gas giant without Earth and Venus assists. Once there, repeated encounters with three of Jupiter's major moons will allow a comprehensive tour of the system.

1 *One of the two facsimile cameras on the deck of the Viking 1 lander took this 110 degree color panorama of its landing area on February 17, 1977. The spacecraft's meteorology boom extends upwards at right. The picture was one taken to document the trenching activities down to a depth of about 12 inches (30cms) below the Martian surface, from which samples were taken for processing in the onboard laboratory. The scoop marks can be seen clearly at lower right.*

2 *A false color image prepared by the US Geological Survey showing Mons Olympus as recorded by the Viking 1 orbiter. The colors exaggerate the very subtle differences in the volcano's lava flows that have accumulated over time.*

generation, the dusty surface conditions required the landers to be equipped with small, nuclear-power generators. The Surveyor landers had been controlled from JPL almost in real time, but the much greater and varying distance of Mars could introduce time delays of more than 20 minutes one-way even at the speed of light, requiring all operations to be controlled by the onboard computers working on blocks of instructions relayed in advance.

As if the missions were not complicated enough, the completed landers also had to be baked at over 230°F (110°C) for almost four days to avoid terrestrial contamination of an unknown biological environment. Building equipment to endure such treatment proved very costly. The lander and its protective atmospheric shell were then sealed inside a 12ft-diameter (3.7m) "bioshield" for mounting on the orbiter, the casing being ejected once the spacecraft was safely away from Earth. The combined weight of more than three metric tons required the use of the powerful Titan-Centaur launcher.

1

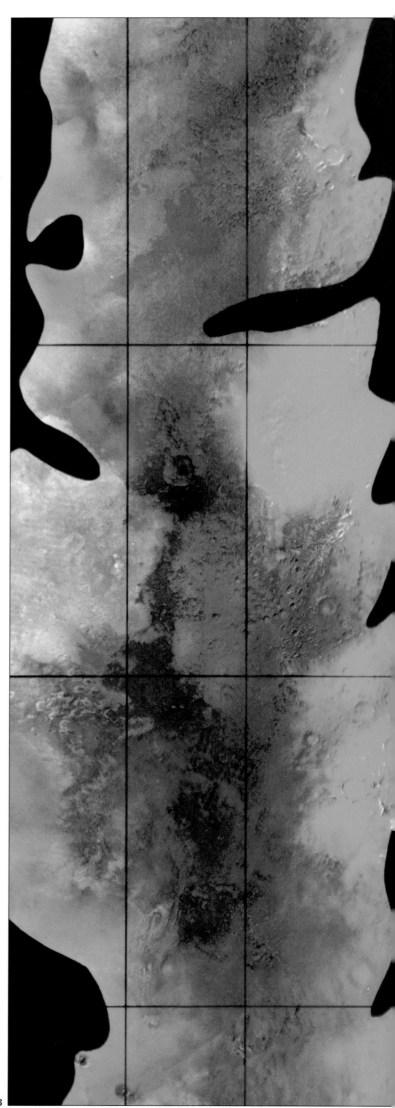

2

3 *As it approached Mars in August 1976, the Viking 2 orbiter took over fifty images of the planet that were used by the USGS to make this mosaic illustrating variations in the planet's surface chemistry. Red depicts rocks containing a high proportion of iron oxides; dark blue less oxidised materials; bright turquoise surface frosts and fogs; and brown, orange and yellow different types of sand and dust deposits.*

Viking 1 arrived at Mars on June 19, 1976, settling in to a 930x20,370-mile (1500x32,800km) orbit that matched the planet's day – more than 40 minutes longer than Earth's. The goal was a July 4 landing, to celebrate America's Bicentenary, at a northern site in the Chryse Planitia (Golden Plain) region. Geologists had spent years studying Mariner 9 pictures to come up with an interesting area that would present few landing hazards to a costly craft, but they were in for a shock. Viking's orbiter carried improved cameras and they showed the preferred site to be considerably rougher than expected. The landing was postponed for two weeks while the scientists and engineers argued over the new photographs pouring in. At least it demonstrated the value of attempting a landing from Mars orbit: the Soviets' less-capable Mars 2 and 3 probes had been forced to come down during the raging storms of 1971 because they were designed to head straight in from their interplanetary trajectories.

The first lander separated on July 20 and hit the atmosphere protected from the 2,732°F (1500°C) heat of entry by its aeroshell. Even that carried two scientific instruments. One analyzed the upper atmosphere down to 62 miles (100km), its discovery of small quantities of nitrogen providing a boost for the scientists searching for life. (A backup unit of this same spectrometer later flew on the Space Shuttle Columbia to monitor the constituents of Earth's high atmosphere during re-entry.)

At this stage JPL's controllers were relying on Viking's computer, because all they could see on their screens 3 was telemetry describing events that had taken place 19 minutes earlier. The lander's radar began the final landing sequence just below 3.7 miles (6km), throwing off the aeroshell and unfurling the 52ft-diameter (16m) parachute. Mars' atmosphere is too thin for a safe landing by parachute alone, so the canopy was discarded with more than a kilometre still to go and three liquid thrusters flared to life, controlled by the radar and computer. Each comprised 18 tiny nozzles to spread the exhaust and avoid disturbing the surface while Viking lander 1 settled down on its three stubby legs at a gentle pace, the first successful landing on the Red Planet.

With its Earth-bound controllers still unaware of their success, Viking went into a pre-programmed routine that would return basic information as soon as possible in case some unknown hazard was about to overwhelm it. Reminiscent of Surveyor 1 a decade earlier, the first task was to photograph the area around a landing pad. It showed a dusty, rock-strewn area not unlike the Moon's. However, the first full-color panorama of the next day left no-one in doubt of the location: a ruddy, rocky desert stretched towards the horizon and the pink sky. In fact, that initial view was at first wrongly processed by JPL's computers to yield a blue sky instead of the pink created by sunlight scattered by airborne dust. The planet's first weather report showed a balmy -27°F (-33°C) in mid afternoon, accompanied by a breeze gusting up to 31mph (50kph). Both landers suffered numerous dust storms during their careers, including two global events.

However, everyone was waiting for the principal question to be answered – is there life on Mars? The robot arm dropped a soil sample into the self-contained biological laboratory on July 28 for distribution among the three separate experiments. The first behaved as though micro-organisms were present, but it was far from conclusive on its own. The other two initially seemed to agree, but as they progressed the suspicion grew that unpredicted chemical processes were occurring. The sampling was repeated several times, with the same results.

Lander 2 touched down on September 3 on the more northerly Utopia Planitia where, it was hoped, the greater moisture found nearer the pole would increase the likelihood of life. It was not to be: there were the same results. In the years since then, scientists have generally agreed that chemical effects were responsible for the results, although there have been dissenting voices. But everyone agrees that Viking was only a very preliminary attempt at finding life and that chances are still good. It could be that such an arid planet harbors life in isolated oases, but this would require an extensive search.

Both landers exceeded their designers' expectations. Not only did they operate throughout a Martian year (almost two Earth years), but they pressed on into the 1980s. In fact, they became something of an embarrassment to NASA as they continued to transmit data year after year. Lander 2 was turned off in April 1980 and, to save money on staff and information processing, the agency proposed shutting down its predecessor, until an outcry from the international scientific community forced a re-think. Lander 1 was then re-programmed to return weekly weather reports and pictures to monitor surface changes at least until 1994, but in November 1982 an erroneous command sent up from JPL killed off the craft anyway.

And what of the orbiters? While their landers grabbed the headlines, they methodically set about mapping Mars at 1,000ft (300m) resolution. Both eventually fell to attitude control propellant exhaustion:

Orbiter 2 in the summer of 1978 and Orbiter 1 two years later. Their collection of more than 50,000 images remains a treasure trove for planetary scientists. Resolution of as good as 65ft (20m) and better enabled NASA scientists to conclude there to be a vast store of water ice just below the surface. If this proves correct, then the future of Mars as a second home for mankind is assured. The planet was not visited by American craft in the 1980s, but JPL's Mars Observer was launched by a Titan 3 in September 1992 to probe the planet with a new generation of instruments for at least a full seasonal cycle following arrival in fall 1993. It will provide a global inventory of Mars' basic resources, including hidden water, and map the surface to a resolution of 5ft (1.5m), better than many areas of Earth. At mission's end, the planet will be better known than the Moon, notwithstanding the six manned Apollo expeditions.

4

4 During NASA planetary encounters there was always pressure on the scientists and technicians to produce their results quickly, and nowhere more so than in the case of the images. But the early images, while frequently of high quality and of a clarity never seen from Earth before, were nonetheless capable of considerable improvement by computer processing techniques. This USGS image of much of the disc of Mars, in which the large volcanoes at left and the enormous Valles Marineris at lower center are the most prominent features, can be compared with the dramatic but raw Martian crescent image on the following page.

5

5 The Viking 1 orbiter images Mars' Noctis Labyrinthus (Labyrinth of the Night) as the Sun rises. Bright clouds composed of water ice can be observed in and around the tributary canyons of this high plateau region of Mars. The scene covers about 4,000 square miles (10,000 sq kms) centered at 9°S, 95°W: the crater at lower right is called Oudemans.

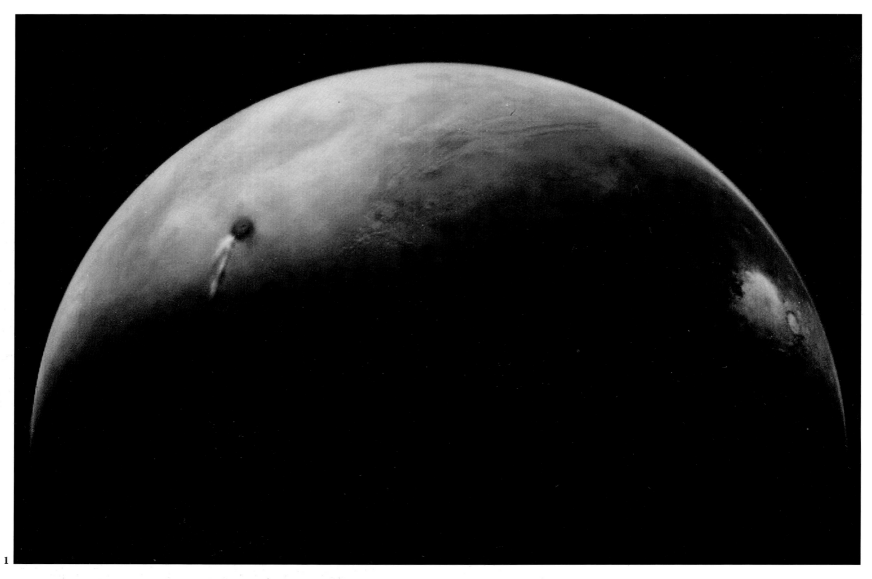

1

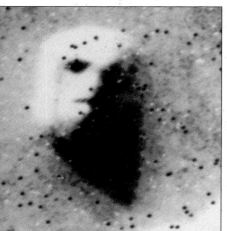

2

1 *August 1976: the Viking 2 orbiter approaches Mars as dawn breaks. At left water ice clouds cover the western flanks of Ascreaus Mons, one of the giant Martian volcanoes. Valles Marineris fills the center of the crescent planet and at right is the large, frosty crater basin Argyre.*

2 *A face on the surface of Mars? This Viking image aroused much speculation but the feature resulted from natural physical processes and system problems; part of the eyes and the nostrils, for example, resulting from missing imaging data.*

The high resolution imagery referred to is also required for planning rover missions. NASA has been working on mobile landers since the early 1980s, but funding restrictions, even with international collaboration, will delay such a mission until at least the 1998 window. Even then it might be sacrificed to wait until a sample return venture is approved; a rover being required to scout out a variety of samples for dispatch to Earth. That mission will be so technically demanding and expensive that 1998 is the first realistic window, with 2001 more likely, should funding allow.

There is also one other plan. The Jet Propulsion Laboratory was working in 1992 on a Mars lander concept based on a program emphasising smaller, lower cost and more frequent planetary missions. Sixteen landers weighing less than 220lb (100kg) each and launched four at a time were envisaged over a period from 1999 to 2005 in the Mars Environmental Survey (MESUR) at a total project cost of $2,000m – with a goal of limiting annual spending to under $150m. Approval for the project was to be sought for a start in fiscal year 1994.

USSR: JOURNEYS TO THE MOON, MARS AND VENUS

Three years were to pass before the Soviets resumed their unmanned lunar missions. The probes carried a rough landing capsule containing a television camera. Just before impact, a retro-rocket would fire, slowing it down to a survivable impact speed. The system was more advanced than the early Ranger capsules but far less sophisticated than the Surveyor landers. Because of their larger size, the second generation Luna probes used the A-2e booster. This was an A-2 (like that used for the Voskhod) fitted with a small "e" (for escape) stage. After the payload reached orbit, the e-stage would fire to send the probe on to the Moon. This was more accurate than the direct ascent launch used for the 1958-60 Luna flights.

The first launch attempt was made on January 4, 1963. The A-2e booster reached orbit but the e-stage failed to fire. A second mission failed during launch soon after. Luna 4, which was launched on April 2, 1963, was sent towards the Moon but a mid-course maneuver failed and it missed the target by 5,283 miles (8,500km). Probably there were major problems on this mission because it was nearly a year before the Soviets tried again, but both the next attempts, in the spring of 1964, ended in launch failures.

These twin mishaps caused another year's halt in lunar flights. The next launch took place on March 12, 1965, but this time the e-stage again failed to fire, leaving the probe stuck in Earth orbit. To hide the failure, the Soviets gave it the name Cosmos 60. The run of failures continued with Luna 5 on May 9, 1965. It was launched out of Earth orbit and successfully placed on a lunar trajectory. As it neared the Moon's surface, however, an instrument failure prevented the retro-rocket from firing and the spacecraft impacted and was destroyed.

Luna 6 was launched on June 8, and the next day a mid-course correction was made. But the rocket engine continued to fire, causing Luna 6 to miss the Moon by 100,000 miles (161,000km). Both Luna 7 and 8 (October 4, and December 3, 1965), also failed: their retro-rockets fired when the spacecraft were too high above the Moon, causing the landers to hit the surface at too high a speed to survive.

The string of failures was finally broken by Luna 9. At 9:44pm Moscow time on February 3, 1966, the spacecraft's radar altimeter showed that it was 47 miles (75km) above the Moon. This triggered the retro-rocket, which began slowing it down. About one minute after ignition, a 16ft-long (5m) probe contacted the surface and the egg-shaped lander was ejected from the retro-rocket module, coming to rest on the surface of the Oceanus Procellarum. The four petal-shaped covers unfolded and it began transmitting a panorama of the lunar surface, showing a rolling landscape with small rocks. Luna 9 also eliminated many exotic theories about the nature of the lunar surface, which certainly could support the weight of a man.

The next Soviet lunar launch was made on March 1, 1966, but once more the e-stage failed to fire and it was named Cosmos 111. The next three successful Soviet Moon missions were orbiters. Luna 10 (March 31) and Luna 11 (August 24) studied the Moon's magnetic field and radiation environment, while Luna 12 (October 22) carried a camera system to photograph the surface.

Soviet lunar activities for 1966 were brought to a close by Luna 13, an improved lander, on December 21. It carried two booms. One had a radiation-measuring device to determine the composition of the lunar surface, and the other carried a soil penetrator to measure the strength of the lunar soil. Both showed it was a loose, granular material. Several panoramas of the lunar surface were also transmitted, revealing a flat, powdery surface with a few rocks.

With this achievement at the end of a generally-successful year, Soviet lunar missions, perhaps surprisingly, halted for 16 months until the Luna 14 orbiter mission in April 1968, about which the Soviets have to date revealed little, and which may have been a TV orbiter of the Luna 12 type that enjoyed scant success. Subsequent Luna missions may best be analyzed in the context of the manned race for the Moon.

Early Interplanetary Flights

The Soviets expressed an early interest in planetary missions. On January 10, 1959, Dr. Anatoli Blagonravov said: "Perhaps we shall try as early as June this year to send a payload weighing from 550 to 750lb on a journey to planet Venus, taking 151 days ...." No such attempt was made. The A-1 booster would have been limited to a payload of around 507lb (250kg). Although this was slightly larger than the early US Mariners, it was over 220lb (100kg) less than the Luna 1 and 2 payloads, and about 110lb (50kg) less than Luna 3. The demands of a deep-space mission, given the state of Soviet technology, could not be met under these tight weight constraints.

To conduct planetary missions, the Soviets developed a larger third stage for the A-booster. Unlike the direct ascent profile of the first generation Luna missions, the Soviet planetary missions would use a "parking" Earth orbit. At the proper time, a small fourth stage (the escape "e" stage) would fire to send a spacecraft on its interplanetary journey. The complete rocket, which was ready in late 1960, was called the A-2e and would serve as the basis for both the Soviet planetary and unmanned lunar missions during the 1960s. It would also launch the Voskhod and Soyuz manned spacecraft.

The launch window for Mars opened in October 1960. Two spacecraft (the basic model could be modified for either Mars or Venus missions) were prepared. Their weights have been estimated to be around 1,422lb (645kg) each, with an intended mission of a photographic flyby. The first was launched on October 10, 1960, but the booster failed before reaching Earth orbit. A second

attempt on October 14 also ended in failure. According to one account, the twin failures were caused by the fuel feed system of the third stage. It was the start of the Soviets' long, frustrating experience with Mars.

Apparently, they had intended to gain a propaganda coup from the Mars launches when Nikita Khrushchev attended a U.N. meeting in New York. A seaman who defected from his ship reported that there were replicas of an "entirely new type of spaceship" aboard. Presumably, had the launches succeeded, the mockups would have been put on display. As it was they remained aboard ship, unseen.

Soviet efforts proved only slightly more successful during the next attempt. On February 4, 1961, their first Venus probe was launched. It reached Earth orbit but the small, e-stage failed to fire. To hide the failure, it was named Sputnik 7, which was described as being for development of "interplanetary ships" and a test of placing heavy payloads into precise orbits. Just over one week later, a second probe was launched. This time the e-stage fired, sending the 1,418lb (643kg) Venera 1 probe toward the planet. Success was short lived, however. Ten days after launch contact was lost, and attempts by Soviet scientists using the huge Jodrell Bank radio telescope in England to pick up any weak signals proved unsuccessful.

3 *Despite repeated (and hidden) failures, the Soviet Union persevered with unmanned missions to the Moon. After three years of failure, Luna 9 in February 1966 ejected a lander which came to rest in the Ocean of Storms, from where it began transmitting pictures. Like the US Surveyors that were to follow, Luna 9 demonstrated that lunar surface dust would bear the weight of men and machines.*

The first Soviet planetary attempts were disappointments. Of four tries, only one was sent on the proper trajectory and it failed before returning any data. Over the next 18 months a new, heavier and, it was hoped, more reliable planetary probe was designed. It was ready in time for the next launch windows. A new information policy was also developed: the "Sputnik 7" failure had apparently been an embarrassment, so in future any probes that failed to leave Earth orbit would not be acknowledged.

The late summer and early fall of 1962 saw intensive Soviet activity, but little success. The first target was

4 *Soviet leader Nikita Khrushchev, seen here giving one of his addresses to the United Nations in New York, used world gatherings as a stage on which to promote the alleged superiority of the Soviet system. There is some evidence that in the fall of 1960 he was planning to announce triumphantly at the U.N. that probes had been launched to Mars, but failures denied him the opportunity.*

1 *Zond (Probe) 3, launched in July 1965, was an interplanetary test vehicle launched towards Mars. During its fly-by of the Moon it imaged much of the lunar farside not covered by Luna 3, but the pictures were not transmitted back to base until it was at a far greater distance from Earth, thereby simulating such a procedure at Venus. This picture was identified only as showing "equatorial and northern parts" of the farside of the Moon.*

Venus and the first launch took place on August 25. The booster reached orbit but the probe was left stranded and US radar detected several pieces in orbit. Two more Venus attempts were made on September 1 and 12, but they, too, were left stranded in Earth orbit.

A month later, the Mars launch window opened and a probe was launched on October 24, 1962. It reached Earth orbit but then exploded, creating a dangerous situation. The incident occurred as the Cuban Missile Crisis neared its climax. The cloud of debris was picked up by US Early Warning radar stations and for a brief time it looked similar to an ICBM attack. Fortunately, however, it was quickly realized that the objects were in orbit and not incoming missiles.

1

The next launch came on November 1. This time the e-stage fired and Mars 1 was sent on its interplanetary trajectory. The probe weighed 1,971lb (894kg) and carried a camera system for photography of the surface as it flew past Mars. The system was similar to that on Luna 3: film would be developed on board and scanned by a television camera for transmission to Earth. Also on Mars 1 were instruments to measure the Martian and interplanetary magnetic fields and to detect any Martian radiation belts, as well as organic compounds on the surface. Cosmic ray detectors and a radio telescope for cosmic radio emissions completed the payload. (It is believed the Venus probes of August/September were similar.) A second launch to Mars followed on November 4, but once again the probe was left stranded in Earth orbit, with NORAD tracking five major pieces of debris that soon re-entered.

The Soviets had made a major planetary effort but had little to show for it. Contact with Mars 1 was lost on March 21, 1963, when its attitude control system failed and the dish antenna lost its lock on Earth. Silenced, it flew past Mars three months later at a distance of 620 to 6,700 miles (998 to 10,783km).

The failures also sparked political controversy. The US Satellite Situation Report listed the August 25, 1962, debris. The US then had second thoughts and

2

2 *Mars 1 was launched to Mars by the Soviets on November 1, 1962. It had an automated camera/film processing system aboard with processed images being scanned and transmitted back to Earth. Unfortunately the mission failed, like the previous Soviet attempts on Mars.*

stopped listing Russian launches (rather than reveal the unannounced failures). All objects were assigned a sequential astronomical designator, however, and it quickly became apparent that some launches were not being listed. In early September, press accounts claimed the US was making unannounced military launches and the Soviet ambassador to the U.N. repeated these charges. Rather than reveal the secret Russian failures, the US representative simply denied the charges, stating that the stories "were not wholly accurate." Finally, on June 6, 1963, US Ambassador Adlai E. Stevenson sent a letter to the U.N. Secretary General, listing the five unannounced planetary failures and a similar unannounced Luna failure on January 4, 1963.

To avoid such embarrassments in the future, the Soviets decided that the "Cosmos" label would be used to "bury" any future planetary failures among the scientific and military satellites launched by the Soviet Union. This procedure was used when the 1964 Venus window opened. The first launch attempt was on March 27, but the spacecraft was left stranded in Earth orbit and re-entered the following day. It was given the name Cosmos 27.

The second launch took place on April 2, 1964, and was successful. Rather than use the "Venera" label, the spacecraft was called Zond "Probe" 1. Over the next several weeks, it made two course corrections but then, in mid-May, the now-familiar communications problems appeared and contact was lost. Presumably the "Zond" label was used to make the mission less obvious should a failure occur.

The Soviets' disappointing record continued with the November 1964 Mars window. Zond 2 was sent towards Mars on November 30. This was late in the launch window and no second attempt was made. Three days after the launch, the Soviets announced that the power output from the solar panels was only half the expected value, which was probably caused by the failure of a solar array to unfold or to function properly. To make up for the loss, the on-board experiments were probably shut off, probably in the hope that enough power could be censerved until Mars was reached. The hope proved in vain and contact was lost on May 4-5. The dead spacecraft flew past Mars on August 6, 1965, at a distance of 932 miles (1,500km). There has been some question about Zond 2's true mission. It is generally believed to have been a flyby, but may have also carried a small landing capsule. Certainly the trajectory would have allowed a low entry speed, minimizing heat shield problems.

The Zond 2 failure caused a shift in the Soviet planetary program. They had launched 13 missions and not one was able to reach its intended target. Although the booster and e-stage problems seemed to have been corrected, the probes themselves had major shortcomings. After Zond 2, Mars missions were abandoned pending development of a more advanced second generation spacecraft, which would be an orbiter carrying a rough landing capsule. Launched by a D-1e booster, the spacecraft would not, however, be ready until the end of the decade. Venus missions would continue, the shorter flight times apparently being seen as within Soviet capabilities.

In the meantime, the Soviets would make a planetary test flight to build confidence for subsequent missions. Zond 3 was launched on July 18, 1965. Although sent on a trajectory that would take the spacecraft out to the orbit of Mars, it would not pass near the planet. The Soviets described its mission as "testing of the station's systems in actual conditions of a prolonged space flight and scientific research in the interplanetary space."

Some 33 hours after launch, Zond 3 flew past the far side of the Moon. During the 68-minute photo session, the camera covered most of the 30 percent of the far side not photographed by Luna 3. It continued on into deep space, but when it was at a distance of 1.37m miles (2.2m km) the photos were transmitted back to Earth; an exercise repeated when it was at a distance of 19.6m miles (31.5m km). This tested out the photo system at a distance equivalent to Venus. Contact with Zond 3 was continued until March 1966 when it was 93.2m miles (150m km) away. This was the last Soviet planetary mission to carry the Zond label: subsequently, it was used for unmanned tests of the Soviets' L-1 lunar spacecraft.

The next series of Venus probes were to carry landing capsules. Despite centuries of ground-based observations and the Mariner 2 flight, the composition of the atmosphere of Venus was unclear. Radio astronomy data indicated it had an extremely thick atmosphere with a surface temperature of 600-800°F (315-427°C). Other astronomers felt the atmospheric pressure on Venus was less than 10 times that of Earth. The radio emissions were blamed on a variety of sources: an ionized layer in the atmosphere or electrical discharges between droplets in the clouds or radiation belts around the planet. This was called the cold-surface model, which was adopted by the Soviets in designing their landing capsules.

Venera 2 was launched on November 12, 1965. The A-2e booster placed it on the correct trajectory for a 14,916-mile (24,000km) flyby, and no mid-course correction was needed. It was the last photographic flyby mission, for the next launch, Venera 3 on November 16, carried the first capsule lander. It was spherical and 35.4in (90cm) in diameter. As the probe neared the planet, the capsule would separate, enter the atmosphere and deploy a parachute, radioing back temperature and atmospheric pressure data as it descended through the swirling clouds of Venus. With luck, it would reach the surface but, unlike the Luna 9/13 capsule, there was no camera system. (It appears a second capsule-carrying probe was planned for the 1965 Venus window. The launch was made on November 23, but the final stage failed and eight pieces of debris were tracked in orbit. It was given the name Cosmos 96 and the debris re-entered 16 days later.)

Venera 2 and 3 continued towards the planet, the latter making a course correction to ensure it would hit Venus. Everything went well during the three-and-a-half-month flight. But as the two probes neared Venus, the heat from the Sun grew and this caused loss of communication with both probes. Venera 2 flew past the planet on February 27, 1966, while Venera 3 impacted the night side of the planet on March 1. Both were silent.

To have come so close, yet failed to gain any planetary data, must have been a major disappointment for Soviet engineers and scientists. But they pressed on: the basic Venera spacecraft was improved and its weight increased from 2,116lb (960kg) to 2,438lb (1,106kg). The work was completed by the time the 1967 Venus window opened.

Venera 4 was launched on June 12, 1967. (A second probe was launched on June 17 but failed after reaching orbit and was given the name Cosmos 167.) Venera 4 reached the planet on October 18 and, as it neared, no radiation belts could be detected. Just before hitting, the 3.2ft-diameter (1m) landing capsule was released, and during entry was subjected to 350Gs of deceleration. (The Venera 4 "bus" burned up.) As it slowed in the thick atmosphere, the parachute opened and radio signals were received for 94 minutes. The Soviets announced that the capsule had reached the surface,

but when the Venera 4 data was combined with that from Mariner 5, this was seen to be not the case. The capsule had begun transmitting at an altitude of 34 miles (55km) and ended at 17 miles (27km); an atmospheric pressure of 15 to 22 times that of Earth's having crushed the capsule.

The data on the composition of the atmosphere came as a surprise. It had been thought Venus would have a nitrogen atmosphere with little (perhaps 1-10 percent) carbon dioxide. But Venera 4 found that it was 90-95 percent carbon dioxide with no nitrogen detected and only traces of oxygen and water vapor. (Ironically, the capsule was designed to float if it came down in an ocean.)

3

At long last the Soviets had achieved their planetary goal. Despite the Mariner 5 data, Soviet scientists still clung to the idea that the Venera 4 capsule had reached the surface, in one case suggesting the capsule had landed on a 15-mile-high (24km) mountain. Venera 4 data clearly indicated a much more rugged capsule was needed, but how much more rugged was not yet clear. In the 19 months before the Venus window opened again there would not be enough time for a complete redesign of the capsule. Instead, the Soviets strengthened it to withstand 25-27 times Earth atmosphere's pressure and the area of the parachute was cut by two-thirds to allow a faster descent. This meant the capsule would descend farther into the atmosphere before it was crushed or its electronics were overwhelmed by the heat. An improved radar altimeter was also installed to prevent any confusion about its height above the surface. In addition to the atmospheric instruments, a photometer was added to measure any lightning flashes.

Venera 5 was launched on January 5, 1969, at the beginning of an eventful year in space. Venera 6 followed five days later. The cruise to Venus was uneventful for both. Venera 5 arrived on May 16, coming down on the night side of the planet. Because of the less-favorable launch window, the G loads were 50 percent higher than those experienced by Venera 4. The parachute opened and the instruments began sending readings. The descent lasted 53 minutes, and when the capsule had descended to 15 to 16 miles (24 to 26km), the 27-times Earth pressure crushed it. The temperature was 608°F (320°C). Four minutes before signals were lost, a single flash of light was detected.

3 Venera 4 being prepared for launch: a still from the film Hallo Venus! *This spacecraft was the first major interplanetary Soviet success and was the reward of great perseverance. When Venera reached Venus in October 1967 a landing capsule was ejected, and it was subjected to 350Gs of deceleration as it plunged into the planet's atmosphere. In due course a parachute opened and data was transmitted back to Earth for more than one-and-a-half hours. The Soviets announced that the capsule had reached the surface, but later evidence indicated that it had been destroyed by the enormous pressure at an altitude of 17 miles (27km). A major surprise was that the Venusian atmosphere was not composed mainly of nitrogen, as expected, but of carbon dioxide.*

Venera 6 followed the next day. It also came down on the night side, some 310 miles (500km) from Venera 5's impact point. The descent lasted 51 minutes. The capsule was crushed at a pressure 26 times that of Earth's, but it was only six to seven miles (10 to 12km) above the surface. It was suggested that the Venera 6 capsule had come down over mountainous terrain, and it is true that later radar data would indicate there were continent-sized, elevated areas on Venus. No flashes of light were detected so the Soviets tended to dismiss the earlier flash as a telemetry error. The data from the two capsules indicated that the atmosphere was composed of 93-97 percent carbon dioxide and 2-5 percent inert gas, with oxygen making up less than 0.4 percent.

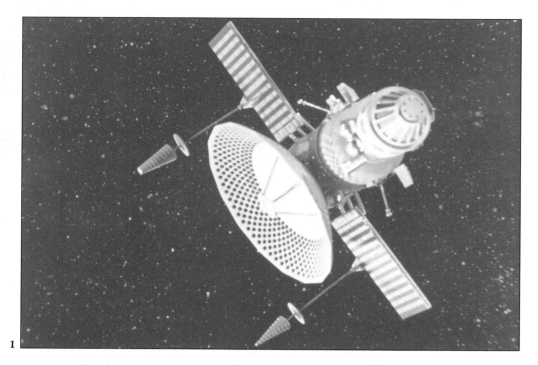

1

1 Encouraged by the success of Venera 4, the Soviets determined to take advantage of the next launch window to Venus occurring in under two years. The gap was not sufficient to permit a major redesign of the basic lander, but it was strengthened and the parachute system modified to allow a faster descent to the planet's surface. Venera 5 was launched on January 5, 1969, and Venera 6, an impression of which is shown here, followed five days later. Both vehicles reached Venus: Venera 5 got to within about 15 miles (24km) of the surface before it was crushed and Venera 6 descended to six or seven miles (10-12km) before suffering the same fate. Just before Venera 5 was destroyed a single flash of light was detected, and there was speculation that this might have been from lightning. Although only partially successful, it was Venera 7 that was the first spacecraft to survive down to the surface of Venus – on December 15, 1970.

There remained only the requirement to build a capsule strong enough to reach the surface. The Soviets took no chances, with a capsule designed to survive 180 times Earth pressure and temperatures of over 1,000°F (540°C) for 90 minutes. The new capsule was smaller than earlier ones and, since heat was seen as the biggest problem, there were fewer openings in the egg-shaped heat shield. Before separating, the capsule's interior would be cooled to below freezing. Its weight was increased to about 1,102 lb (500kg), with the total weight of the spacecraft being 2,601lb (1,180kg).

Venera 7 was launched on August 17, 1970, and the second of the pair followed on August 22. The e-stage of the latter fired but burned for only 25 seconds before shutting down. "Cosmos 359" was left stranded in an elliptical orbit and it re-entered the atmosphere on November 6.

Venera 7 reached Venus on December 15. As it neared entry, power was switched to the capsule's batteries and the capsule's interior was cooled down. The capsule apparently remained attached to the bus until atmospheric entry began, whereas earlier capsules had been separated several thousand kilometres before entry. This may have been to keep the capsule cool as long as possible, or the separation system may have failed, the straps holding it to the bus burning through during entry.

The entry and parachute deployment followed the pattern of earlier capsules, but there was a failure in the telemetry system so only temperature data was sent. However, the atmospheric density could be determined from the capsule's descent rate. The descent lasted 26 minutes. When the capsule reached the surface it

apparently rolled over, possibly because of winds which, in the high atmospheric density of Venus, would have the power of ocean waves. Signal strength dropped to one percent of the normal value, which was just over the level of the background noise. The signals from the surface lasted 23 minutes and the data indicated a surface temperature of 887°F – plus or minus 68° – (475°C, plus or minus 20°) and a pressure of 90 times (plus or minus 15 times) that of Earth was derived. It was the culmination of a decade of effort.

To the Red Planet

With the new decade of the 1970s, Soviet attention switched to Mars again. With the D-1e booster a larger probe/lander could be used. The first two attempts may have been made in the spring of 1969, but neither launch was successful. It was not until the 1971 window that the Soviet's second generation planetary probe made its debut. Although launch opportunities to Mars occur about every 25 months, in 1971 Mars made its closest approach to Earth in 17 years. This meant a greater payload could be sent to the planet, and both the US and the Soviets made use of the opportunity.

The first Soviet launch was made on May 10, 1971. The D-1e booster reached Earth orbit but failed to re-start: it was named Cosmos 419 and the failure came only a day after Mariner 8's Atlas Centaur was lost. However, Mars 2 and Mars 3 followed in quick succession. The probes each weighed just over 10,250lb (4,650kg) and carried a 992lb (450kg) landing capsule. The new generation of Mars spacecraft was four times heavier than the Venera probes and was designed by a team whose average age was under 30. The control system was fitted to the base with the mid-course correction engine, resulting in a more compact and efficient design. This was attached to a cylindrical fuel tank, which also carried the two solar panels and the dish antenna. At the top was the landing capsule under its "coolie-hat" heat shield.

The lander was similar in design to the Luna 9/13 capsule. Because of the thin atmosphere on Mars (about one percent that of Earth) a more complex system had to be used compared with that of the Venera capsules. The capsule would be separated before the bus reached Mars and it would then fire a small, solid-fuel rocket to put it on an impact course with the planet, the bus meantime maneuvering into an elliptical orbit around Mars to begin taking photos and measurements. About four-and-a-half-hours later, the capsule would hit the Martian atmosphere. Once the entry heating was over, the main parachute would deploy and the heat shield would separate. The atmosphere was too thin for a parachute to be used for the complete descent (it was said that the parachute would have had to be as large as Red Square!) so a long probe hung down from the capsule. When this touched the surface, a retro-rocket would fire to slow the capsule and a second retro-rocket would carry the separated parachute well away from the lander.

Once the capsule had landed, four "petals" would unfold, bringing it to an upright position. It was roughly spherical and rested on a pad-shaped base/shock absorber with a pair of stereo cameras at the top. Four antennas were also on the upper part of the lander. It carried instruments to measure soil composition, and the design resembled several ideas for rough landing capsules developed in the early 1960s for the US Voyager program (a Mars landing program cancelled in 1967).

As Mariner 9 and Mars 2 and 3 neared the planet, a dust storm sprang up and it soon completely covered Mars in a reddish-orange blanket. On November 27,

Mars 2 reached Mars and separated the lander, which entered the atmosphere but suffered a malfunction, being destroyed on impact.

Mars 3 followed on December 2, 1971. This time the landing sequence went as planned. Ninety seconds after landing, the capsule began transmitting television signals up to the orbiter, and Soviet scientists eagerly awaited their first view of Mars. For 20 seconds part of a panorama was transmitted, but it was dark and showed no detail. Then contact was lost.

Several possible reasons have been offered for the failure. Since Mars was enveloped in a planet-wide dust storm, high winds and blowing grit and dust could have damaged the lander. Another cause might have been the failure of relay aboard the Mars 3 orbiter. In any event, it was a frustrating end to a promising mission, even though the Mars 2 and 3 orbiters continued to circle the planet, taking pictures. Their imaging system was similar to that used on earlier Soviet missions: the planet's surface was photographed on film which was developed and then scanned with a light source for transmission to Earth. The two orbiters continued to operate until March 1972.

By that time it was again the turn of Venus. Venera 8 was launched on March 27, 1972. The second of the pair followed four days later, but the burn of the final stage lasted only 125 seconds, leaving Cosmos 482 in a highly-elliptical Earth orbit, from which it did not re-enter until May 1981.

Venera 8 was aimed at a target area only 311 miles (500km) wide on the day side of the planet: the first such landing. It was the last of the first-generation landers and one of the mission's goals was design information for the heavier, second-generation missions. The capsule was cooled down and released on July 22. After the parachute opened, the two radar altimeters were used to estimate the density of the soil and the data indicated a loose material. As Venera 8 descended, the composition of the atmosphere was measured: 97 percent carbon dioxide, less than two percent nitrogen and less than 0.1 percent oxygen. Small traces of ammonia were also found. At high altitudes, wind speeds were found to range from 328 to 131fps (100 to 40 m/sec) but below six miles (10km) fell to only 3.2fps (1 m/sec). As the new landers would carry cameras, light measurements were made as the capsule descended. These indicated only 1.5 percent of the Sun's light reached the surface, which meant that "day" on Venus was like an Earth night just before dawn. (This proved to be misleading, in fact, because the Sun was only five degrees above the horizon.)

The capsule landed on an upland plains area; a type of surface comprising about 60 to 70 percent of the total area. Analysis of the soil indicated that it was similar to granite. The surface pressure and temperature were found to be 90 plus or minus 1.5 Earth atmospheres, and 878°F – plus or minus 15° (470°C plus or minus 8°). The data were transmitted through two antennas: one on the top of the capsule and the other on a cable lying a short distance away. The second antenna had been added to prevent communications problems such as had occurred on Venera 7. Venera 8 lasted 50 minutes after landing (its pressure capability had been cut to 105 atmospheres and the weight saved used for increased heat protection), and its demise brought the first stage of Soviet exploration of Venus to a close. After years of hard work and many failures a picture had finally emerged of the planet. It was wrapped in a thick atmosphere of carbon dioxide and the clouds were composed of sulfuric acid droplets. The surface pressure was akin to that on the floor of a shallow

ocean and the temperature would melt lead. Gone were the theories of a planet-wide ocean or steaming, tropical jungles. Venus proved to be very much like the popular idea of hell.

The following year saw the next Mars window. The Soviets made an intensive effort: their last chance to beat Viking to the first surface photos of Mars. The 1973 launch window was not as good as the previous one. Even with the large D-1e booster, it was not possible to launch a combination orbiter/lander like Mars 2/3, so this time the functions would be split: two orbiters and two landers.

The first pair were the orbiters. Mars 4 was launched on July 21, 1973, with Mars 5 following four days later. The capsule-carrying spacecraft followed soon after: Mars 6 on August 5 and Mars 7 on August 9. The trajectories of the orbiters meant that they would arrive a month ahead of the landers. Mars 4 was first on February 10, 1974. The retro-rocket failed and it flew by the planet instead of going into orbit, but it did manage to photograph the Martian surface as it sailed past. Mars 5 proved more successful, going into orbit on February 12. It had been intended that each orbiter would handle a separate lander, but because of the Mars 4 mishap, Mars 5's orbit was adjusted to cover both landing sites.

The landers arrived in early March. Mars 7 had pulled ahead of its twin. The capsule was separated on March 9 but missed Mars by 808 miles (1,300km), although the flyby bus took measurements as it went past Mars. Mars 6 reached the planet on March 12. The capsule entered the Martian atmosphere but contact was lost seconds before landing and an ambitious effort had ended in failure.

Disappointment at Mars

The Soviet Mars program had cost around $3,000m. All there was to show for it were the orbital photos obtained by Mars 2, 3 and 5, and the flyby data from Mars 4, 6 and 7. The attempts to gain the first successful landing on Mars had all ended in failure and it would be 15 years before the Soviets would try Mars again.

The Soviet planetary program of the 1960s and early 1970s enjoyed very mixed fortunes. Once the initial failures were overcome the Venus program was impressive. Mars was a great disappointment. On the technological level the situation was ironic. In the manned area, the Soviets' large boosters were an advantage. With unmanned efforts this advantage was offset by the limitations of Soviet technology. Soviet electronics were air cooled. This meant the systems were in containers at sea level pressure: virtually armor-plated. The net result was that, feature for feature, Soviet unmanned probes were heavier than their US counterparts yet far less versatile. Viking, for example, represented a capability far beyond anything the Soviets could hope to build. Another difference was the reliability of the probes. Several US unmanned missions were lost in launch accidents, but none failed to leave Earth orbit or "died" en route once that hurdle was overcome. Even as late as 1972, the Soviets' e-stages were still failing and the spacecraft themselves were inconsistent in performance.

But there was one major difference between the "superpowers" that worked in the Soviets' favor. Despite its very considerable achievements, the US abandoned manned spaceflight between 1975 and 1981 and NASA launched no planetary probes between 1978 and 1989. While the US sank into isolation and self-doubt, the Soviets began a long and determined campaign of consolidation and gradual progress.

2

2 *Soviet attempts on Mars began again in the early 1970s. A larger orbiter/landing probe had been developed, and with Mars at its closest to Earth for 17 years in 1971, Mars 2 and 3 were launched, with NASA doing similarly with its Mariner 9 orbiter. The Soviets intended that their orbiters would go into elliptical orbits around the Red Planet while the landers would descend to the surface using a combination of heat-shield protection, parachute and retro-rocket for the final stage. On November 27, 1971, Mars 2 reached the planet but the lander was destroyed on impact, although the orbiter continued to operate. Mars 3 followed on December 2, 1971 – the spacecraft is shown here being checked out in the laboratory – and the lander descended to the surface. It started to transmit the first TV panorama to the orbiter above, but after some 20 seconds contact was lost. The cause could have been damage resulting from the severe dust storm which covered much of Mars at that time, or a failure aboard the orbiter. The Mars 2 and 3 orbiters conducted an imaging program based on film, but the failure of the landers was obviously deeply disappointing. Obtaining the first view from the surface of Mars would be a prize beyond compare!*

7

SPACE STATIONS IN EARTH ORBIT: 1969-1981

1 *If manned missions were to be increased significantly in length, much more information about the body's response to prolonged weightlessness was required. Soyuz 9 made a significant contribution in June 1970, when the two-man crew spent 18 days in Earth orbit. Shown here in the spacecraft is flight engineer Vitali Sevastyanov.*

2 *Released in Moscow the day after the Soyuz 9 mission began, this picture records Sevastyanov undergoing a medical test.*

The Apollo 11 moon landing was a tremendous blow to Soviet self-respect. The Soviet people had been led to believe they were still far ahead of the Americans, and among the Soviet leadership there remained a commitment to "beat the Americans." The Soviets' initial response followed their past pattern. They denied they had ever tried to reach the Moon, and this propaganda campaign to deny the American triumph would last 20 years. Their other effort was a space spectacular.

On October 11, 1969, Soyuz 6 was launched. Its crew was Georgi Shonin (commander) and Valeri Kubasov (flight engineer). In the orbital module was the Vulkan smelting furnace. The following day, Soyuz 7 went into orbit: its crew was Anatoli Filipchenko (commander),

Vladislav Volkov (flight engineer) and Viktor Gorbatko (research engineer). The flight's goals were described as group flight with Soyuz 6 and observing the Earth and stars. The third Soyuz launch in three days followed on October 13. Soyuz 8's crew was Vladimir Shatalov and Aleksei Yeliseyev (from Soyuz 4 and 5 of the previous January). The mission's goal was described as joint operations and scientific observations with Soyuz 6 and 7. It was the first time three manned spacecraft had been in orbit simultaneously and each spacecraft had specific research areas: Soyuz 6 medical and biological tests; Soyuz 7 photography of Earth and sky in different spectral bands; and Soyuz 8 investigation of the polarization of sunlight reflected by the atmosphere.

On October 15, all three spacecraft maneuvered and approached each other. The following day was Soyuz 6's last in orbit. The Orbital Module was depressurized and welding experiments were conducted using the Vulkan equipment, which was operated by Kubasov from inside the Descent Module. Once these tests were completed, the Descent Module was repressurized and Kubasov stowed the samples. This brought Soyuz 6's mission to a close. It re-entered after five days in orbit. Soyuz 7 and 8 followed at one day intervals.

The purpose of the propaganda "three ring space circus" was unclear. It was reported that Soyuz 6 was originally planned for an April 1969 launch but was delayed to make way for the N-1 test launch preparation. It was then simply combined with the dual Soyuz 7 and 8 missions. It has long been suspected these two were to dock, but apparently a malfunction prevented it.

The Beginning of Salyut
As in the US, the issue of post-Apollo space goals faced the Soviets. The N-1 failures and the shortcomings of the L-1 and L-3 spacecraft eliminated the Moon (although the Moon program was not formally cancelled until early 1974). However, it was within Soviet capabilities to launch a small space station into Earth orbit on the D-1 booster, with the Soyuz acting as a ferry spacecraft. Initial flights would be short and limited by the space station's lifetime as well as onboard supplies. Later, separate space station modules could be assembled in orbit, with unmanned re-supply spacecraft bringing up fuel, oxygen and food. All this could be done with existing Soviet boosters, spacecraft and technology.

The project was given the name "Salyut" and formally

1

began on January 1, 1969. Actually, it was two related programs. The civilian Salyut was to be built by the Korolev Bureau, while a military photo reconnaissance Salyut (called Almaz) was to be developed by the Chelomei Bureau. Both used the same basic technology and structure but with major modifications for the intended mission.

One of the major unknowns was how the human body would respond to prolonged weightlessness; which in the late 1960s meant anything over two weeks. So far Soviet medical information had been limited to five day missions, and biomedical data in extended flights was the goal of Soyuz 9, which was launched on June 1, 1970, with Andrian Nikolayev (commander) and Vitali Sevastyanov (flight engineer) aboard. Nikolayev had flown on Vostok 3, while Sevastyanov was a rookie. The launch took place at night (the first manned launch in darkness), and this would allow an afternoon landing at the end of the 18-day mission.

As Soyuz 9 orbited, the crew conducted medical and psychological tests. There were two strenuous exercise periods per day. The working day, which included eating and exercise, lasted 16 hours: the other eight hours were for sleep. It was a demanding mission for the crew. The Gemini spacecraft had been allowed to drift to conserve attitude control fuel during long-duration flights. Soyuz 9 could not do this. It was powered by solar cells on two wing-like panels, and these had to be kept pointed at the Sun. Accordingly, the spacecraft was spin-stabilized. This set up centrifugal forces, and the cosmonauts reportedly could feel the different levels of the force between the Orbital and Descent Modules. Some areas of the spacecraft were more unpleasant than others.

In addition to the medical experiments, the crew observed cloud formations, performed navigation tests using bright stars to orient the spacecraft, saw small meteors enter the Earth's atmosphere and photographed the Earth's surface. The cosmonauts were also given one day (their tenth in orbit) off, when they played chess and read books.

Effect of Re-entry on Cosmonauts

Re-entry came on June 19. The re-entry profile would limit the forces to only 3Gs, because after such a prolonged exposure to weightlessness, minimizing the stress on the crew was important. When the ground recovery crew reached the Descent Module, they found that Nikolayev and Sevastyanov, who said that they felt as though they were still under several Gs, were unable to get out of their seats. The crew was moved into isolation at Tyuratam, since in their weakened condition they would be vulnerable to infections. It was only on the tenth day after their return that they again began to feel normal.

This reaction was more severe than that on the Gemini flights. The Gemini 7 crew was able to walk immediately after 14 days in orbit. One is tempted to blame Soyuz 9's spin-stabilization for the problems. Despite this, the Soyuz 9 data indicated that humans could survive weightlessness for perhaps a month without permanent damage and, moreover, the Salyut space station would not be spin-stabilized.

The way was now clear for the launch of the first space station. It was hoped that this would erase the Apollo triumph and return the Soviet Union to the position of space pre-eminence it had enjoyed in the early 1960s. Instead it ushered in years of disappointment, failure and death.

Salyut 1 was launched on April 19, 1971. This was only a few days after the tenth anniversary of Gagarin's flight. (The name "Salyut" was meant as a "Salute" to

the anniversary.) The station was made up of three cylindrical sections, looking like a telescope. At the forward end was the Transfer/Docking Compartment. This contained the docking hatch (eliminating the need for a spacewalk transfer as on the L-1/L-3 Moon ship and Soyuz 4 and 5) and an EVA hatch. The next unit was the Work Compartment. This housed the controls for the station. Behind this was a funnel-shaped housing for a solar telescope. At the end was the propulsion unit – the Soyuz Instrument Module. Much of Salyut 1 was based on Soyuz technology: the four solar panels (two on the forward section and two on the propulsion unit) were from Soyuz, as were the control and instrument panels.

Soyuz 10 was launched on April 22, 1971. The crew was Vladimir Shatalov (commander), Alexei Yeliseyev

2

3

(flight engineer) and Nikolai Rukavishnikov (research engineer). The docking was successful but the crew was unable to transfer over to Salyut. (According to one rumor, they were unable to equalize the pressure between the two spacecraft.) The Soyuz undocked and had another try, again without success. The crew therefore returned to Earth, making the first night manned landing.

Whatever failure prevented the transfer was evidently seen as correctable, for Salyut 1 was boosted into a higher orbit to await the next mission.

The original crew for Soyuz 11 was Alexei Leonov (commander), Valeri Kubasov (flight engineer) and Pyotr Kolodin (research engineer). However, a week prior to the launch Kubasov was found to have a lung condition. Initially, the Soviets thought about replacing him only. Then, four days before the launch, it was decided to replace the whole of the original crew with the backup crew of Georgi Dobrovolski (commander), Vladislav Volkov (flight engineer) and Viktor Patsayev (research engineer). The Soyuz 10 crew was recycled as the new backup crew. Kolodin was very unhappy about being passed over and spent the next several days arguing for a reversal. He was unsuccessful and was subsequently taken off cosmonaut status, medical problems being given as the reason.

Soyuz 11 was launched on June 6 and docked the following day. After boarding Salyut, the crew spent two days activating its systems and shutting down those on Soyuz 11. Medical tests took up much of the crew members' time. Experiments included geological

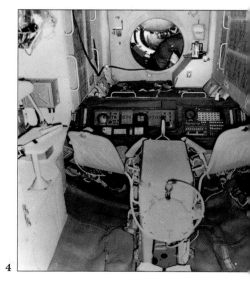
4

3 *"In a few minutes the hatch will open and the cosmonauts ... will fall into the warm embrace of the people meeting them." said the official caption to this picture of the returned Soyuz 9. In fact, initially the crew members were unable to get out of their seats unaided.*

4 *A photo taken in Salyut 1 looking toward the docking compartment.*

1 Andrian Nikolayev, the commander of Soyuz 9, was one of the original cosmonauts and flew the Vostok 3 mission in 1962.

2

3

2 The crew of Soyuz 11, who flew the first mission to a space station – Salyut 1 – but whose triumph turned to tragedy: from left to right Viktor Patsayev, Georgi Dobrovolski and Vladislav Volkov. After breaking the spaceflight endurance record, all three crewmen died because of a spacecraft malfunction during the return to Earth.

3 Moscow, July 6: relatives and friends mourn the dead cosmonauts.

observations, astronomy and cosmic ray studies, navigation tests, cloud observation and atmospheric studies. The crew also spotted aircraft contrails and ship wakes. On virtually every other day the crew made television broadcasts to Earth: they joked, did weightless acrobatics and showed off Salyut 1. Two of them grew beards and Patsayev celebrated a birthday in orbit. The Soviet media highlighted the flight, stressing the superiority of space station missions over those of the US Apollo program in contributing to science, medicine and the national economy.

Propaganda aside, the Salyut 1/Soyuz 11 flight was a genuine first; a real accomplishment achieved well in advance of NASA's Skylab missions. Dobrovolski, Volkov and Patsayev became national heroes and it seemed that the glory days had returned.

But behind the scenes there were problems. During the mission, a cable caught fire. The crew wanted to come home early but Designer Vasili Mishin was able to reassure them. (An electrical fire on June 17 had long been suspected, but it was not confirmed by the Soviets until 1990.) On June 24, the crew broke Soyuz 9's flight record of 18 days, and next day announced they were beginning preparations to return to Earth. In addition to checking out Soyuz 11, they exercised longer and harder. Finally, on June 30, 1971, and after 23 days in orbit, Soyuz 11 separated from Salyut 1. The spacecraft was oriented and retrofire commanded: the cosmonauts were 40 minutes from home. However, in the words of an official announcement issued later: "When recovery forces reached the Soyuz, the men were found in their seats without signs of life."

The impact of this tragedy on the Soviet people was devastating. The only comparable events for Americans were the deaths of President Kennedy and of astronauts Grissom, White and Chaffee. It was the grief of crushed hopes and dreams.

Many questions followed the deaths. For example, had they resulted from the effects of prolonged weightlessness? In fact, the Soviets soon identified a hardware failure as the cause of the tragedy. After the retro-rocket burn, the Orbital and Instrument Modules separated. As the explosive bolts holding the Orbital Module fired, the shock prematurely opened a vent that was later intended to equalize the pressure inside the Descent Module once the parachute had opened in the atmosphere. But now it was venting into space. The crew were not wearing spacesuits, as these had been considered unnecessary. Patsayev unstrapped himself from his seat and tried to block the vent with his finger. There was a manual closing system but it took too long to operate.

The air was gone in about one minute. The crew lost consciousness about 15 seconds after all pressure was lost. The Descent Module made a normal re-entry and landed under automatic control. Once it had re-entered the atmosphere, air flowed back into the capsule but it was too late: Dobrovolsky, Volkov and Patsayev were already dead.

The Soyuz spacecraft underwent a major redesign. Now, two crewmen would wear spacesuits for use during launch and re-entry – one of the seats being removed to make room for the spacesuit controls. On flights to the Salyut the two solar panels were removed: batteries would now supply electrical power. (A separate variety of the Soyuz for solo flights retained the solar panels.) This lightened the spacecraft and improved orbital maneuverability. There was a price, however, in that battery power limited a flight to about two and a half days. Thus, any prolonged docking problem would force a return to Earth. This was the form the Soyuz spacecraft would take until the early 1980s.

Recovery from Tragedy

The first unmanned test of the modifications, a six day mission, was Cosmos 496, which was launched on June 26, 1972. A month later, on July 29, a Salyut reportedly failed during launch, leading to a further, lengthy delay. This put pressure on the Soviets because Skylab would be launched in May 1973. After the Soyuz 11 tragedy, the Soviets felt they deserved the first successful space station mission. So, in the spring of 1973, they prepared two Salyuts for launch.

The first was Salyut 2, which was launched on April 3, 1973, only six weeks before Skylab. It was noticed in the West that the spacecraft transmitted on radio frequencies previously used only for Soviet photo reconnaissance satellites. In fact, Salyut 2 was the first military space station, and to this day the Soviets have refused to release a single photo of the military Salyuts. From "leaks" and a study of the later space station modules, however, it is possible to make educated guesses as to its configuration.

The Soviets apparently removed the Transfer/Docking Compartment, which was replaced by a large re-entry capsule able to carry up to 1,103lb (500kg) back to Earth. This would be used to return exposed film at the completion of manned operations, and it is thought that it looked similar to the Gemini capsule.

In the Work Compartment was a 33ft (10m) focal length reconnaissance camera that, from an altitude of 150 miles (240km), could resolve objects as small as 12 to 18in (30 to 40cm) across. The military Salyut also used a different attitude control system. A spherical gyroscope was spun inside a magnetic chamber: as it spun, the Salyut would move in the opposite direction. This both conserved fuel and held the station in position much more precisely; a factor which could contribute to reducing the "smear" in images caused by movement.

Electrical power was provided by two solar panels mounted on the wide part of the Work Compartment. They could be rotated to face the Sun even when the camera was pointed towards the Earth. (The four panels on Salyut 1 were fixed.) The solar cells were on both sides because reflected sunlight from Earth provided

4

about 20 percent of the normal power level.

Since the forward end of the military Salyut was occupied by the return capsule, the docking hatch was moved to the rear and the maneuvering engines were located on the rim of the Work Compartment. The Soviets themselves likened the military Salyut's appearance to that of a bird in flight.

From Salyut 2's orbit a manned Soyuz launch on April 4 could have been expected. No such launch took place, however. Instead, Salyut 2 was put into a slightly higher orbit. This changed the manned Soyuz's launch window to the morning of April 10. On April 8, however, Salyut 2's onboard engine boosted it into a high parking orbit, well above the orbit used for rendezvous. On April 14, its orbit was again raised. The station then failed for reasons that are not clear.

Nonetheless, the Soviets still hoped to upstage Skylab. On May 11, 1973, (only three days before Skylab's launch), a civilian Salyut was launched. On its very first orbit, a control problem appeared. As a spacecraft orbits, it tends to drift. When the amount (i.e. speed) of this drift becomes too great, control rockets fire. As these cannot exactly cancel out the motion, there is always some drift. On this launch, the limits were set too tightly. When one thruster fired, a second would fire to cancel its residual motion. This, in turn, would cause a third firing, and so on. The nearly-continuous firing of the thrusters consumed the station's entire fuel supply and it was rendered useless. To hide the failure it was called Cosmos 557.

Both Salyut 2 and "Cosmos 557" re-entered within a few days. The Soviets had hoped to operate the two stations simultaneously. Now they claimed that Salyut

2 was never intended to be manned but was designed for automatic operation. Cosmos 557 was only a scientific satellite. In the West, analysts were not fooled. They also began to realize the differences in the Salyut program. Unlike Salyut 2, "Cosmos 557" transmitted on the same radio frequency as Salyut 1. This was the first indication of the civilian/military split.

Fuel Tanker Concept

The twin failures of 1973 had a long-term impact on the Soviet space program. "Cosmos 557" gave birth to the idea of unmanned fuel tankers for resupplying the stations. The impact went far beyond that, however. The twin failures basically resulted from the attempt to upstage Skylab. In the future, the quest for "firsts" would increasingly give way to a careful buildup.

Thus the years between 1967 and 1973 had been disappointing. They were described by the Soviets in glowing terms that tried to hide the true emptiness. However, there was another factor: after Skylab was completed, US manned spaceflight would end effectively until the late 1970s and the launch of the shuttle. (In reality, it would be 1981 before this would occur.) During that time, the Soviets would have space to themselves. There was now time and opportunity for the step-by-step program that had been lacking up to now. It was the road that would lead to success.

The first step on that road was the launch of Cosmos 573 on June 15, 1973, only a month after "Cosmos 557" failed. Cosmos 573 was a two-day test of the Soyuz transport spacecraft. This was followed on September 27, 1973, by the launch of Soyuz 12, crewed by Vasili Lazarev (commander) and Oleg Makarov (flight

4 An artist's impression of Soyuz 11 docking with the Salyut 1 space station. The crew spent 23 days in space, during which time maximum publicity was given to their achievements, which were contrasted favorably with the US Apollo missions which were still continuing in 1971. There were problems during the mission but nothing that occurred then presaged the tragedy to come, when a fault on separation of the descent module from other units of Soyuz allowed the atmosphere of the cabin to vent into space. Within about one minute all of the air in the module was exhausted and the crew quickly lost consciousness. Inevitably, the Soyuz spacecraft underwent a major redesign and no cosmonauts flew in space again until September 1973 – more than two years later.

1

1 *Following some lack of success with Salyut and other missions earlier in 1973, cosmonauts returned to space with the Soyuz 12 mission in September. This was a two-day checkout mission, followed three months later by the eight-day Soyuz 13 mission. Shown here during training are Soyuz 12 commander Vasili Lazarev and flight engineer Oleg Makarov. In April 1975 these two cosmonauts flying together once again were to endure the world's first manned launch abort.*

1 *Following some lack of success with Salyut and other missions earlier in 1973, cosmonauts returned to space with the Soyuz 12 mission in September. This was a two-day checkout mission, followed three months later by the eight-day Soyuz 13 mission. Shown here during training are Soyuz 12 commander Vasili Lazarev and flight engineer Oleg Makarov. In April 1975 these two cosmonauts flying together once again were to endure the world's first manned launch abort.*

2

2 *Shown here training in a Soyuz simulator is the Soyuz 15 crew: commander Gennadi Sarafanov (right) and flight engineer Lev Demin. They flew to the Salyut 3 military space station which was launched on June 24, 1974, but were unable to dock and were forced to return to Earth prematurely.*

engineer). This was a two-day checkout of Soyuz systems followed three months later by Soyuz 13. Launched on December 18, 1973, its crew of Pyotr Klimuk (commander) and Valentin Lebedev (flight engineer) operated a set of telescopes mounted on the nose of the spacecraft. Carrying solar panels, the mission lasted for eight days and the crew was able to undertake some of the research planned for the lost Salyut missions of the previous summer.

Two unmanned Soyuz tests were also made before the way was clear for a further Salyut mission. Cosmos 613 was launched in November 1973. It maneuvered into a high orbit and was then powered down. This simulated a Soyuz docked to a Salyut. After nearly two months it was revived, showing that the Soyuz systems could last for a prolonged period in space. Cosmos 613 returned to Earth after 60 days. The second unmanned flight, another two day Soyuz test, was Cosmos 656 in May 1974. Now the way was clear for the launch of a Salyut.

Salyut 3 lifted off on June 24. Western observers quickly noted that it was transmitting on military radio frequencies like Salyut 2. Soyuz 14 followed on July 3, with a crew of Pavel Popovich (commander) and Yuri Artyukhin (flight engineer). The following day they docked and boarded Salyut. There, the crew conducted medical tests, solar and atmospheric studies, hardware testing, housekeeping and "Earth resources studies." The last was, in fact, manned, military photo reconnaissance. According to one report, objects were put on the ground near Tyuratam as photo resolution targets and the crew used code words to hide the nature of their activities. They also learned to live in weightlessness, at one point Popovich engaging in a wrestling match with a vacuum-cleaner hose.

By July 19 the work was completed: Soyuz 14 separated and landed successfully at the end of a fifteen day mission. The crew suffered no ill effects and the cosmonauts were able to walk from the capsule.

The second launch to Salyut 3 followed six weeks later with the lift-off of Soyuz 15 on August 26, 1974, for what was probably planned as a 30-day mission. The crew was Gennadi Sarafanov (commander) and Lev Demin (flight engineer). Demin was 48 and was then the oldest man to go into space, as well as the first grandfather. The day-long rendezvous and docking sequence went normally until Soyuz 15 was within 100 to 165ft (30 to 50m) of the Salyut. At this point, the automatic rendezvous system failed, causing the engine to burn for too long. The crew made several docking attempts, but each time the approach speed was too high. With the docking aborted and electrical power in the batteries limited, the cosmonauts were forced to return home. To conserve power, voice and telemetry transmissions were stopped: re-entry and landing were successfully accomplished on August 29.

The results of the flight to Salyut 3 were mixed: the plans for back-to-back flights had failed, but Soyuz 14 had succeeded and Soyuz 15's crew had survived. It would take several months for the system's problems to be fixed, and the 90-day lifetime of Salyut 3 would be exceeded by then. On September 23, therefore, the re-entry capsule (containing film exposed by the Soyuz 14 crew) was separated and recovered.

Preparation for Apollo-Soyuz

In parallel with the Salyut developments, the US and Soviets were working on the Apollo-Soyuz Test Project. Part of the ASTP planning envisioned the launch of rehearsal missions. Two unmanned flights, Cosmos 638 (April 1974) and 672 (August 1974), were made. The

manned rehearsal was Soyuz 16, launched on December 2, 1974, with Anatoli Filipchenko (commander) and Nikolai Rukavishnikov (flight engineer) aboard. During the six-day flight they tested the reduced atmosphere inside the Soyuz, simulated the ASTP maneuvers, and tested the universal docking system. All went well and Soyuz 16 landed on December 8.

The year that followed saw both intensive Soviet activity and the beginning of the six-year-long abandonment of space by the US.

Salyut 4 was launched on December 26, 1974, and Salyut 1. Most prominent were the new solar panels: three were fitted, each larger than the earlier version and they could rotate to face the Sun. Soyuz 17 followed into space on January 11, 1975, with a crew of Aleksei Gubarev (commander) and Georgi Grechko (flight engineer), both of whom were rookies. They docked the following day. As the crew entered Salyut 4, they were surprised by a "wipe your feet" sign.

The crew's activities fell into several broad categories. The main scientific research instrument was the OSF-1 solar telescope, but Gubarev and Grechko had to carry out a difficult repair job before it would work. This was accomplished and they were able to take spectrographs and photos. Other astronomical equipment included the Filin-2, an X-ray telescope used to observe the Crab Nebula and other X-ray sources, as well as the ITS-K infrared telescope to observe Earth, Moon, planets and stars. One use of this telescope was to measure the water vapor and ozone content of the Earth's upper atmosphere, measurements of sunlight being taken as it passed through the atmosphere during sunrise and sunset.

Biological tests took up much of the crew's time, and the growth of plants, bacteria, fruit flies and frog embryos in weightlessness were studied. The effects on the cosmonauts' own bodies were also studied, with analysis of blood, bone-tissue density, lung ventilation and muscular condition. The equipment used included a treadmill, a bicycling machine and a special suit that reduced the pressure on the lower body to improve circulation.

The flight continued for 29 days, surpassing Salyut 1/Soyuz 11, Soyuz 9 and Skylab 1. Re-entry took place on February 9, 1975, and extreme weather conditions were encountered during the landing itself when the capsule was buffeted by 45mph (72kph) winds. The landing site had falling snow, low cloud cover and 1,640ft (500m) visibility. However, the crew emerged unharmed and Gubarev and Grechko adapted quickly to life in normal gravity.

Their replacements aboard Salyut 4 were to be Vasili Lazarev (commander) and Oleg Makarov (flight engineer) with launch taking place on April 5, 1975. This would allow a landing after 51 to 63 days, doubling Soyuz 17's flight time and completing the mission before the ASTP launch on July 15.

The launch went normally until separation of the four strap-on boosters and core stage shut-down. At this point, the third stage rocket engines started. Normally, explosive bolts also fired to separate the core stage, but this time an electrical failure caused only half the bolts to fire. So the core stage remained attached to the third stage, even as its engine fired. The added weight caused the booster to go off course and the Soyuz's retro-rocket automatically fired to blast it clear of the booster. (A Soyuz crew has no means of triggering an abort on their own: it must be done by the ground or automatic systems.)

But the crew's troubles were not yet over. The Descent Module underwent a severe re-entry, with a maximum load of 20.6Gs, causing both breathing and vision difficulties for the cosmonauts. Slowly the G forces eased, but the capsule's landing point was uncertain; it was heading towards China and might land on the wrong side. In fact, the crew landed 978 miles (1,574km) downrange in western Siberia, some 199 miles (320km) short of the border. Lazarev and Makarov's troubles had still not ended: the capsule landed on a mountain ledge and caught on a tree at the edge of a cliff. The injured crew crawled out with difficulty and lit a fire. The helicopter recovery crews had great difficulty in lifting them from the nearly-inaccessible ledge, and great skill and courage were shown by all involved.

The Soviets made only a brief announcement of the world's first manned launch abort, reporting only that the failure had occurred and that the crew was alive. Behind the scenes, what began with near tragedy ended in farce. Soviet cosmonauts are paid a flight bonus for specific achievements (e.g. 5,000 rubles for a 90-day flight) and the payment for the successful test of the Launch Escape System was 3,000 rubles. But some bureaucrats did not want to pay the two crew members, and the problem had to go all the way up to Communist Party General Secretary, Leonid Brezhnev. He ordered that they should be paid. The Soviets initially referred to the aborted flight as the "April 5 Anomaly." Later they called it "Soyuz 18-1."

Two Missions in Orbit

It took six weeks to correct the booster failure. Soyuz 18 was launched on May 24 with Pyotr Klimuk (commander) and Vitali Sevastyanov (flight engineer) aboard. Both were veterans, having flown on Soyuz 13 and 9 respectively; they had also been the backup crew for the April 5 Anomaly. They successfully docked with Salyut 4 the next day.

The delay caused by the abort generated problems because the planned two-month-long mission would overlap with the ASTP flight. This required the Soviets to control two separate manned missions; something seized upon by US opponents of manned spaceflight generally to raise a "safety" issue. The Soviets, however, indicated that they would operate two separate ground control teams. Only twice would both ASTP and Salyut 4 be in communication with the same ground station: once for 60 seconds and once for 90 seconds.

Meantime, the Soyuz 18 crew conducted their experiments. The schedule was different from that used on Skylab; the cosmonauts spending most of each day on specific areas of study. The first few days were taken up with biological studies of the growth of onions, peas, beetles and flies. Earth resources and X-ray astronomy received much more attention than on the Soyuz 17 flight. During a 10-day program, some 2,000 photos of the Soviet Union were taken in both single and multispectral wavebands. The images were used to monitor crops, mineral deposits and atmospheric pollution. By the end of the flight some 3.3m square miles (8.5m sq km) had been photographed.

Intensive X-ray astronomy studies were made. These included studying previously detected X-ray sources and searching for new ones. The constellations of Scorpio, Virgo, Cygnus and Lyra were examined, and Soviet scientists felt they had enough data to prove that Cygnus X-1 was a black hole (a star with gravity so strong that not even light can escape). The solar telescope was also in heavy use. By June 4, over 100 spectrograms of the Sun had been taken, and by the end of the mission about 600 images of the Sun had been secured. On June 18, the cosmonauts were requested by the Crimean

3

4

Astrophysical Observatory to record a solar disturbance just detected by ground telescopes, and the Soviet press gave extensive coverage to the crew's activities.

One type of activity that received less press attention was repair and maintenance. According to some Western reports, Salyut 4 had major life support system problems. The humidity control system failed and the windows became fogged. Green mold started to grow and spread through the station. Klimuk and Sevastyanov gave regular reports on the struggle with the green mold, mixed with requests to come home, which were refused.

3 Baikonur Cosmodrome – May 1975: the Soyuz 18 module and its booster is transported to the launch pad. The mission had a special significance because six weeks before a launch had gone badly wrong when a Soyuz with its two-man crew had to be rocketed free of a malfunctioning launch vehicle. The crew of what came to be called Soyuz 18-1 had to survive extreme G forces and a landing in a mountainous region close to the Chinese border, from where they were rescued with great difficulty.

4 A Soyuz spacecraft being assembled. At top is the orbital module; at center the descent module which is occupied by the cosmonauts during launch and landing; and below is the instrument module containing the main propulsion system.

5 A dramatic view of the clustered engines of the Soyuz booster.

As the Salyut 4/Soyuz 18 crew endured their flight, the ASTP mission took place, and the return of the Apollo command module from that marked the last splashdown of a US expendable spacecraft. Ahead lay the shuttle, but prefaced by a six-year abandonment of space by the US. There was no joint follow-up to ASTP: with the fall of southeast Asia to communist forces, Soviet moves in Africa and the stalled SALT II treaty, detente was fading.

Soviet space activity for 1975 ended with the return to Earth of Soyuz 18 on July 26 after 63 days in orbit. This was second only to Skylab 4's record 84 days. Salyut

imized resolution of reconnaissance photos and the higher orbit reduced the amount of fuel needed to correct orbital decay.

It was expected the Salyut 5/Soyuz 21 mission would last 90 days, and the Soviets had discussed such a duration. More importantly, it would fit the step-by-step pattern of durations: 16 days, 30 days, 63 days and then 90 days. But at noon Moscow time on August 24, it was announced that Soyuz 21 was returning home. Volynov and Zholobov hurriedly loaded equipment aboard the Soyuz, powered down the station and separated. They landed that evening after 48 days in space. The flight

4 was used for one final test: on November 17, the unmanned Soyuz 20 was launched. It docked with the station and remained attached for 90 days. Its successful recovery proved the Soyuz systems were safe for a 90-day mission.

With the civilian Salyut 4's mission completed, it was again the military program's turn. Salyut 5 was launched on June 22, 1976, with Soyuz 21 following on July 6. The crew of Boris Volynov (commander) and Vitali Zholobov (flight engineer) set to work. Although they conducted some scientific research – medical and biological studies,materials processing and pollution monitoring – this was only window dressing. The crew's main activity was photo reconnaissance. One possible target was Operation Sevier, a large air and naval exercise east of Siberia, and photographs of the US and China could also be taken. One indication of the crew's secret activities was the lack of Soviet press coverage; two or three days would pass without any mention of the mission and, as with the Salyut 3/Soyuz 14 mission, voice commu-nication was scrambled.

By this time, the differences between the civilian and military Salyut were clear to Western analysts. On military Salyuts, both crewmen were military officers, whereas on civilian space stations the flight engineer was a civilian. In addition to the radio frequencies noted earlier, another visible difference was in the orbit. Military Salyuts went into a 155-mile (250km) orbit: civilian Salyuts had a higher, 218-mile (350km), orbit. The low orbit max-

had clearly been ended abruptly. According to one report, an acrid odor had begun coming from the life support system. The crew was unable to fix it and was forced to leave.

While the Soviets pondered the problems that had cut short Salyut 5/Soyuz 21, another manned mission was launched. Soyuz 22 (September 15, 1976) was not directed towards Salyut 5. Instead, it was the last "solo" Soyuz and was, in fact, the backup Soyuz for ASTP. During the eight day flight, the crew of Valeri Bykovsky (commander) and Vladimir Aksyonov (flight engineer) was to test the MKF-6 camera: a multispectral (six waveband) system built by Carl Zeiss in East Germany. Soyuz 22 returned to Earth on September 23, having successfully tested a major experiment for future Salyuts.

Only two months after Soyuz 21 was called home early, another manned launch was made to Salyut 5. Soyuz 23 was launched on October 14, 1976, with the crew of Vyacheslav Zudov (commander) and Valeri Rozhdestvensky (flight engineer). When they turned on the automatic rendezvous system it failed: they were unable to attempt a manual docking and were forced to return to Earth, where their troubles were not yet over. The spacecraft splashed down in Lake Tengiz. Divers tried and failed in the cold and dark to put a line on the floating capsule. Amphibious vehicles were also unable to reach it. Finally a helicopter got a line on the capsule and dragged it towards the shore and across a frozen swamp. Only then, six hours after

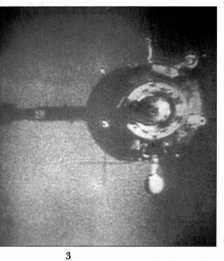

1 *Valeri Bykovsky – commander of Soyuz 22. The mission did not dock with the Salyut 5 space station but conducted independent activities, including a detailed test of a new multispectral camera system produced by Zeiss in East Germany.*

2 *Moonrises seen from Earth orbit were common although always attractive pictures released by NASA. But this moonrise was taken from Salyut 6 in January 1978, when it was occupied by the crews of Soyuz 26 and 27.*

splashdown, was the crew able to get out. (Zudov later said the flight would have lasted 15 days had the docking been successful.)

Military Program Ends

It was not until the new year that another flight to Salyut 5 was made. Soyuz 24 was orbited on February 7, 1977. Its crew of Viktor Gorbatko (commander) and Yuri Glazkov (flight engineer) was to conduct a relatively brief mission. They docked successfully, but it was not until 11 hours later that they entered Salyut. This suggested that the report of life support problems was correct; another indication being the venting of part of the station's atmosphere to space on February 21. Gorbatko and Glazkov returned to Earth on February 25 after 18 days in orbit. The following day, the recovery capsule from Salyut 5 was de-orbited, bringing both the military Salyut program and the post-Apollo rebuilding effort to a close.

Looking back on the years 1971 to 1977, it is clear that the Soviet manned space program lacked consistency. One mission might be a success while the next would be a cliff-hanging adventure. Looking beyond this, however, the outlook was more hopeful. The pointless quest to "beat the Americans" had been replaced by a more careful, step-by-step development program. The ultimate goal was the permanent human habitation of space, and 1977 would see the next step towards that target.

Salyut 6 was launched on September 29, 1977. It was the first of the stations to have two docking ports and had been long awaited. Three long-duration flights were planned: 90 days, 120-140 days and 180 days. Soyuz 25 with Vladimir Kovalyonok (commander) and Valeri Ryumin (flight engineer) aboard was launched on October 9, 1977. As this was only a few days after the 20th anniversary of Sputnik 1, the Soviets made much of this beginning of the third decade of space-flight. However, while Soyuz 25 "soft" docked with Salyut 6 it was unable to make an airtight "hard" docking. The cosmonauts, like others before them, had to hurry home, but this was the last time that two rookies would make up a crew.

It was not clear if the docking equipment on Soyuz 25 or on Salyut 6 had failed. If it was Salyut 6's equipment then the Soviets' ambitious plans would be in danger. To check the situation, Soyuz 26 was launched on December 10, 1977. The crew was Yuri Romanenko (commander) and Georgi Grechko (flight engineer), who had been paired together at short notice. Soyuz 26 docked at the rear port and the cosmonauts spent the next several days checking out the station. On December 19 they made a spacewalk using new, self-contained spacesuits. When Grechko checked the docking equipment, latches, and electrical and hydraulic connections he found them to be in perfect working order. Romanenko was to have stayed in the airlock, but Grechko later said that his crew mate left the airlock to take a look. They then realized that Romanenko's tether had not been fastened and that he was drifting away, causing Grechko to grab the end of the tether. (Grechko has told this story many times in different ways so the details are not clear.)

With the forward docking port checked out, the way was clear for the next step. A Soyuz could spend only 90 days in space before the systems began to deteriorate, so this meant replacement Soyuzes would have to be launched to support long-duration flights. Although they could be flown unmanned, sending a crew along would ease the space station crew's isolation.

The procedure was tested by the Soyuz 27 crew of

4

Vladimir Dzhanibekov (commander) and Oleg Makarov (flight engineer), who lifted off on January 10, 1978. After docking at the forward port, they spent about four days aboard Salyut 6 conducting experiments and testing systems. They loaded the Soyuz 26 Descent Module with film and experimental data, switched their couches to the same spacecraft and landed on January 16. Left behind was the "fresh" Soyuz 27.

Launch of Progress 1

The next step was taken on January 20, 1978, with the launch of Progress 1, an unmanned resupply spacecraft. It was basically a Soyuz but with fuel tanks replacing the Descent Module. New equipment, repair parts, mail and newspapers were carried in the Orbital Module. A Progress could support a two-man crew for a month or more. Once docked at the rear port, fuel was pumped from the Progress into the Salyut 6 tanks. (The Progress' own rocket engine could also be fired to raise Salyut's orbit.) The crew unloaded the equipment and put trash aboard Progress 1 which separated on February 6 and was de-orbited over the Pacific Ocean two days later. This became standard practice, with any debris falling into the sea. The Soviets now had the technology to keep humans in space on a permanent basis. The unknown factor was man himself, and it would take nearly a decade before enough biomedical data was available to resolve this problem.

Another use for the Soyuz "swap" was to bring up guest researchers for brief stays aboard a Salyut. In September 1976, the Soviets selected two cosmonauts each from Czechoslovakia, Poland and East Germany.

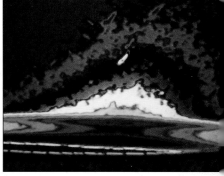

5

3 Soyuz 26 seen from the Salyut 6 space station. The picture was probably taken as the Soyuz 27 crew left the station on January 16, 1978, in the "old" Soyuz spacecraft flown up by the crew still aboard the station.

4 The Nile Delta and northern Gulf of Suez photographed from Salyut 6.

5 A color enhancement of an image of the zodiacal light obtained by Soyuz 26 cosmonauts Yuri Romanenko and Georgi Grechko from Salyut 6.

1

2

2 Alexei Gubarev, commander, and Czech air force officer Vladimir Remek during water survival training prior to their Soyuz 28 mission in March 1978. Remek was the first national from another country to fly on a Soviet spacecraft in the Interkosmos program.

eight days. Six days later, Romanenko and Grechko undocked Soyuz 27 and landed after 96 days in orbit, thereby surpassing the Skylab 4 record. The US and Soviets had traded the manned spaceflight duration record since the mid-1960s, but after Salyut 6 it would belong to Russia into the next century.

The second long-duration flight to Salyut 6 began on June 15, 1978, when Vladimir Kovalyonok (commander) and Alexander Ivanchenkov (flight engineer) rode into space aboard Soyuz 29. During their 139-day mission, they played host to two Intercosmos flights: Soyuz 30 (June 27, 1978) with Polish cosmonaut Miroslaw Hermaszewski and Soyuz 31 (August 26) with East German Sigmund Jahn. Both lasted just under eight days and Soyuz 31 was the "switch" spacecraft. To support orbital operations, Progress 2, 3 and 4 brought fuel, film, mail and even a guitar for Ivanchenkov.

The crew aboard Salyut made use of the MKF-6M Earth resources camera and the KATE-140 mapping camera. Astronomical observations were carried out with the BST-1M telescope, which could observe in infrared, ultraviolet and other wavelengths. It had to be cryogenically cooled and required large amounts of power, and because of this it was used less than the Earth resources equipment. The BST-1M's targets included the Earth's atmosphere, the Moon during eclipses, Venus, Mars, Jupiter, stars such as Sirius and Beta Centauri, galaxies and the interstellar medium. Experiments to manufacture exotic materials which could not be produced in gravity were conducted, and, finally, there were biological studies on flies, plants and the crew members themselves.

Kovalyonok and Ivanchenkov returned to Earth on November 2, 1978, thereby completing the second of the three planned long-duration flights.

Failure of Soyuz 33

The third mission began on February 25, 1979, with the launch of Soyuz 32. Its crew of Vladimir Lyakhov (commander) and Valeri Ryumin (flight engineer) was to stay in space for six months, and plans called for visits by three Intercosmos crews including Bulgarian, Hungarian and Cuban cosmonauts. (The additional guest countries were added in March 1978). Events proved otherwise. After only two weeks, Progress 5 docked with the station and at the same time the Soviets announced that one of Salyut's three fuel tanks had suffered a failure. Although the tank was isolated and fuel was transferred into the other two, the station's main rocket engine could not be used. Any boosts to overcome orbital decay would have to be made by the Progress or Soyuz spacecraft.

The first Intercosmos flight to the new crew began on April 10. The Soyuz 33 crew was Nikolai Rukavishnikov (commander) and Georgi Ivanov, the Bulgarian flight engineer. As they neared the rendezvous with Salyut 6, a six-second engine burn was to be made. The engine fired erratically for three seconds, then shut down with abnormal vibrations: the crew's main engine, the rocket they would fire to bring them home, had failed. They would have to rely on the Soyuz's backup engine. If it had been damaged and fired for less than 90 seconds, then they would be trapped in space. Rukavishnikov thought briefly about the novel *Marooned*.

On April 12, the engine was fired, and did so for 213 seconds. Rukavishnikov and Ivanov endured an 8-10G ballistic re-entry and the capsule was still glowing as it descended under the parachute, but they were safe.

The failure of Soyuz 33 had several consequences. The subsequent Intercosmos flights were put off: Lyakov and Ryumin would not have visitors. More importantly,

All were Air Force officers. The first to fly in the *Intercosmos* program was Vladimir Remek of Czechoslovakia aboard Soyuz 28, the commander of which was Alexei Gubarev. They were launched on March 2, 1978, and, following the docking, Remek conducted Czech experiments. All four cosmonauts on the space station made speeches about socialist solidarity and Czech-Soviet friendship. The Soyuz 28 mission ended on March 10 after nearly

Soyuz 33 was to have been switched. Now Soyuz 32's three-month safety lifetime would soon expire. To make the necessary switch, the Soviets launched Soyuz 34, unmanned, on June 6. Experimental data and film was loaded onto Soyuz 32 and it returned to Earth.

Progress 6 and 7 were also launched to Salyut 6. The latter delivered the KRT-10 radio telescope which was deployed from the rear docking port as Progress 7 separated. After a month of operations, the crew attempted to jettison it but found that the 33ft (10m) dish was caught on Salyut 6. Reluctantly, because it was thought that the crew might be too tired, ground control agreed to the cosmonauts' suggestion that they make a spacewalk to free it. This they did, and Ryumin successfully used wire-cutters to free the antenna. Four days later, on August 19, the crew returned to Earth after six months in orbit. The Soviets had completed the three long-duration flights. More importantly, they had also overcome difficulties by improvisation. It showed a maturity that the Soviet program had previously lacked.

Despite the propulsion problem mentioned earlier, Salyut 6 was still operational and it was decided to launch additional missions. The first of these was an unmanned test of the improved Soyuz T spacecraft, which had solar panels, improved systems and a three-man crew. Soyuz T-1 was launched on December 16, 1979, and remained docked for over three months.

Additional Missions to Salyut 6
The first of the manned additional flights began on April 9, 1980. The crew of Soyuz 35 was a surprise: the flight engineer was again to be Valeri Ryumin. The originally selected crew was Leonid Popov, the rookie commander, and Valentin Lebedev, who had flown aboard Soyuz 13. But Lebedev was injured in a trampoline accident, and as the backup flight engineer also had never flown (and the Soviets required one experienced cosmonaut on a mission) a replacement was needed. Ryumin volunteered to go.

Popov and Ryumin played host to four guest crews. Soyuz 36 (May 26, 1980) carried Hungarian cosmonaut Bertalan Farkas, while Soyuz T-2 (June 5, 1980) was a two-man Russian crew for a three-day test of the new spacecraft. The third flight, Soyuz 37 launched on July 23, 1980, was a particularly bitter experience for the US. The guest cosmonaut was Pham Tuan of the Vietnamese Air Force. As the flight took place during the Moscow Olympics (which the US boycotted to protest against the invasion of Afghanistan), it had an anti-American tone. Tuan was described as "the only man to have shot down an American B-52," which was incorrect since all those shot down had been lost to SAM missiles. On the flight he would: " ... study the effects on the Vietnamese countryside, plants and forests of the enormous amounts of defoliants and firebombs dropped during the war." Although the flight was described as being "timed" to coincide with the Olympics, this was press ignorance. Beginning with Soyuz 33, the guest countries' cosmonauts were being launched in (Russian) alphabetical order. The specific timing was dictated by the political need for Salyut 6 to be visible in the night sky above the guest cosmonaut's homeland while he was on board; constraints set by the Salyut 6's orbit and the launch window.

The final Intercosmos flight of this mission was Soyuz 38, on September 18, 1980. It carried a Cuban cosmonaut, Arnaldo Mendez, who was also the first black in space. As with all the Intercosmos flights, the duration was just under eight days. In addition, supplies were brought up by Progress 8, 9, 10 and 11.

Popov and Ryumin remained aboard Salyut 6 for another month, returning to Earth on October 11 aboard Soyuz 37. The flight had lasted 184 days and Ryumin's total flight experience was now 362 days and 150m miles (241m km).

There was still life in Salyut 6, so more flights were scheduled. Soyuz T-3 (November 27, 1980) was a 12-day repair mission. Its three-man crew of Leonid Kizim (commander), Oleg Makarov (flight engineer) and Gennadi Strekalov (research cosmonaut) was the first since Soyuz 11/Salyut 1 nearly a decade before.

This cleared the way for Soyuz T-4 on March 12, 1981. Vladimir Kovalyonok (commander) and Viktor Savinykh (flight engineer) spent 74 days aboard the station. They also played host to two more Intercosmos flights: Soyuz 39 (March 22, 1981), which had

3

Jugderdemidyin Gurragcha from Mongolia, and Soyuz 40 (May 15) with Dumitru Prunariu from Rumania aboard. (Soyuz 40 was the last of the original Soyuzes.) Supplies were brought up by Progress 12.

When the Soyuz T-4 crew came home on May 26, manned operations on Salyut 6 came to a close. In three-and-a-half years, it had been home for 16 crews and had been occupied for 676 days. Some 13,000 photos of Earth had been taken and 1,310 astrophysical, technological and biomedical experiments had been conducted. The final test came on June 19, when Cosmos 1267 docked with the station. It was a space station module built by the Chelomei Bureau and was believed to be based on the military Salyut. Similar modules would be used to expand the Salyut into a modular space station. The two continued to orbit until July 1982 when, like the earlier Salyuts, they were de-orbited over the Pacific.

By this time, the US had made its thunderous return to space.

1 *Toward the end of the 1970s and in the early 1980s, the number of actual mission images released by the Soviets increased and the quality improved. This picture is of Soyuz 29 commander Vladimir Kovalyonok, taken by his crew colleague Alexander Ivanchenkov during an EVA conducted from Salyut 6. The two cosmonauts were in space for 139 days and received visits from two other crews: those of Soyuz 30 and 31.*

3 *Leonid Popov – commander of the Soyuz 40 mission – floats at top left while an experiment is conducted aboard Salyut 6 by long-stay occupants Kovalyonok and Viktor Savinykh. These two cosmonauts spent 74 days aboard the station and manned occupation of Salyut 6 was ended with their return to Earth in May 1981. An unmanned docking took place with a test module during the next month and about a year later they were commanded to re-enter the atmosphere, where they burned up.*

8

AFTER APOLLO: SKYLAB AND ASTP

1

1 *President Richard Nixon (left) announces the appointment of Dr. Thomas O. Paine as Administrator of NASA in March 1969. Man had not yet walked upon the Moon at that time, but for a number of years some top management within NASA had been looking at the projects that would follow. Paine associated himself strongly with the Apollo Applications Program – in due course to be renamed Skylab – which had been promoted strongly by George Mueller, who was chief of the Office of Manned Space Flight at NASA HQ.*

For many of the early thinkers and writers on spaceflight, as well as those participating in turning the thought into reality, the natural progression appeared to be extended stays in Earth orbit and the creation of space stations before venturing further afield. President Kennedy's challenge changed all that, but while Apollo progressed NASA management was already looking towards the late 1960s and pondering what would follow, assuming success was achieved. Bearing in mind the long lead times in program development, George Mueller (chief of the Office of Manned Space Flight at NASA HQ) saw it largely in terms of how to keep the Apollo team and its hard won experience and expertize together for the next challenge.

During the 1960s there were numerous ideas on Earth orbit activity. Many advocates favored the creation of a rotating station that would create at least a partial gravity in furthering the generally-agreed objectives of greatly extending the length of man's stay in space and conducting scientific research in Earth orbit. Others favored pressing Apollo equipment into use, and as early as 1962 engineers from the Douglas Aircraft Company proposed converting obsolete S-IV third stages of the Saturn V booster into orbiting laboratories. At this time some were thinking in terms of extended Apollo missions in Earth orbit and to the Moon, beginning in 1968, with custom-built space stations, and with journeys to Mars following in the 1970s and 1980s respectively. The reality was both a good deal harsher and less clear cut.

In 1965, a rival to NASA (particularly in terms of winning Congressional support) appeared in the form of the USAF's proposed Manned Orbiting Laboratory. Within NASA itself there was dissension. Administrator Webb tended to regard post-lunar landing plans as a distraction from the immediate task, while rivalry developed between the Manned Spacecraft Center at Houston and the Marshall Space Flight Center in Alabama. So far as Apollo was concerned, Marshall was in charge of booster development and Houston of spaceflight systems generally, as well as the astronauts: the responsibilities were evident and agreed. Once the lunar landing was achieved, management at Marshall under Wernher von Braun could foresee the end of the Saturns, and it would therefore need a new role. It saw the planning and construction of space stations as the new task and Houston interpreted this possibility as a

challenge to its hard-won position as the lead center for manned spaceflight.

As various plans were proposed, Marshall and Houston were usually found on opposite sides, and it was not to be until 1970 that the situation changed substantially.

In the meantime, Mueller sought to develop a coherent program, and in the summer of 1965 announced the formation of the Apollo Applications Program Office. At this stage the AAP (cynically defined as *Almost A Program* by some) was regarded as embracing extended activities both in Earth orbit and on the Moon, and Mueller at least had ambitious plans for it, contemplating no less than 29 Saturn launches between 1968 and 1971. But while there were some high points as well as low in the years ahead there was a sign of the problems to come when, for the fiscal year 1967, the AAP secured funding of only $42m compared with a request for $250m. Nonetheless, with many fits and starts the shape of a project started to emerge. An S-IVB third stage would be launched into orbit and subsequently cleaned of its residual propellants and modified by astronauts to form a laboratory. In the spring of 1966 the Office of Space Science and Applications had its Advanced Orbiting Solar Observatory cancelled for budgetary reasons and began serious discussions with Mueller's office on adapting the telescopes for flight in the AAP. OSSA found a warm welcome, and in due course it was proposed that the Apollo Telescope Mount (ATM), as the concept came to be called, would be launched aboard modified lunar modules on possibly two AAP missions. Since a series of flights was envisaged, another piece of hardware which was proposed early on was a docking adapter with multiple ports and with an airlock attached.

Early in 1967 Mueller was still looking forward optimistically to an expansive first stage AAP program in the period 1968-71, and envisaging the build up of Earth orbital flights from an initial 28 days in 1968 to as much as one year by 1971. Reality indicated otherwise. This was the period when American society seemed to be turning against "technology" as it faced what it considered to be many more pressing problems. NASA's budgets tumbled, and the tragic Apollo fire of January 1967 made the task even harder, to say nothing of NASA's apparent inability to present a clear and convincing post-Apollo concept that could gain the

2 *Initially, AAP was seen by George Mueller as a far-ranging development of the early Apollo thrust to the Moon, which would see tens of missions going both into Earth orbit and back to the Moon as early as 1968. A major experiment to study the Sun was added in 1966. However, the original plans had to be scaled back greatly and by 1971 – when AAP had become Skylab and with the Moon removed from the equation – the project was recognized as America's first space station, with man due to spend months at a time in Earth orbit. This is an early artist's impression of Skylab. The main workshop is at right, with the Apollo command module ferry vehicle at left. Between are the docking adapter and airlock module, and extending above is the solar studies Telescope Mount with its own solar array.*

support of American politicians and the public. NASA's problem here should not be underestimated, for on the one hand it was advised by politicians that there was merit in a major new goal (that is, something that could be interpreted as dramatic) rather than loosely related scientific experiments, and on the other it was counseled that it should become more responsive to considerations of "specific pay-offs," for example the remote sensing of Earth resources.

The years 1967 and 1968 were the low point of the Apollo Applications Program. It shrank in size and its first launch dates were postponed as economy measures. For example, out of a requested $439m for the fiscal year 1969 it obtained $253m. But if the wide-ranging plans had to be curtailed a hard shape to the program became more discernible. The success of Apollo 8 in December 1968 and the arrival a short time before of a new Administrator, Thomas Paine, who identified himself strongly with AAP, provided a powerful impetus to morale. By the middle of 1969 ideas for the launch of a "wet" (that is, fueled) third stage and multiple facility

launches had been abandoned in favor of a "dry" stage fully constructed and modified on the ground, which would be launched with other units as a cluster. The cancellation of the USAF's Manned Orbiting Laboratory (on the grounds that its unmanned satellites were sufficient) also helped. The going was still tough – NASA's fiscal year 1971 budget was the lowest for nine years – but in February 1970 AAP (shorn of any lunar associations) was renamed Skylab and recognized in its own right as America's first space station. Its mission was still evolving and the budget cuts (besides reducing the number of Apollo lunar missions) had put its launch back to 1973. But the Manned Spacecraft Center and Marshall had worked out most of their differences and Skylab had survived. Now it was time to get down to detailed planning and building.

3 *Another early artist's impression of Skylab. The enormous space available in the orbital workshop – which housed both experiments and living quarters – can be gauged by the comparative size of the astronauts.*

4 *But when actually in orbit Skylab looked significantly different: this picture was taken by the last crew to visit it when they returned to Earth in February 1974. The major difference is obvious: the workshop only has one solar array and the events leading to this placed the entire project in jeopardy.*

1 *The first Skylab crew – (left to right) Joseph Kerwin, Pete Conrad and Paul Weitz – at work in the orbital workshop simulator at the Johnson Space Center, Houston. Mission commander Conrad is riding the bicycle ergometer while Kerwin (a physician) is on a rotating chair which was used in a (trying) experiment which tested astronauts' reactions to body and head movements in micro-gravity. Study of man's ability to work for long periods of time in space was a prime mission objective of Skylab, along with the remote sensing of Earth resources, study of the Sun and materials processing in micro-gravity.*

As finally envisaged, Skylab – complete with one of the docked Apollo CSMs which would bring three successive crews of three astronauts up to the station – comprised a cluster of units weighing a little under 200,000lb (90,000kg), with a total length of 118ft (36m). By far the biggest single element was the orbital workshop (OWS), a modified S-IVB stage with a staggering working volume of almost 10,500 cubic feet (300 cubic metres) compared with the Apollo CSM's 210 cubic feet (almost six cubic metres). With a waste disposal tank at its furthest end, the OWS was divided into two major compartments. The lower level was given over to crew quarters for sleeping, hygiene and "waste management," food preparation and eating, as well as a number of experiments. The upper level was devoted to a working and storage area which housed water tanks, food freezers and scientific airlocks, besides additional experiment equipment.

At the forward end of the OWS was the airlock module, which was Skylab's communications, electrical and environmental control center. The airlock had a tunnel section through which crew members could move from the workshop, and also a port through which they exited when performing EVAs – an essential part of station activity. One end of the airlock had a shroud on its outside which was the same diameter as the OWS (22ft – almost 7m), and high pressure containers of oxygen and nitrogen which provided the station's atmosphere were housed in the annular space between the outside of the tunnel and the inside of the shroud.

Attached to the airlock was the multiple docking adapter (MDA), which was the control center for the solar, Earth-remote sensing and metals/materials processing experiments and which also housed other experiments. At its forward end was the primary port where the Apollo CSM docked, but there was an alternative port (should a rescue mission be necessary) located on the side of the module between the optical window and the control and display panel for the remote sensing experiment. The MDA was one of the prime work areas and its volume of almost 1,200 cubic feet (around 32 cubic metres) made it larger than earlier NASA spacecraft. Astride the adapter at a right angle in orbit was the support truss and cylindrical canister of the solar observatory, or ATM. During launch the ATM structure was located at the forward end of the MDA, where it was protected – like the rest of the station apart from the OWS – by a shroud. Once in orbit, the shroud was jettisoned and the ATM rotated through 90 degrees to its operational position.

Skylab had two separate solar power generation systems which were connected (the result of an early, fortunate decision) so that power could flow in either direction. The ATM had a four-wing solar array capable of generating 4,000 watts and the OWS had two solar wings on either side with the same capacity. Each system was connected to a complex battery, regulator and distribution system, and as designed they were jointly more than enough to meet an electrical demand of about 6,000 watts when Skylab was fully activated. Maneuvering the station in orbit and control of its

2 *In this training picture, Conrad is operating one of the materials processing experiments (one devoted to studying the characteristics of molten metal in the vacuum and micro-gravity environment found in space) located in the space station's multiple docking adapter. The spherical object is a work chamber in which many of the experiments were conducted.*

3 *This picture again shows a work area in the multiple docking adapter, but actually in orbit. Skylab 3 scientist-astronaut Owen Garriott is seen at the console controlling the solar physics telescopes. There is one significant feature of the picture which may not be initially obvious: Garriott appears to be sitting down – but there is nothing there on which to sit! In conditions of weightlessness – more accurately described as micro-gravity – adopting such a position without support presents no problems provided the astronaut is securely anchored, in this case with foot restraints.*

attitude was achieved either by six thrusters at the aft end of the OWS using nitrogen gas or, more usually, by three "control-moment gyroscopes." Attitude was measured by a combination of Sun sensors, rate gyroscopes and star trackers, which fed data to attitude control computers, all of which equipment was located on the structure of the ATM.

The atmosphere provided was three parts of oxygen to one of nitrogen at 5psi (0.352kg/cm²) – the proportions, while the reverse of those on the surface, being neces-

sary to maintain the correct sea-level equivalent oxygen pressure. Losses due to leakage and crew consumption of oxygen were automatically replenished. One result of the low pressure was that normal conversations between crew members beyond a few feet were impossible, and so intercommunication stations were located at thirteen points within the OWS.

The environmental and thermal control systems aboard the station were much like those providing heating and air conditioning in the home, although with the twin objectives of providing not only comfort for the astronauts but the correct temperature conditions for the many experiments and items of equipment aboard Skylab.

"Habitability" (as it came to be called) was a major concern of NASA managers. A brief mission of less than two weeks to the Moon was one thing, but they realized that much thought had to be given to the conditions under which astronauts would work and live when missions extended to several months. The astronauts themselves tended to adopt a "right stuff" lack of concern for such matters, but fortunately others did not – the original appearance of the quarters being described as "austere" and "forbidding." In 1967 Mueller enlisted the services of a top design consultant who made recommendations about color schemes, illumination and general planning in the OWS, and it was this same designer who insisted (against the firm wishes of engineers) that a window be placed in the wardroom area so that crew members could enjoy the view from orbit as a form of relaxation – something which in fact was enormously appreciated by those who later flew. The window was inserted on George Mueller's explicit order. The engineers continued to be reluctant on such matters, but managers at the Manned Spacecraft Center and elsewhere persisted, and in 1970 Chris Kraft, the deputy director of the Center, stated that astronauts should not have the last word on this because they were too prone to accept a "make-shift situation."

Thus, much attention was paid to food which could be eaten in the galley area from a tray which heated the individual items (although once again food was not

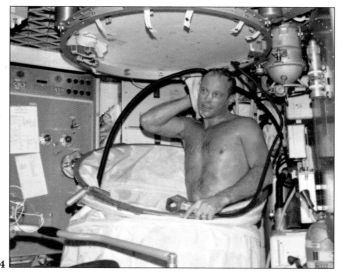

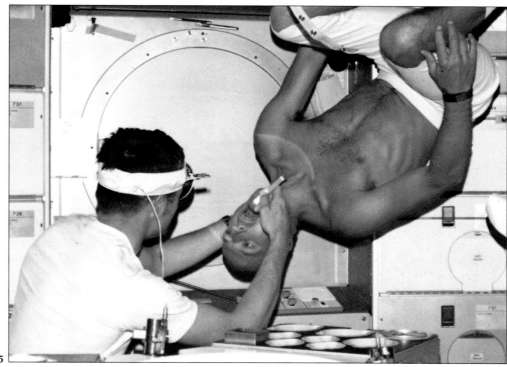

to prove a great success with the crews, "blandness" being a continuing criticism); privacy was provided in curtained sleeping cubicles complete with locker and multilayered sleeping bag and personal entertainment (usually in the form of music tapes) was provided; outer clothes in a pleasant, golden-brown color were provided as well as T-shirts and shorts and a variety of other garments – the astronauts having the pleasure of no laundry chores as the clothes were deposited in the large trash collector in the OWS when changed; and there was maximum provision for overcoming microgravity when an astronaut needed to stay in one position – handrails and handholds colored blue were located throughout, while in the workshop the walls and ceilings comprised a metal triangular gridwork at any point of which a crew member could locate himself by inserting a triangular cleat fastened to the sole of his shoes through the mesh.

"Waste management," a euphemism for defecation and urination, had always been a tedious and difficult function in the close confines of earlier spacecraft, but aboard Skylab it was made as easy as possible consistent with the absence of gravity, which normally separates liquid and solid wastes from the body. A fecal-urine collector used airflow as a substitute for gravity, but the unit also had to allow for the demands of medical experiments, which required that urine samples be collected and frozen and solid waste be dried before

4 *The volume of the orbital workshop made many luxuries possible – such as a shower. However, in a spacecraft a jet of water is not coherent as on Earth but breaks down into globules which float around the available space unless restrained in some way. The Skylab shower necessitated a curtain being pulled up from the floor and attached to the ceiling. Water was ejected from a push-button shower head attached to a flexible hose – which is seen clearly in this picture of Skylab 3 astronaut Jack Lousma wiping his face after showering. The water was drawn off from the curtained area by a vacuum system.*

5 *Space can be fun as well as daunting and dangerous: here Skylab 2 commander Conrad facilitates a dental examination by his medical colleague Kerwin by simply floating upside down – but notice the knee restraint to hold him in position.*

6 *Time for a haircut, but in microgravity shorn hair would float around the spacecraft and possibly cause equipment failures. As his hair is cut by Conrad, astronaut Weitz draws the clippings straight into a vacuum hose.*

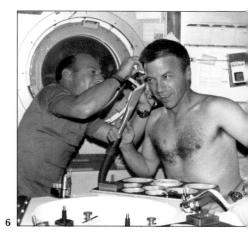

1 *Oceanic processes in remote areas were a particular target of Skylab 4, and shown here is part of the Falkland Current. Plankton, a source of food for many creatures of the sea, appeared as a fluorescent green in the Current, "like a dye marker" according to crew commander Gerald Carr. Also seen as a reddish color at bottom right in this view is a plankton "bloom."*

2 *Colorful twilight layering seen from Skylab 4.*

1

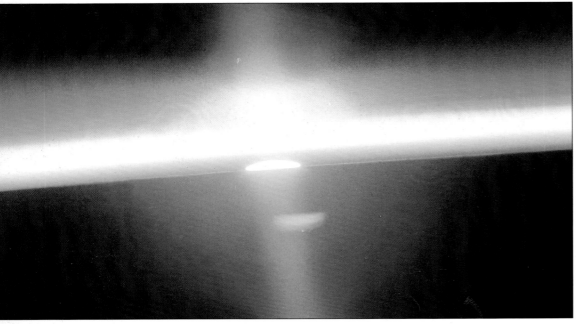

2

3

3 *A star of the Skylab 3 mission: Arabella, one of two common cross spiders taken into orbit.*

4, 5 *The Skylab missions revolutionized our knowledge of the Sun: these two color-enhanced images taken by a US Naval Research Laboratory ultraviolet instrument show two different types of prominence and much detail in the chromosphere.*

return to Earth for examination and analysis of mineral balance. Since water could not be run in the normal way crew members usually washed by means of moistened washcloths, but a shower facility was also provided. A cylindrical curtain was raised around the astronaut and clipped to the ceiling, with water being discharged through a hand-held showerhead. The floating water droplets were then drawn into a water collection system by movement of air. Crew reactions were mixed, partly because of the length of time required, but as one said "you come out smelling good." Spaceflight had come a long way.

But Skylab's serious purpose was to test man's prolonged exposure in orbit to micro-gravity and to conduct experiments. Medical experiments were one of the major justifications for the workshop from the beginning, and ultimately there were nineteen life sciences experiments. Ideally the doctors would have liked a regime of eating, drinking, work and exercise so closely monitored and controlled as to make crew members' lives intolerable, so compromises were inevitable. The most important groups of medical experiments concentrated on measuring the rate of energy expenditure during controlled exercise, using a highly-instrumented exercise bicycle; on cardiovascular function as revealed when the heart was stressed by subjecting the astronaut's lower body to a partial vacuum; and on establishing bone and muscle changes by mineral balance measurements which depended for their success on sampling body wastes and returning them to Earth in the required manner. At the end of July 1972, three astronauts (who did not fly on the missions) were isolated in an altitude chamber for eight weeks in the Skylab Medical Experiments Altitude Test (SMEAT), which proved of great value to the mission, for there were various equipment breakdowns and a basic fault in the urine collection system was discovered which, if it had gone undetected, would have ruined one of the most important experiments.

The eight telescopes mounted in the ATM – if not of such long standing as a Skylab payload as the medical experiments – were regarded as of major importance and made heavy demands on NASA's construction, flight planning and training resources. Six experiments were to study the Sun in UV and X-ray wavebands and to image the solar corona (or outer atmosphere) from above the obscuring layers of the Earth's atmosphere, recording the results in all but one case on photographic film. This gave better spectral and spatial resolution than other forms of recording but it meant that regular EVAs would need to be made by crew members to change film magazines. The ATM had to be pointed within 2.5 arc seconds of a target area on the Sun's disc and held without drifting more than the same angular distance over a 15 minute period. This made considerable demands on Skylab's attitude control and fine pointing systems: the solar control and display panel in the MDA was described as "probably the most complicated ever put into a spacecraft," and it had three times as many controls as the Apollo command module. While long-term schedules for solar observations could be drawn up, the operation of the ATM led to a revolution in NASA procedures, because in the fullness of time the solar physicist experimenters (with the willing cooperation of the astronauts in orbit) greatly increased the scientific harvest from the telescopes by reacting to solar events as they happened – on frequent occasions holding discussions direct with the astronauts and not via the capcoms, an unprecedented change.

Study of the Earth and Sun

The six major Earth remote sensing experiments – referred to as EREP, the Earth Resources Experiment Package – were by comparison late additions. The '60s saw an upsurge of interest in the remote sensing of the environment of Earth in the widest of contexts, and EREP was a reflection of that. In 1967 an earlier AAP plan for a dedicated, remote-sensing mission had been cancelled for lack of funds, but within a year or so NASA was being urged by the politicians to do more work in the area, and one senior NASA manager believed that an Earth resources project might be the salvation of the space program. "Whether or not justified, earth resource sensing from aircraft and space has been widely advertised as promoting great economic returns." It was too good an opportunity to miss. Over $40m was found from somewhere in the budgets to build a package which included a multispectral camera battery and high resolution, large format camera; a spectrometer and ten-band multispectral scanner; and a microwave radiometer to which a passive L-band radiometer was added later. NASA cannot be blamed for capitalizing on the opportunities of the moment but, in a reversal of normal scientific procedures, having decided unilaterally what instruments to fly it then invited potential experimenters and investigators to demonstrate the usefulness of the data.

Like the solar experiments, EREP had a major effect on the Skylab missions, because to maximise the return the station's orbital inclination was increased to 50 degrees, taking it over about three-quarters of the Earth's surface. This meant that Skylab would be in contact with one of the 13 stations of the tracking and data network for about one-quarter of each orbit only. As a result data were recorded and had to be "dumped" when Skylab reached the next ground station – the telemetry system eventually developed needing only five minutes to transmit data that had taken two hours to collect. In most cases the attitude required by the solar telescopes was incompatible with that for the EREP instruments, so solar and Earth resources experiments had to be operated at different times, another planning complication in an already complex program of experiments.

Other experiments covered stellar astronomy and space physics; eighteen were devoted to the examination of materials science and manufacturing in space (a potentially very attractive application of conditions resulting from micro-gravity); and a further nineteen

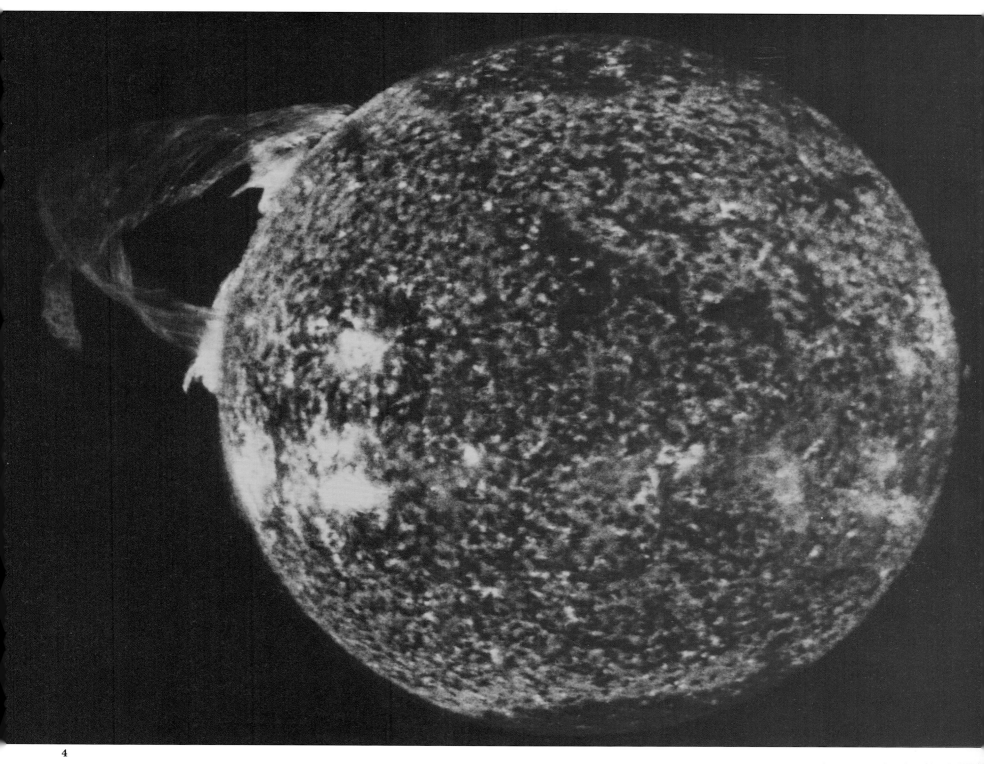

4

resulted from NASA's invitation to American students to develop experimental projects that could be flown on Skylab. The attraction of this to both sides was evident, and while the accepted experiments embraced astronomy, space physics and the biological sciences, the one that ultimately attracted the most attention was "web formation in zero gravity," which in due course was to create something approaching star status for two spiders called Anita and Arabella.

A New Challenge

Skylab demanded many changes in NASA practices and procedures. At the Kennedy Space Center, one major decision concerned the launch of three Saturn IBs that would take three crews to the station. For reasons of efficiency and economy managers wished to launch these missions from Complex 39 on Merritt Island, from which the Saturn Vs were launched. However, the two-stage Saturn IB was more than 140ft (43m) shorter than the Saturn V, resulting in a mis-match of support structures, work platforms and access arms. The solution was the building of a four legged pedestal

weighing 250 tons (nicknamed the milkstool), on which the IBs would sit and which would bring the top of the launchers to the height of a Saturn V. There was some concern about proceeding to manned launches with a modification that had not been tested beforehand, but studies showed that there should be no problems, and such was the case.

KSC's responsibilities for checking out launch vehicles, spacecraft and all systems before a launch had been difficult and complex during Apollo lunar missions, but Skylab and its experiment equipment presented a new dimension, not least because of the complex manner in which spacecraft and experiment systems interfaced. During check-out more than 60 percent of experiments had to be removed because of faults and late design changes, and it was estimated that without such checks one third of Skylab's experiments would have failed in space.

Controllers and technicians at Houston, which in February 1973 changed its name to the Lyndon B. Johnson Space Center, also faced major changes. Although endeavoring to foresee possible emergency

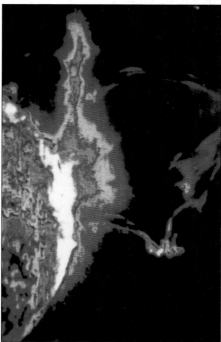

5

situations, Apollo missions had been organized in a precise and rigid manner. Controllers received data from the spacecraft all the time (save when the command module went behind the Moon), and responsibility for the scientific experiments could be located within a specialist group in Mission Control. Overall, the lunar missions led to an intense but relatively brief period of activity. Skylab would be very different. It would be long drawn out – a matter of months rather than days or weeks. It demanded flexibility in the conduct of experiments, with frequent and close participation on the part of investigators, and also of the engineers from the Marshall Space Flight Center, who had built so much of the station and experiment equipment. Data-taking runs would need to be changed at short notice: for example, because of weather conditions in the case of remote sensing of Earth resources, with all that that implied for schedules, alterations to Skylab's attitude and so on. And, far from being in contact with the station and crews at all times, the controllers faced high intensity dumps of data only when the station was in contact with a ground station.

With numerous changes to flight plans and a mission of such different character, training the controllers for Skylab was difficult, to say the least. By the end, the books, maps, cue cards, checklists and charts used for planning each of the three visits came to more than 10,000 pages.

Scientists on Skylab

What of the astronauts? It was inevitable that the role of the scientist-astronauts in Skylab missions would be raised. In 1970 Homer Newell, chief of the Office of Space Science and Applications, pressed for two scientists to fly on each of the missions to the space station. Houston management agreed to one but not to two: it stressed that maximum attention needed to be paid to "systems management and malfunction procedures," based on a concept of "maximum cross-training," giving each crewman the same degree of proficiency on the various experiments. Thus, any specific academic background was unimportant. (In the event, there was a tendency to specialize. The commander took responsibility for the Apollo spacecraft, the scientist concentrated on EVAs as well as solar and medical experiments, and the pilot devoted his attention to the remote sensing experiment and workshop systems generally.)

When the crews were announced early in 1972 there were indeed only three scientists and, perhaps a little

more surprisingly, only two of the nine astronauts had been in space before. They faced an intense training schedule which totaled 2,200 hours spread over eighteen months, compared with 1,200 hours for an Apollo lunar mission. Early on the crews faced a ten week course on solar physics provided by the ATM investigators, and some 450 hours in all were planned for experiment work. Training in simulators was to take a planned 700 hours and EVA training in the neutral buoyancy tank at Huntsville was regularly undertaken. As launch came closer, individual training began to give way to team training in "mini-sims" located in the workshop simulator at Houston. The work did not let up, and early in 1973 the first crew, who were scheduled to lift off in the spring, were working a 60-hour week.

But at long last machines and men were nearing the point of readiness. However, Skylab had a major surprise in store.

Crisis During Launch

In the early afternoon of May 14, 1973, the Saturn V carrying Skylab as its third stage lifted off from the Kennedy Space Center. Its 270 miles (436km) orbit was achieved and within minutes the shroud had been jettisoned, the ATM rotated into position and its solar arrays extended. But it quickly became apparent that all was not well: at just over forty minutes into the mission the workshop's solar arrays should have unfolded but there was no confirmation that this had taken place. Moreover, an earlier signal had indicated that the meteoroid shield protecting the workshop had deployed prematurely. Within hours controllers were wrestling with the near certainty that the shield had been torn off during the launch, and that one or both of the workshop solar arrays had gone with it.

Initially, most media attention was directed towards the possible loss of half Skylab's electrical power. But of more immediate concern was the loss of the shield. While fulfilling the protective role after which it was named, the meteoroid shield was of major importance in protecting the workshop underneath from the heat of the Sun. Fitted tightly to the OWS during launch, once in orbit the release of fold-out panels should have raised the shield to stand some 5in (13cm) from the workshop, thereby creating the correct thermal control within. Its destruction left the gold foil of the OWS wall totally unprotected and rendered conditions within uninhabitable: engineers calculated that temperatures there could rise to as high as 170°F (77°C). (An enquiry was later to attribute the shield's loss to the build up of pressure between it and the OWS wall during launch, caused by an inadequate fit: earlier, KSC engineers had reported considerable difficulties in fitting the shield, which was described as "like fastening a corset around a sleeping elephant.")

Controllers in Houston faced a demanding task during these first hours and days: exposing the ATM's solar arrays to the Sun (in the solar inertial position) created maximum power, but this caused temperatures to soar. Experimentation showed that the best compromise was reached by pitching Skylab up at about 45 degrees toward the Sun, when temperatures stabilised at around 108°F (42°C). In the first three days after launch almost one-quarter of the attitude thruster gas was used up in finding this compromise attitude and in overcoming problems with the gyroscopes; there was a possibility that plastic insulation lining the walls of the workshop had given off toxic fumes (the workshop being vented to space and repressurised four times as a precaution); and studies needed to be made of how the extremely high temperatures would have affected food, films and other supplies loaded aboard the station.

The first crew had been scheduled to launch one day after Skylab. At the beginning of an 11-day period that came to be called "the 11 years in May," this was postponed and involved engineers in many locations throughout the US striving to find solutions. Telemetry indicated that one of the solar arrays might still be on the workshop, but possibly jammed by debris from the meteoroid shield. Two tools, a cable cutter and an implement with prongs for prying and pulling, were modified for use attached to a pole around 10ft (3m) long. Possible techniques for freeing the wing if still there were practised in the neutral buoyancy tank.

Protecting the workshop from heat was more involved. After intensive study two ideas were put into practice. An engineer at the Johnson Space Center proposed a parasol which would be unfurled after it had been passed through one of the scientific airlocks in the upper level of the workshop. Springs and telescoping rods enabled the relatively small port to be used; the four rods or arms of the parasol extending to 21ft (6.5m) when fully deployed. Various ideas for covering or coating the nylon fabric, which deteriorates quickly on exposure to strong sunlight, came up against the problem of the small size of the airlock, and NASA management finally decided to send the parasol with the first crew, with a second design to be sent up later. The second idea originated at Huntsville. During an EVA, two astronauts would fix a bracket base plate on the ATM structure, and two 57ft (17.5m) rods, assembled from sections, would extend out over the workshop. A continuous loop of rope would run the length of each of the poles through eyelets at the far ends. With the shade attached to the ropes, it would be pulled out much as a flag is raised.

With parasol, tools, replacement film and other items the Skylab 2 crew (the station was Skylab 1) lifted off on May 25. The commander was Gemini and Apollo

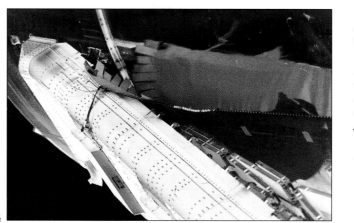

4

5

1 *The Skylab 2 crew lifts off aboard a Saturn IB booster on their delayed mission on May 25, 1973.*

2 *That launch had been preceded by the Saturn 5 launch of the station itself on May 14. However, damage to the outside of the workshop for some time appeared to make a total failure of the project possible.*

3 *One of the proposed sun shades developed by NASA engineers to bring down temperatures inside the workshop.*

6

4 *A close up of the remaining main solar array wing which had become jammed when the workshop's meteoroid shield was ripped off during launch. The strap which was part of the problem is plainly visible.*

5 *Joe Kerwin (foreground) and Pete Conrad perform the EVA which freed the solar array and helped to save the mission. Conrad is carrying the cable cutter used to cut the strap. The solar array was subsequently forced open partially and it then continued to open on its own until fully deployed.*

6 *As the Spacelab 2 crew returned home, they took pictures of the battered station – with the parasol looking distinctly inelegant. However, despite the demands made on them in making the workshop habitable, the astronauts completed between 60 and 100 percent of their scheduled tasks.*

1 *Film was used to record the images obtained by the solar telescopes aboard Skylab, and at intervals the astronauts had to perform EVAs to retrieve exposed film and to load magazines of fresh film. In this splendidly observed and exposed picture, a Skylab 3 crew member stands at the top of the telescope mount while (at left) a colleague passes film up using a pole. One of the solar arrays is visible by the astronaut's helmet and the support truss of the mount fills the foreground.*

2 *Jack Lousma of Skylab 3 performs an EVA with colleague Owen Garriott to fit a replacement sunshade over the existing parasol. The space station and the Earth beyond is reflected in Lousma's visor.*

veteran "Pete" Conrad, accompanied by group five astronaut Paul Weitz as pilot, and physician Joe Kerwin as group four scientist-astronaut. When the crew sighted Skylab, they confirmed both the bad news and good: only remnants of the meteoroid shield remained and they were jamming one of the solar array wings. There was no sign of the second wing. Paul Weitz – his legs held by Joe Kerwin – performed a stand-up EVA as Conrad brought the command module in close, but no amount of wrestling (and four letter words) would free the wing. Neither was docking with Skylab smooth: the crew had to don pressure suits again and partially take the docking mechanism to pieces before a hard dock was achieved at the end of a very long day. Next day the crew entered the station. They found a high, dry heat "like the desert" and returned to the command module at intervals for relief. In due course the parasol was extended but it wrinkled and did not deploy fully: nonetheless it covered about two-thirds of the intended area and over the next few days temperatures steadily dropped, with engineers predicting that they would stabilize at about 78°F (26°C); above the planned level but more than bearable.

1

The crew then proceeded to activate experiments, with power consumption running close to the maximum available. In subsequent days some of the batteries gave problems and an early attempt to free the workshop solar wing appeared imperative. The procedures were worked out in the neutral buoyancy tank by Rusty Schweickart, commander of the first mission back-up crew, and Ed Gibson, who was to fly on Skylab 4 as scientist-astronaut. Having received guidance from Schweickart and having discussed the situation with Mission Control, Conrad and Kerwin went outside the workshop on June 7 to do what could be done, but without much optimism. Their doubts, however, were confounded. While the absence of restraints resulted in difficulties, in an EVA lasting 3.5 hours an aluminum strap across the solar array was cut and a tether was successfully used to force the array partly open. Solar

2

heating accomplished the rest: the wing sections unfolded and within a few hours Skylab's power capability was up to 7,000 watts.

Dealing with the results of the launch damage inevitably had an effect on the attention that Conrad and his colleagues could devote to experiments. Nonetheless, by the time they returned on June 22 at the end of their 28-day mission it was estimated that they had conducted virtually all the required medical experiments, over three-quarters of planned solar observations and about 60 percent of the scheduled remote-sensing work. Conrad was the type to tell Mission Control when it was pushing too hard, and that was valuable in enabling the whole crew to settle down to work despite the alarms. Not surprisingly, each astronaut found different ways of doing things, for example working the bicycle exerciser, and with Joe Kerwin the only medical doctor to fly on Skylab the crew brought back to Earth the valuable advice that time for more exercise must be fitted into the program of the later crews to maintain fitness levels. All in all, they reported favorably on the "house in the sky," even though there was the customary lack of enthusiasm for the food.

A few days before returning to Earth, Conrad and Weitz performed a final EVA to change ATM film magazines. They also hoped to repair a power module. Tests had shown that a relay was probably stuck and that a blow to the battery housing in the right place could well free it. Conrad was told precisely where to strike, and a smart blow with a hammer on the housing did indeed set electricity flowing again. Somehow it all seemed quite in character for Skylab.

Skylab 3 Sets the Pace

The Skylab 3 crew lifted off for Skylab on July 28, 1973, on a mission that lasted for 59 days. The commander was Alan Bean, the fourth man to walk on the Moon, who was joined by Owen Garriott as scientist-astronaut and Jack Lousma, a major in the US marines and selected in 1966 in group five, as pilot. They got off to a slow start since all three suffered from space sickness. Activating the station presented them with a very busy schedule and Bean stressed that for future flights rest (and meals if the crew felt like it) should be given priority over activation, even if a day or two longer needed to be taken. Another early problem was the leaking of propellants from two of the command module's thruster units. Since control of the spacecraft was vital for a safe return this was a worrying development, which resulted initially in talk of a rescue mission being launched. However, analysis of the problem indicated that it was not as serious as first thought.

From that point on Alan Bean and his colleagues built up to a workload which extended the nominal working day from eight to twelve hours. They basically ignored the distinction between working and off-duty hours, and by the end of the mission on September 25, 1973, they had exceeded their experiment goals by 50 percent. During a record breaking EVA of 6.5 hours on August 6, Garriott and Lousma deployed the twin-pole sunshade from the center station of the ATM and temperatures inside the workshop fell close to the originally-planned level. Over the weeks that followed, the crew worked on all experiment groups, but the solar physicists in particular were delighted – the recording of a massive loop prominence nearly three-quarters the size of the Sun itself ("the most significant [solar] event since the launch") earning high marks and reflecting the close liaison that steadily grew up between the astronauts and experimenters.

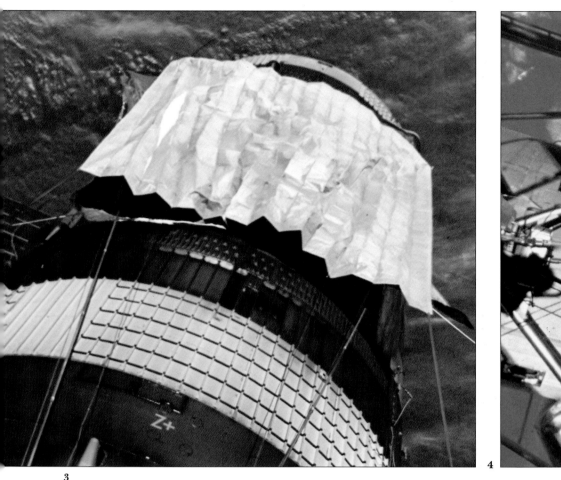

3

4

Although working hard, the crew appreciated the value of humor and variety in presenting their life aboard Skylab to the public in film, TV and sound transmissions. Lousma delighted in showing his unseen audience around the workshop, singing the praises of the Marine Corps and entering into jokes, such as seeming to strain at lifting heavy weights and then suddenly soaring up with the weights held easily above his head. Owen Garriott concentrated on demonstrating good science, while Bean, perhaps the keenest exerciser of the three, after one hard session on the bicycle claimed that he was surely the first person ever to have cycled around the world non-stop.

Repair work included attending to the erring coolant system and gyroscopes aboard the station, but nothing happened to prevent the mission running its full length. Workload schedules (sent up overnight via the teleprinter) could easily fill 6.5ft (2m) lengths of paper, but the crew took it all in their stride. Their main quibble seemed to be with the food, which once again was criticized as "bland," and the menu for being "too regimented." As the conclusion of the mission drew near, Bean requested an extension but this was turned down on the grounds that the doctors wanted more data (i.e. to examine the Skylab 3 crew) before agreeing to a longer stay: besides, supplies of film and food were dwindling. Back on Earth, Bean recommended the continuation of a 70-hour work-week to achieve. maximum returns from the time left aboard Skylab, and the addition of new experiments.

A Difficult Start

The pace set by Bean's crew (once their initial difficulties were resolved, and that point needs stress-ing) created problems for the final crew. It also led to a situation which demonstrated that insufficient attention had been given by NASA to human factors in the context of spaceflight. None of the Skylab 4 crew had been in space before: the commander was Gerald Carr, a marine test pilot, with William Pogue, a USAF officer, as pilot: both

were group five (1966) astronauts. The scientist-astronaut was Ed Gibson. They had been scheduled to launch on November 11 but stress cracks had been found in the fins of the Saturn IB launcher and these had to be replaced. Five days later the crew lifted off, with additional solar, remote sensing and medical experiments to carry out as well as studies of Comet Kohoutek, which had aroused great excitement since its discovery in March 1973.

The mission got off to a bad start when Pogue became space sick and vomited while still aboard the command module. After discussion with Gibson, Carr decided not to report this fact to the ground, although he did inform Mission Control that the pilot was suffering nausea. Carr forgot that the on-board tape recorder was running: in due course the contents were transmitted to Houston, and the deception was revealed. Carr readily admitted that his decision was "dumb," and this error of judgement meant that when flight controllers started to pressure the crew, almost certainly too early into the flight, he was reluctant to react as Conrad and Bean certainly would have done. Thus, despite accomplishing much in activating the station again, finding their space legs, making repairs, for example, to the antenna on the EREP microwave sensor, and conducting experiments, the crew found themselves being bustled in a manner that they resented.

At various times all of them commented on this adversely to the ground. Carr talked about the "frantic" pace of the first two weeks and of being "driven right into the ground." Somewhat later Gibson talked of their time on Skylab since they had arrived being nothing but a "33-day fire drill," and that henceforth he intended to get the tasks done "right, rather than on time." The crew had not established close contacts with the flight controllers prior to the mission, and in striving to meet experiment and work schedules it seemed that the ground was not listening to the astronauts. Towards the end of December, Carr decided to send a considered statement to the ground, and this

5

3 A view of the twin-pole sunshade deployed by the Skylab 4 crew. The first shade was deteriorating as a result of exposure to strong solar radiation and needed to be replaced. Temperatures in due course fell to around the originally-planned level.

4 Scientist-astronaut Edward Gibson photographed during the final Skylab 4 EVA on 16mm movie film by mission commander Gerald Carr. Carr's umbilical and tether stretches away from the camera and on into the station.

5 All was not work: Skylab 4 astronauts Carr and William Pogue have fun and use micro-gravity to pose as supermen.

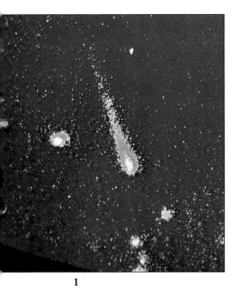

1 A special far ultra-violet sensitive camera designed by the US Naval Research Observatory was used aboard Skylab 4 to photograph Comet Kohoutek. This image (the result of subsequent color enhancement) was obtained on Christmas Day 1973, and shows the comet against background "hot" stars in the constellation of Sagittarius.

1

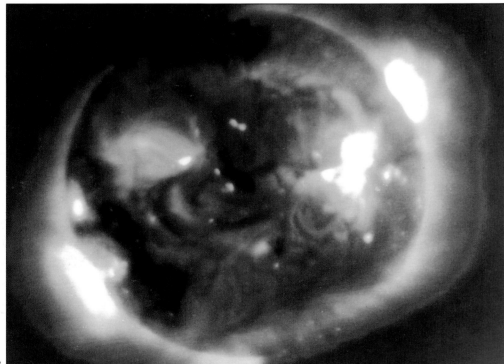

2

2 An X-ray image of the corona – the tenuous outer atmosphere of the Sun. Structures with temperatures higher than one million degrees Centigrade can be observed in X-rays, and the loops, arches and other features seen here are produced by the interaction of the Sun's magnetic field with the ionized gas of the corona. The Skylab solar physics experiments played a major role in elaborating features such as large dark "holes" in the corona and much smaller "bright points" – both phenomena being clearly visible in the image.

eventually led to a constructive exchange of views which cleared the air to a considerable degree. Crew morale improved and, when the ground studied the performance so far, the controllers found to their surprise that between the 15th and 30th mission days there was "no significant difference" between Skylab 3 and 4.

So the crew worked on toward the conclusion of their 84 days in space. From inside and outside the station they photographed Kohoutek, and Gibson achieved one of the goals of the entire solar observation program by recording the development of a flare from beginning to end. Conferences with experimenters greatly improved the crew's insight into their work and over four EVAs (which included dealing with a jammed filter wheel on one of the ATM telescopes) the crew amassed a total of over 22 hours outside the station. In January 1974 Skylab enjoyed its mid-summer, with 46 revolutions in continual sunlight, and Carr could show how far the atmosphere had improved by reacting to the transmission of instructions filling almost 50ft (15m) of teleprinter paper by remarking to the capcom: "I understand you are going to [send] up the Old Testament tonight."

On February 8 Carr and his colleagues left the space station and returned to Earth. The media had already queried whether there would be any more visits, even though they had not been planned. It was pointed out that the workshop systems could be expected to deteriorate as time went on, and that some essential elements, such as atmosphere to breathe, would be exhausted. (Besides, there was no money.) So, as the real work for the scientists in analyzing data began, the controllers in Houston placed Skylab in a "gravity gradient" stabilized attitude (in which the multiple docking adapter pointed to Earth) and switched off the power. Four years passed before they reawakened it, and then it was near its end.

The Skylab Achievement

The results of the Skylab experiments can be presented in terms of statistics – 18,000 blood pressure measurements, 127,000 film frames of solar activity and 46,000 Earth resources remote sensing images, for example – but of course these were but the basic data. What was the meaning and value of the vast flow of information?

In terms of examining man's future role in space the medical experiments were clearly the most important of the entire program. Motion sickness remained an unpredictable problem, five of the nine astronauts suffering from some symptoms in the early parts of the flights, but mineral loss from bones, the loss of muscle mass and the deconditioning of the cardiovascular system were extensively elaborated, although numerous questions remained. The conclusion was that none of the physiological changes observed would preclude continuous stays in space of nine months, but beyond that no predictions could be made. After Skylab, in the 1970s and on into the next decade, the vital work in this area would be done by the Soviets as a result of the record breaking, long duration flights aboard the Salyut and Mir space stations.

Both in the immediate and longer terms, the solar observations were greeted with universal acclaim. The telescopes had revealed new phenomena or important information about previously inadequately explored solar features, such as coronal transients, bright points revealed in the X-ray waveband in both the chromosphere and corona, and coronal "holes" which were believed to be closely associated with the origin of the so-called solar wind. While the data was taken over a limited period of the Sun's eleven year cycle of activity, when, despite all of the phenomena observed, it was close to the minimum, the investigators were in no doubt as to the achievements. One referred to "one of the most important events in the history of solar physics," and there was general agreement that Skylab had yielded the best observations ever obtained from space. Another investigator pronounced the observations "extraordinarily valuable, perfect and complete ... staggering in quantity and quality ... Skylab has vindicated the use of man in space to perform scientific experimentation, notwithstanding opinions still voiced to the contrary."

The value of the EREP experiments was less clear cut. Much of the data was of high resolution, and the Skylab 4 crew in particular had demonstrated man's ability to observe and report on surface phenomena in a manner beyond the capability of unmanned spacecraft. Numerous studies were published subsequently, but this group of experiments had tended to suffer most from an operational clash with the demands of the solar observations, and many in the fast-growing body of remote sensing investigators valued regularity as well as frequency of data gathering highly, and this was provided by unmanned satellites. At a NASA sponsored symposium held in mid-1975, for example, only about one-eighth of the papers stemmed from the Skylab results, with almost all the rest interpreting data derived from the unmanned Landsat satellite that had been launched in 1972.

While perhaps attracting little attention at the time, a group of experiments with much significance for the future was concerned with materials processing. In conditions of micro-gravity, such laboratory and industrial activities as crystal formation, diffusion in liquid metals, solidification of molten metals and alloy formation from components of different densities are often considerably enhanced. Thus Skylab produced crystals of greater size and purity than would normally be produced on Earth. This work (the program of experiments being overfulfilled by the crews by no less than 220 percent) was highly exploratory in nature but it was still continuing many years later, and in the early 1990s materials processing was regarded as a major function to be carried out on NASA's proposed Freedom space station.

The $2,500m Skylab project was undoubtedly

successful, both in yielding valuable information about man's performance during long spaceflights (Skylab 4's endurance record of 84 days stood until the Soyuz 26 mission to Salyut 6 in 1977-78) and in demonstrating the manner in which he could perform science. In an ideal world, a more logically conceived and planned program would perhaps have resulted in flights where each was devoted to a major experiment group without compromising the quality of other groups. But NASA was struggling hard with budget cuts in a country which appeared to have fallen out of love with space and, understandably, it determined to make the most of the three flights that it could afford. And lack of funds also had a major impact on the nature of Skylab's demise.

Before the Skylab 4 crew undocked from Skylab in February 1974 Gerald Carr fired the Apollo spacecraft thrusters to nudge the workshop into a higher (433x455km) orbit. It had been realized early in the project that some parts of the 165,000lb (75,000kg) space station would survive re-entry through the atmosphere. In 1970, the NASA Administrator Thomas Paine ordered a study which showed that adding retro-rockets and systems to enable controllers to bring surviving Skylab fragments down to a pre-selected area of an ocean would exact a weight penalty (entailing considerable redesign of the station) and also cost a minimum of $10m. The chances of a fragment striking anyone were calculated at 55 to 1. Faced by financial problems, and the likelihood of lengthy delays caused by modifications if a retro package were to be added, NASA management decided to accept the small risk. Meanwhile, the final Skylab crew did what it could to prolong the station's life.

Early, over-optimistic estimates were that Skylab re-entry would take place around 1983, by which time it was hoped the space shuttle would be operational and could fly a mission to the workshop, where it would

3

4

3 *In this hand-held image Israel (with the Sea of Galilee and the Dead Sea prominent), Jordan, Lebanon, Syria and Cyprus (top left) are the main features.*

4 *Growing crystals was an important element of the materials processing work aboard Skylab, and such experiments still continue in space today. Shown here is a one-inch-long crystal of Germanium-Selenium, which was of excellent quality: it was reported that these crystals grown conventionally were rarely longer than one-tenth of an inch (2-3mm) and had significant defects.*

5 *Plans for the shuttle to fly a mission that would boost Skylab into a higher orbit came to nothing.*

5

1 *In the Kremlin on May 24, 1972, President Richard Nixon and Soviet Premier Alexei Kosygin signed a five-year science and technology agreement which included plans for cooperation in space. The major proposal was for a joint mission in which an Apollo and Soyuz spacecraft would rendezvous and dock in Earth orbit. This became known as ASTP (Apollo Soyuz Test Project.) By the late 1960s, the intense political rivalry of earlier years had tended to ease and by 1970 some leading NASA managers, such as George Low, were keen on promoting cooperative ventures with the Soviet space authorities. Later exchanges of visits between the two sides proved successful and the formal 1972 agreement was the result.*

2 *Soviet Academician Alexander Vinogradov (left), in Houston for a lunar science conference in January 1971, examines a lunar rock with Dr. Robert Gilruth, Director of the MSC (right). By this time US-Soviet space contacts were becoming closer.*

attach a propulsion module to boost the station into a higher orbit. Two things conspired against the idea: high levels of solar activity as the next maximum of the cycle (1980-81) approached had the effect of increasing the density of the atmosphere at orbital altitude and thus the drag on Skylab. In addition, low budgets, delays and constant problems meant that the shuttle would not be available. Early in 1978 the Soviet Cosmos 954 satellite re-entered the atmosphere, scattering radio-active debris over a large area of northern Canada. There was a justified outcry – and the media in the US and elsewhere which, for the first time in almost ten years, had decided in 1974 not to televise live the return of a NASA crew to Earth, realized that it had a major news story on its hands, despite the fact that there was no radioactivity threat from Skylab.

NASA was subjected to considerable criticism as the months went on, and an inter-agency team was formed with other US government departments to keep world governments informed on re-entry. The 50 degree inclination chosen to increase the scientific return from Skylab worked to NASA's disadvantage now, for Skylab flew over areas occupied by about 90 percent of the world's population. Contact with the space station was re-established and teams of controllers in Houston experimented with the most suitable low drag attitude control, keeping a watching brief from the summer of 1978 onwards. Once re-entry was known to be imminent from information supplied by the USAF radar tracking system, they would be able to exert a limited control over the probable area of impact by setting the station in an end-over-end tumble. In the event Skylab did not break up as quickly as expected once this maneuver was commanded, and on July 11 or 12, 1979, depending

It was indeed, and over the years ahead that resolve was to prove but a pale shadow of the early days of Apollo.

THE APOLLO-SOYUZ TEST PROJECT

After Skylab there was one final Apollo mission before the "long winter" of US absence from Earth orbit began while the space shuttle was developed. The joint mission flown in 1975 by Apollo and Soyuz crews was the culmination of several years of hard, detailed work devoted to the mission, and of many more years of preliminary exploration and discussion set against a shifting political background.

The late 1950s and the early 1960s were a period of Soviet "firsts" in spaceflight and of much political tension. Even so, despite the inevitable effect of events such as the Bay of Pigs invasion in Cuba, and the erection of the Berlin Wall in 1961, as well as the Cuban missiles crisis in October 1962, there were contacts, albeit halting, about possible spaceflight cooperation, particularly as the US program began to develop strength. Between 1962 and 1964 meetings between Hugh Dryden, NASA's Deputy Administrator, and Anatolii Blagonravov, chairman of the Soviet Academy of Sciences' space exploration committee, led to technical cooperation which was described subsequently as taking the form of an exchange of information rather than true cooperation. As the US drive to the Moon developed, talk of initiatives for the most part died, though in September 1963 President Kennedy, in an address to the U.N. General Assembly, somewhat quixotically floated the idea of a joint US-Soviet mission to the

1

2

on one's location on the Earth, instead of fragments falling 800 miles (1300km) east of Cape Town in the Indian Ocean, well south of the shipping lanes, they fell near Perth in Western Australia. There were no reports of injuries to persons nor of damage to property.

In the short term at least the controversy over Skylab's re-entry tended to overshadow the achievements of the project. While some NASA management decisions could be criticized, inadequate funding was a major problem of the later 1960s and 1970s when the Soviet Union once again pulled ahead of the United States in Earth orbital activity. Instead of flying in space, back-up Skylab hardware found its way into a museum. A year or two earlier, in evaluating the success of Skylab, the NASA program director commented that the only limiting factor was not man's ability to work in space nor his technical capabilities, but the US nation's resolve.

Moon, the technological difficulties of which brought a predictable reaction from Robert Gilruth.

But with the appointment of Thomas Paine as NASA Administrator in the autumn of 1968 and the success of Apollo, the attitude on the US side changed. After the Apollo 11 landing, and with President Nixon's approval, Paine began contacts with leading Soviet scientists about the possibility of a joint rendezvous and docking mission. His overtures were received in a friendly and constructive manner and, despite Paine's own departure from NASA in September 1970 when he was replaced in top level negotiations by (acting) Administrator George Low, who was equally as keen on possible cooperation, a first, major meeting was held in October 1970, when Gilruth led a small party of NASA managers to Moscow. Although exploratory in nature, the discussions did go into considerable detail on the

3

4

3 *An artist's impression of the Apollo and Soyuz spacecraft about to dock. Of particular importance is the newly-developed docking tunnel which the NASA crew carried into orbit. By strange coincidence, the highly dramatic lighting effects from the Sun which the artist painted in this scene were to some extent actually created during the mission in an experiment conducted by the two crews (page 126 picture 3).*

4 *Soviet engineer V.S. Syromyatnikov is framed by part of the docking mechanism during discussions between engineers of the two partners at Houston in April 1972.*

5

6

5 *There were an immense number of difficulties to be resolved before the ASTP mission could take place. Many of these were of an institutional nature; for example, the need for even senior Soviet managers to defer to Moscow before releasing information or making decisions. Matters did improve slowly, and the fact that this was so resulted in no small way from the relationship struck up between the technical directors leading the two sides. Shown here is Konstantin Bushuyev in a picture released at the time of the summit meeting in Moscow.*

6 *Dr. Glynn S. Lunney was the ASTP technical director for NASA. The picture was taken at a joint press conference in Houston in April 1974.*

President of the Soviet Academy of Sciences, promised to consider the idea, and intensive work on both sides eventually led to the project being formally approved at the Nixon/Kosygin summit meeting that took place in May 1972. It was, in the words of James Fletcher, the Nixon appointee as NASA Administrator, the "most dramatic" of the items included in the overall space cooperation agreement signed at the summit. Moreover, unlike the SALT agreement on the limitation of nuclear strategic arms, it had a firm timetable. Given the lengthy development periods typical of space hardware there was little time to spare to meet a 1975 deadline for what was officially called, from June 1972 onwards, the Apollo-Soyuz Test Project (ASTP).

Docking Mechanism

From the earliest contacts between the two sides it was realized that there were many problems of compatibility to be solved between spaceflight philosophies which had developed quite separately. These obviously included vital details such as radio, optical and other guidance and communications systems, and wide-ranging issues such as overall flight control procedures for bringing about the successful rendezvous and docking of two manned spacecraft from the world's leading space powers – one of which, the Soviet Union, had favored automated and ground-controlled rendezvous whereas NASA had pursued a more manual, astronaut-controlled system.

Notwithstanding all of these many issues, the absolutely vital matters to be agreed were the means by which the two craft would dock, and how crew members would transfer between Apollo and Soyuz. Up to this time, both American and Soviet docking techniques had been based on the use of probe (male) and drogue (female) units, though of different designs. There would obviously have to be a new system devised for a joint mission, and at an early stage it was agreed that this system should be "universal," in the sense that the unit fitted to each craft could be used either in an active or a passive mode depending on the agreed procedure for a particular docking. This clearly had important implications for missions further into the

major problems that needed to be solved before a joint mission involving rendezvous and docking could take place. Working parties were set up and plans laid for future meetings.

Subsequently the US side became convinced that the proposed project should feature *current* spacecraft and take place in the near future, rather than be regarded as a vague, unspecified development far into the future. The attraction for NASA was the chance to fly another Apollo mission in the period after the conclusion of Skylab and before the shuttle flew. George Low took the opportunity of a visit to Moscow in January 1971 – to initial an agreement which involved among other things an exchange of samples from the lunar surface – to suggest in confidence the development of compatible systems for use with Apollo and Soyuz spacecraft. The Soviet side, led by Academician Mstislav Keldysh,

1 *A group portrait of the ASTP crews. Standing (left to right) are Apollo commander Tom Stafford and Soyuz commander Aleksei Leonov. Seated (left to right) are Deke Slayton; "rookie" astronaut Vance Brand; and experienced flight engineer Valeri Kubasov. Both Stafford and Leonov went on to occupy important posts in their respective countries' aerospace establishments.*

future and for possible rescue missions, although the latter tended to be played down as time went on. The life-sciences term "androgynous" was applied to this new system. Both sides contributed much to the eventual design adopted, which allowed each of them to construct their version of the docking mechanism in their own country - but with certain agreed modifications which still rendered one unit compatible with the other.

Transfer between the two spacecraft faced the major difficulty of the different environmental systems that had been adopted for US and Soviet spacecraft. Soyuz crews breathed a nitrogen/oxygen mixture at 14.7 psi ($1.03kg/cm^2$), much like that breathed on Earth, while Apollo crews breathed pure oxygen at 5psi ($0.352kg/cm^2$). With these conditions unaltered, anybody passing from Soyuz to Apollo during the mission by an intermediate tunnel would have to spend several hours there prior to entering Apollo, otherwise they would risk suffering the "bends." The solution was for the pressure in Soyuz to be dropped to 10psi ($0.703kg/cm^2$) during the docked phase, which permitted transfer to take place in a much shorter time.

this was done, that craft's hatches would be opened.

But the construction of the docking systems and the module itself (contracted to North American Rockwell – later to become Rockwell International) lay some time in the future as the US and Soviet technicians, engineers and managers began the long round of traveling to the meetings and discussions needed to thrash out the many issues to be resolved. Time was needed to get used to one another's way of working, but the tone of what might be called "cooperative frankness" was set by the men chosen as technical directors by the two sides – Konstantin Bushuyev and Houston's Glynn Lunney. Bushuyev was later quoted as saying that there had only been one difference between them: that Lunney drank his coffee black while he himself preferred it with cream. This demonstrated the good spirit which had developed between the two sides, but it was not always easy.

A Slow Beginning

At first, information had to be drawn item by item from the Soviet engineers – a process described ruefully by their opposite numbers as a "tooth pulling competition." Worried by the slow pace at which information was being exchanged and decisions were being taken, the US side had to insist strongly that conferences by telephone, "telecons," be held when necessary. Even then, during the 1973 "Mid-Term Review," Lunney complained to Bushuyev that in some cases there had been delays of up to nine months in receiving information. The need for Professor Bushuyev to refer decisions back to Moscow was recognized as a side effect of a highly centralized bureaucratic system, and occasionally communication failures appeared to result from pure oversight, as when the US side found out almost by chance that the Russians were prepared to launch a second Soyuz should the first one encounter problems.

Any Soviet equipment limitations, for example on the orbital height Soyuz could attain, tended to be wrapped in obscure technical explanations which served little purpose, because the NASA engineers realized that the Apollo and Soyuz craft had been developed with very different tasks in mind and that no question of "inferiority" arose. Besides, there were areas where the Soviets led the way. Early on Lunney recognized that the standard Soyuz docking system, to which the Soviet docking specialist Vladimir Syromyatnikov, who played an important part in the development of the ASTP docking system, had contributed, was more durable and flexible than the Apollo system. In addition, when the Soviets were informed that they could not enter the Apollo command module during the mission wearing flammable suits they proceeded to develop a suit that was superior to that of the Americans, which burned slowly in an oxygen atmosphere, in that it "self-extinguished."

1

The means of transfer was the docking module, which was to be launched into space occupying the position in the Saturn adapter previously occupied by the lunar module, and which would be extracted by the command and service module after entering orbit. Essentially it was a cylindrical airlock made from aluminum and placed between the Apollo command module and the Soyuz orbital module, to which it was docked by means of the newly developed system. It was a little over 10ft (3.15m) long with a maximum width of just over 4.5ft (1.4m) and a weight of 4,436lb (2,012kg). In or around the docking module were tanks of oxygen and nitrogen; radio, communications and TV equipment; a small electric furnace to be used in one of the experiments and storage lockers. (It was also eventually to be pressed into service as a spare bedroom.) But its prime purpose was that of an airlock. Hatches at both ends sealed off the Apollo and Soyuz spacecraft, with their different pressures and atmospheres, from each other. Crewmen entering the module would close the hatches to the one craft from which they were exiting and operate pressure equalisation valves to change the atmosphere to that of the craft they were to visit. When

2

2 *Astronauts indulge in snowballing during a break in training at Star City, near Moscow, where they were familiarizing themselves with Soyuz equipment. Host officials and cosmonauts look a little bemused!*

For some time the Soviet side was slow to brief the Americans fully on problems encountered during Soyuz/Salyut missions, which were of obvious concern to NASA since crew safety was the prime requirement of any manned spaceflight. Thus, it took two years for the cause of the tragic loss of the three Soyuz 11 cosmonauts in June 1971 to be revealed, but when it was it demonstrated, in the words of NASA's official ASTP history, that "both sides were reaching the level of trust necessary to build a genuine space partnership." Subsequently, the Soviets became more quickly forthcoming in informing NASA about mission anomalies – there was a safe launch abort involving two cosmonauts at the start of a Soyuz mission on April 5, 1975, for example. NASA was perfectly happy about the safety aspects of the forthcoming joint flight, although any problems on the Soviet side were seized upon by that redoubtable opponent of spaceflight in the US, Senator William Proxmire, who seemed suddenly to have developed a previously undisclosed interest in aerospace safety.

NASA's Open Policies

Media coverage of ASTP was another potential source of trouble between the two sides. The US media was traditionally aggressive in its demands for maximum access to events and, to its credit, NASA from its beginning had carried out its activities in the full glare of publicity. The Soviet experience was quite the opposite. Thus, when the Russians in 1971 requested that the discussions in general should be kept confidential, NASA refused. Even later, on the occasion of the Soyuz 16 dress rehearsal flight for ASTP in December 1974, the Russians offered to give advance notice of the launch (to facilitate a test of tracking procedures) provided NASA withheld the information from the media until the flight had begun. Again Lunney and his colleagues refused.

The NASA public affairs officers were determined to secure maximum exposure for the mission, but they faced a difficult task. Eventually they did secure agreement to extensive in-flight television coverage aboard both Apollo and Soyuz (even if, in the event, the black and white television camera on Soyuz failed during the mission, resulting in the Apollo CSM's approach to the Soviet spacecraft not being recorded). There was also agreement on the observance of normal NASA policy in both countries for coverage by accredited correspondents during *joint* phases of the mission. This meant that NASA could not influence Soviet policy over the admission of western correspondents to the Baikonur launch site, which had never been permitted before and which was not allowed for the mission. Nonetheless the Soviet side did considerably ease its normal complete secrecy. In June 1973 it invited a party of American aerospace journalists to Zvezdny Gorodok (Star City), the Soviet equivalent of the Johnson Space Center. After the successful Soyuz 16 flight in December 1974 the crew members Filipchenko and Rukavishnikov gave a press conference together with several senior Soviet space figures: this was the first time that the press had been able directly to question a crew after a Soviet mission, and the first time most of the western correspondents had been allowed to visit Star City, where the conference was held.

During the mission, the launch of the Soyuz spacecraft and its return to Earth were televised live for the first time, and there was even a media "first" on the US side when color TV pictures were transmitted from the interior of the command module while on the launch pad. Nonetheless, it was perhaps an indication of continuing different attitudes between the two sides that, whereas 24 Soviet journalists were accredited to both Houston and the Kennedy Space Center for ASTP, none chose to attend the Apollo launch, perhaps because it was thought this would lead to adverse comment from the American media about the enforced absence of western correspondents from Baikonur.

A Mission for Slayton

And what of the men who were to fly? The names of the American astronauts were announced in January of 1973 and, in another departure from normal Soviet practice, the Soyuz crew was named four months later. The Apollo commander was Gemini and Apollo veteran Tom Stafford. The command module pilot was Vance

Brand, for whom this was the first flight in space, although he had been a back-up crew member for the Apollo 15 lunar mission as well as back-up commander for two of the Skylab missions. Also new to spaceflight (because of a heart anomaly identified as far back as 1962), was Mercury astronaut Donald "Deke" Slayton, who had been returned to full flight status in 1972. At the age of 48 he was the oldest man yet to be selected for a spaceflight. Both members of the prime Soyuz crew had been in space before. Alexei Leonov, the commander, had performed the world's first EVA during the Voskhod 2 mission in 1965, while flight engineer Valeri Kubasov had flown on the Soyuz 6 flight in October 1969. All members of the two prime crews were 40 years of age or over.

The crews and their back-up and support colleagues on both sides faced the normal hard work of preparing for a demanding mission that would involve performing complicated rendezvous and docking maneuvers as well as numerous experiments. But to these was added the necessity for extensive language training. Clearly no single language could be used on such a symbolic mission, and as training progressed it seemed to be tacitly

agreed that the Russians would speak English and the Americans Russian during both the training sessions and the mission itself. Three major joint training sessions in each country were planned, and at the end of 1973 Stafford reported to Houston that Leonov and Kubasov were having individual English tuition for between six and eight hours a day. The American crew also stepped up their language training, and by the time they flew the mission Slayton's tuition record was typical of the other two: out of total training lasting 3,075 hours, approximately one-third (1,077 hours) had been devoted to Russian, which was about twice the time the astronaut had spent in the simulators. Once again, the astronauts and cosmonauts were but the most visible element in complex organizations, and an enormous amount of work by translators and interpreters on both sides was a vital contribution to the eventual success achieved.

By the time the crews participated in a three-part series of involved inter-control center simulations (held in Houston and Kaliningrad from May 13 to July 1, 1975) they were ready to fly, but there had been fun along the way. Leonov and Kubasov had been to Disney World – of course – while their American colleagues had indulged in a snowball fight somewhere between Moscow and Star City on one occasion and on another had celebrated the Fourth of July by letting off fireworks and a sky rocket before the eyes of an indulgent Soviet policeman.

3 *Although a constant battle, NASA made some gains in stressing that ASTP should be as "open" as possible: this TV still shows Leonov a short time before entering the spacecraft on July 15, 1975.*

4 *A Soviet Academy of Sciences picture released by NASA of the Soyuz 19 launcher on the pad.*

5 *Lift off! Leonov and Kubasov launched first on the morning of July 15, and were followed seven-and-a-half hours later by the Apollo crew.*

1 *International handshake in space: Tom Stafford greets Aleksei Leonov at the Soyuz end of the docking tunnel – a still from a 16mm movie frame. There was slight delay in getting the hatch open. The handshake should have taken place over the south coast of Britain but in fact occurred over France.*

2 *For all the world like an insect, with wings, eyes and proboscis! While Soviet restrictions had become somewhat more relaxed, in 1975 this was probably the best view of the Soyuz spacecraft that had ever been seen in the West. It was, of course, photographed from Apollo and released by NASA.*

In Orbit at Last

But by July 15, 1975, it was time for serious work and, following the Soyuz 19 launch at 12:20 GMT on that day, the joint mission went well, doubtless proving easier to the two crews than the long drawn out training. As with the last Skylab mission, the fins of the Saturn IB launcher for Apollo had to be replaced, but the complex orbital mechanics of Apollo's chase of Soyuz had been proven from the days of Gemini onwards, and a flawless docking took place at 16:10 GMT on July 17, with the first international handshake in space taking place a few hours later high above the French city of Metz, at approximately 6°E 49°N. There then began a lengthy program of crew transfers, memento exchanges, conducted TV tours of the two craft and of the Earth below and press conferences, which were greatly facilitated by the use of the geostationary ATS-6

after information. These tasks completed, the two spacecraft began to move apart. Leonov and Kubasov landed safely in Kazakhstan at 10:51 GMT on July 21. The three Americans still had plenty of experiments to perform, however.

NASA had been concerned from the beginning to derive as much benefit as possible from the mission, and 28 experiments were flown. Five were joint experiments conducted with the Russian crew; two derived from West Germany; and the remaining 21 were unilateral US experiments. From an original estimated cost of $10m the US expenditure on experiments was subsequently raised to $16m. Experiment disciplines included astronomy and earth environment studies, as well as remote sensing and observation of the surface, life sciences and materials processing and applications. Considerable success was achieved, including the first

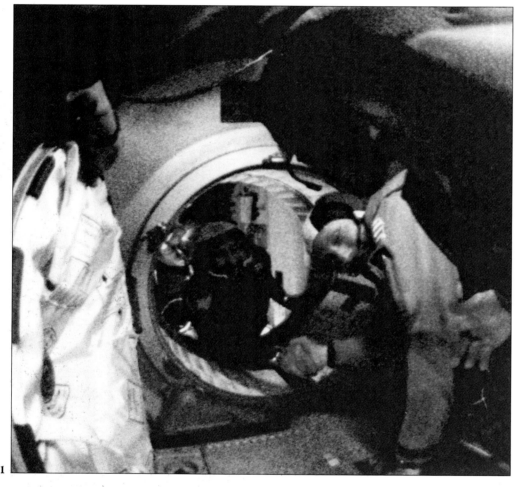

1

2

3

3 *After final separation on July 19, the two spacecraft moved apart and in a joint experiment Apollo occulted the Sun, with the resulting eclipse being photographed from Soyuz.*

communications satellite, which increased the time the docked spacecraft were in direct contact with the ground from 15 minutes to 55 minutes during each 87 minute revolution. The Soviet leader Leonid Brezhnev sent a message to the crews, while President Ford indulged in a lengthy question and answer session with them in person.

On July 19, at just after noon GMT, following nearly 44 hours of docked flight, the two craft separated for a brief period, following which they redocked with the Soviet docking system being active. Final separation took place after several hours, during which two experiments were performed: the creation of an artificial solar eclipse with Apollo acting as the occulting device, and an attempt to determine the quantities of atomic oxygen and nitrogen existing at the orbital altitude. For this experiment, Apollo maneuvered around Soyuz and fired laser beams at retroreflectors on the sides of the Soviet spacecraft, the reflections being received back by a spectrometer on Apollo that recorded the wavelength of the light which, after analysis, would yield the sought-

detection of a pulsar outside the Milky Way galaxy; the detection of extreme ultraviolet radiation from four stars, one of which was the hottest "white dwarf" known; and the successful separation of live biological materials by electrophoresis, a technique which continues to be of great interest in the 1990s. The crew also devoted much time to Earth observations and photography, the veteran Tom Stafford passing a message to Farouk El-Baz, the chief investigator for the experiment, about how much more detail could be seen from the ASTP orbital height compared with the higher altitude at which he had flown during Gemini.

But finally it was time to end the last Apollo mission. Stafford, Brand and Slayton landed in the Pacific not far from Hawaii on the evening of July 24, 1975, close to the planned target point. Pleasure at a safe return was somewhat marred when it was learned that all three astronauts had inhaled highly toxic nitrogen tetroxide fumes from the command module's thrusters, after a failure to operate switches which initiated an automatic landing sequence. Fortunately, no long-term effects resulted, and after a stay in hospital the Apollo astronauts set off with their Soviet colleagues on gruelling public relations tours of their two homelands.

ASTP – A Sign for the Future

The value of the Apollo-Soyuz Test Mission, which cost some $250m, was debated at great length both during the project's lifetime and subsequently. To critics it was a "circus" or a "junket," from which the US gained nothing: on the contrary, there was the risk of "technology transfer" from the Soviet access to Apollo. There was also a doubt about Soviet capabilities – at least some American politicians wishing to fly a full battery of experiments aboard Apollo to make overall mission expenditure worthwhile in case the Soviet Union withdrew for political or technical reasons.

In fact, there was little if any technology transfer because Apollo systems had been extensively described and were no longer "state of the art." NASA managers and engineers did not for the most part regard the Soviet equipment and procedures as in any way inferior, they were *different*. Whereas, for example, NASA simulated missions as a test of new techniques, the Soviet Union tended to fly a mission to achieve the same goal. As such, and despite the undoubted great achievements of the NASA method of doing things, the American engineers and managers were given a very valuable insight or window into a different philosophy and practice of spaceflight, which could only be a benefit. Even the successful may become too insular.

During ASTP the cooperation survived such political stresses as differences over the alleged treatment of Jews in the Soviet Union, the position of the dissident scientist Andrei Sakharov there and, most important of all, the Yom Kippur War in the Middle East. There may have been a technical disappointment in the fact that the universal docking system that performed so well on ASTP did not fly on further missions, which was one of the reasons for its development. To some degree, like Project Apollo itself, ASTP was a technical achievement that took place for political reasons. Ultimately George Low identified the strongest justification for it when, besides pointing out the economic waste of duplication in spaceflight, he said "We live in a rather dangerous world. Anything that we can do to make it a little less dangerous is worth doing ... ASTP was one of those things."

Subsequently the detente that had led to ASTP faded somewhat but the scene changed once again when Mikhail Gorbachov rose to power in the Soviet Union. Expanded space cooperation was approved by the then President Ronald Reagan and Gorbachov in 1987, and although technology transfer doubts once again arose this did not prevent the launch of a NASA Goddard SFC ozone mapper aboard a Soviet Meteor weather satellite in the summer of 1991 and the attachment of US-built cosmic ray detectors on the outside of Mir by spacewalking cosmonauts at roughly the same time. Space figured in discussions between the Soviet leader and George Bush subsequently and it was perhaps not surprising when a Soviet celebration of ASTP was published in the form of a booklet entitled *A Step Towards a Dream*. In addition to a summary account of ASTP and its importance as the first major step in US-Soviet cooperation in space research, it suggested that the joint exploration of Mars in stages between 1991 and 2015 could become the second major step. During ASTP the two sides "proved that by pooling efforts, any organizational, technological or simply human problems can be solved as long as there is enough goodwill in exchanging ideas, knowledge and experience. Space

exploration has advanced tremendously. Humanity can now translate into reality very daring ideas not only in near-Earth space but in deep space as well. What seemed only a dream 15 years ago can become a reality at the turn of the 21st century."

The break up of the Soviet Union led to an inevitable pause in active cooperation between the USA and (primarily) the Russian republic now led by Boris Yeltsin, but as 1992 progressed matters moved very quickly. In April the Bush administration approved the purchase at a total cost of $14m of Russian space hardware, including a nuclear power system and electric thrusters which ironically were to be used in Strategic Defense Initiative projects. More far reaching developments followed. With the collapse of the Soviet Union the 1987 agreement had lapsed but on June 17 at their summit meeting in Washington Bush and Yeltsin signed a five-year agreement on space that promised quite unprecedented cooperation – including cosmonaut and astronaut flights aboard the shuttle and Mir respectively in 1993; the docking of a shuttle orbiter with Mir possibly in 1994; adding US experiments to a hard lander on the Russian Mars 1994 spacecraft; and a study to establish whether the Soyuz spacecraft could be used as an interim solution to NASA's search for a rescue vehicle (Assured Crew Return Vehicle – ACRV – in the jargon) to be used in case emergencies occurred aboard the space station Freedom. Several of these subjects figured in a $1m applications study contract awarded the very next day (June 18, 1992) by NASA to the Russian aerospace company NPO-Energiya.

Three weeks later NASA Administrator Daniel Goldin announced the establishment of a new function at headquarters – that of Associate Administrator for Russian Programs. He then departed for Russia and the Ukraine on a seven day trip on which he was joined by representatives of various US agencies and government departments. This went well and it was significant that the 1975 practice of cooperating through various working parties was confirmed in the discussions that took place. Despite the exciting promise of what had been agreed good sense prevailed on both sides and Goldin was quoted as saying that: "In our relationship with Russia, we need to start slowly and deliberately to build a strong foundation .... In this way we will ensure that what we do together will be successful, both technically and scientifically". His words echoed those quoted earlier from the Soviet publication *A Step Towards a Dream*.

4 *Return to Earth: the Soyuz 19 touch down was televised live, and this image was taken direct from a TV screen at the time.*

5 *The last Apollo mission ends as the command module, after nine days in orbit, floats in the Pacific west of Hawaii. There was danger from the crew inhaling toxic thruster fumes during the descent, but they quickly recovered.*

6 *From aboard the recovery ship USS* New Orleans, *the astronauts talked on the telephone to President Gerald Ford.*

9

THE SHUTTLE: THE YEARS OF PREPARATION

1

Early in 1969, the newly elected President Richard Nixon asked his Vice-President Spiro Agnew to head a Task Group that would examine the space program and make recommendations for the future. The Apollo 11 moon landing had taken place by the time the Group reported in September 1969, and the Vice-President at least was still enthusiastic. He presented a number of options for a long-term goal in space, ranging from a major program of a space station in Earth orbit, a lunar base and manned landing on Mars by the mid 1980s, to a minimum program to defer the Mars landing but develop a space station and a shuttle. The cost for the

was lower than that for 1970 – a decrease which reflected "current financial constraints."

NASA hoped nonetheless to include both station and shuttle in its plans for the post-Apollo period, but as 1970 progressed studies showed that each element would cost around $5,000m and, therefore, that simultaneous development was out of the question. Considerable work had in fact been done on the station, but it was evident that support for it would depend on the availability of a low-cost supply vehicle – the shuttle – which must consequently take precedence.

George Mueller had made what was probably

2

3

1 *NASA Administrator James Fletcher (left) discusses the proposed space shuttle vehicle with President Richard Nixon on the day that formal US Government approval for the project was announced: January 5, 1972. The new system was presented inaccurately and inadvisedly as making space into "familiar territory, easily accessible for human endeavour ...."*

all-out effort would be up to $10,000m a year by 1980 and for the limited proposals up to $5,700m a year.

The report was totally out of step with the feelings of a period when Apollo itself was already under considerable pressure. A manned trip to Mars was denounced as "the utmost folly" by one scientist, and even supporters of space in Congress found it too much. The President caught the general mood in March 1970, when he stated that "space expenditures must take their proper place within a rigorous system of national priorities." The Administration, he went on, was recommending a NASA budget for fiscal year 1971 that

the first public announcement of NASA's interest in reusable space vehicle concepts at a meeting of the British Interplanetary Society in August 1968. That theme had been established early in the writings and proposals of the spaceflight pioneers. In the 1950s, for example, Wernher von Braun had published details of a three-stage rocket ship with the crew and payload housed in a winged reentry vehicle which would glide back to a landing strip on Earth, while the first and second stages would parachute back and be recovered. As time went on, detailed theoretical and practical work on merging the roles of aircraft and spacecraft, on developing "lifting bodies," and on progressing concepts for reusable or recoverable space vehicles was carried out in such programs as that featuring the "X" series of rocket powered aircraft, and the USAF Dyna-Soar vehicle, although the latter was cancelled in 1963.

Concept Changes
The space shuttle as envisaged by NASA in 1969 was a fully reusable, two-stage vehicle. The booster would be about the size of a Boeing 747, and the orbital stage about that of a 707, both being rocket powered and fueled by hydrogen and oxygen carried in on-board tanks. The vehicles would be attached in parallel for a vertical lift off, after which the booster would be piloted back for a horizontal landing and refurbishment. The orbiter would do the same after completion of its mission in space. The initial cost estimates looked reasonable, but continuing studies and the requirements of the Department of Defense (for the President was regarding the new generation of spacecraft as serving all users, both civil and military) caused a considerable uprating in the performance of the orbiter, such that the development cost for a fully reusable vehicle was estimated at around $10,000m.

NASA realized that this level of funding would not be forthcoming and, although some design elements remained to be worked out, by the turn of 1971-72 it had revised the concept to that of a thrust-assisted-orbiter shuttle system (TAOS). This was to consist of a manned orbiter vehicle, an expendable and external fuel tank supplying propellants for the orbiter's on-board engines, and two solid-fuel boosters which would be recovered by parachute.

During this time political support for the shuttle was by no means firm. In fiscal year 1971 NASA requested $170m for feasibility and other studies devoted to the shuttle and received $80m: in the following year it requested $190m and received $100m. During an earlier discussion in the Senate, an amendment to delete funds for the shuttle from NASA's budget entirely was defeated by only four votes. But the President himself had responded to NASA's arguments, and in January 1972 he endorsed the program and ordered that it should proceed at once. He was optimistic on the prospects: " ... the space shuttle will give us routine access to space by sharply reducing costs in dollars and preparation time" ... [and the] ... "resulting economies may bring operating costs down as low as one-tenth of those for present launch vehicles." The intended role of the new vehicle as the primary, ultimately only, way into space was emphasized by the name chosen to describe it: the Space Transportation System (STS) and, some years later, the even more emphatic National Space Transportation System (NSTS). However, just to be on the safe side, the USAF did manage to make sure that it had a few Titan III launchers available in reserve.

In making the case for the shuttle NASA had not in fact regarded its primary justification as being economic; rather it was a way to the future of spaceflight which, like Apollo, would enhance the technological prowess and prestige of the United States. The President agreed with the national importance of manned spaceflight, although it seems likely that he was not convinced entirely about the shuttle's economic prospects.

NASA nonetheless made optimistic noises about shuttle economics. At the end of 1972 it estimated the cost of developing the shuttle at $5,150m, with a cost per flight of about $10.5m. (Orbital flight tests were anticipated in 1978 with operational status being achieved in 1980.) Savings on the use of the shuttle would result

from the frequency of flights, and in predicting eventual economies of $1000m per year NASA took as the basis for its calculation a flight schedule of 580 launches in the twelve years from 1979 to 1990: almost one a week. This was a scarcely credible forecast, which sat ill with the political realities of the time. The President may have given the shuttle his blessing, but his talk of a space policy that "must be bold, but ... must also be balanced" did not inspire. Initially in NASA's plans the shuttle had been clearly linked to the space station, but with the postponement of the latter it promoted the shuttle as a utilitarian, cost-effective system for civilian space science and applications and for defense missions. This was a laudable objective but again it did not have the inspirational or imperative nature of the Apollo challenge. As one space historian has written, "it was a commitment

to technological development without a clear link to an overriding political or other policy justification."

As it set out to make the shuttle reality NASA entered what Robert Frosch, its then Administrator, described in 1979 as an "austere budget environment." In three fiscal years (1974, 1975 and 1977) its budget requests for shuttle development funds were cut by a total of almost $275m by the Office of Management and Budget. This led to schedule delays of more than a year and, NASA claimed, added $70m to the eventual total development costs. During his term of office from 1976 to 1980, President Carter continued the Nixon approach: spaceflight was a means, not an end in itself. It was a place to work in when it contributed to national objectives. While it retained the characteristics of challenge and excitement, there was need for a "short term flexibility to impose fiscal constraints" when necessary on the space program just as on any other program.

Unfortunately, the annual funding process is not well suited to the progress of a major technological project – one which, moreover, was encountering the probably inevitable delays and increased costs resulting from difficult technical problems. NASA could borrow funds from its shuttle production budget and switch them to the development budget and defer work to the next fiscal year for a while, but this way of dealing with "constrained budgets" could not go on indefinitely. Ultimately it had to return to Congress with a request for additional funding every year from 1977 to 1980, with considerable recriminations on both sides. On the one

Early thinking on what became the space shuttle envisaged a fully reusable vehicle in which both a separate booster as well as an orbiter would fly a mission and then return to Earth. This attractive concept was subsequently modified by economic necessity into a system in which a large, external fuel tank was not retrieved.

2 *A very early Grumman representation of an orbiter retrieving a satellite.*

3 *The final configuration.*

4 *Work on the first space-going orbiter proceeding at Rockwell International's plant in California.*

4

5

5 Robert A. Frosch was Administrator of NASA during the later years of development of the shuttle and is shown here (right) during President Carter's visit to the Kennedy Space Center on October 1, 1978. The occasion was the twentieth anniversary of NASA's establishment and the award of the Congressional Space Medal of Honor to a number of astronauts.

1

2

3 Two Martin Marietta engineers inspect aluminum slosh baffles in a section of a shuttle external fuel tank. The baffles limit the movement of liquid propellants which otherwise would induce guidance and balance problems during launch. This imaginatively-conceived photograph gives the scene the overtones of a cathedral.

4 Rockwell technicians repair a thermal protection tile in the rim of one of Columbia's windows. The tiles were a pacing item in the development of the shuttle and proved vulnerable to damage during the early missions.

hand NASA was endeavoring to develop a unique space vehicle within what might be described as a politically inspired financial straightjacket, while even supporters in Congress could criticize it for over-optimistic shuttle development estimates in the beginning and for being less than frank in informing the Congress about its serious and real funding problems as time went on.

By 1980 the total program cost (based on comparable 1971 dollar values) was running at about 20 percent, or $1,000m, above the original estimate. By 1985 the increase was stated to have risen to 30 percent. In addition, the program had slipped two to three years and, while allowances must be made for inflation and other factors, the cost per flight once the shuttle flew bore no comparison to the original estimate of $10.5m. The first flights were costed at over $300m each at current prices. This had reduced to $125m per mission in 1984, but by the end of 1990 the figure had climbed back to well over $200m. NASA's original claims about the economics of the space shuttle were based, as has been mentioned, on a program of almost one launch per week over a period of twelve years. When it failed to meet a schedule of even one launch per month, the basis of its original calculations, no matter how sound or otherwise, fell to the ground.

But there was still the technological challenge of building and operating a complex space vehicle, the like of which the world had never seen before.

Three-Part System

The shuttle vehicle as finally envisaged after the numerous political and technical deliberations consisted of three main elements: the orbiter, the (expendable) external tank (ET) containing the propellants for the orbiter's main engines, and the solid rocket boosters (SRBs). While all were vital to success it was the orbiter itself, and especially those systems such as the main engines and the thermal protection tiles that were pushing the limits of current technology, which had the greatest impact on the time schedule.

While the rocket motors in the SRBs were the largest ever developed for spaceflight (with a diameter of 146in – 371cm), industry had considerable experience with this type of launcher, and the development and building of the boosters by what is now the Thiokol Corporation did not delay the first flights into space. Each of the two boosters to be used for a mission were just over 149ft (45.5m) in length and 12ft (3.7m) in diameter, with a weight of 1.3m lb (589,670kg). Each developed 3.06m lb (1,387,990kg) of thrust at liftoff.

The statistics of the external tank were perhaps more impressive. It was 154ft (47m) in length with a diameter of 27.5ft (8.4m). With the two tanks for liquid oxygen and liquid hydrogen filled before launch its weight was nearly 1.7m lb (760,220kg). It was the largest such structure built for any rocket program, and the Martin Marietta Corporation won the $158m contract for its development in August 1973. The weight of the ET proved to be a continuing problem, as did its thermal protection system, and during its development each tank was taking about two years to build. In 1971 the target cost per flight for each ET was $1.8m, but toward the end of the decade the estimated cost was $2.5m. By that time NASA was already contracting with Martin Marietta for the development of a lighter weight tank.

But it was the orbiter itself which represented the maximum degree of innovation. Dubbed by NASA the first true aerospace vehicle, it was 122ft (37m) long, with a wingspan of 78ft (24m) and a landing weight of some-what over 200,000lb (90,718kg). Its cargo bay was 60ft (18.3m) long with a diameter of 15ft (4.6m): the maximum

payload weight, despite an initial target of 65,000lb (29,500kg), was about 50,000lb (22,680 kg). Competition for the contract to develop and build the vehicle – and to integrate all of the elements in the system – was intense, but in July 1972 NASA selected the North American Rockwell Corporation (later to be renamed Rockwell International). With supplemental agreements the contract by the end of 1975 was worth $2,700m.

While advanced, some features of the vehicle had built on the experience of earlier spaceflight. Three fuel cells were to provide electrical power. The two orbital maneuvering system (OMS), liquid propellant (hypergolic) rocket engines located in pods aft on either side of the orbiter generated 6,000lb (2,722kg) of thrust each, and were recognisably descended from Apollo: these were used for the final thrust into space after the main engines had shut down and also for achieving changes in orbit, rendezvous, orbit circularization and deorbit. Providing more limited power for controlling the orbiter's attitude whilst in orbit and during the earlier stages of re-entry were the primary (870lb thrust – 395kg each) and vernier (24lb thrust – 11kg each) thrusters of the reaction control system (RCS) located in the OMS pods, with an additional set located in the forward fuselage. Another essential system featured three hydraulic pumps, each driven by its own auxiliary power unit (APU), which provided power for gimbaling the orbiter's main engines during lift off and operating its control surfaces, landing gear brakes and steering during re-entry and landing.

Inevitably, the avionics incorporated in the orbiter were a marked advance on those employed during Apollo. Apollo 11 command module pilot Michael Collins, the best writer among the astronauts to date, described the instrumentation of the shuttle as being the equivalent in complexity of perhaps four or five Apollo command modules. Five general purpose computers were at the heart of a system that could provide automatic vehicle flight control for almost every phase of a mission, although manual control was available as an option at all times – that method of control, however, not being direct and being routed through the computers in a "fly-by-wire" system. Overall, avionics aboard the orbiter comprised more than 300 major electronic black boxes connected by over 300 miles (483km) of electrical wiring.

However, while the avionics had advanced enormously and other, more restricted systems, such as the Canadian-built remote manipulator system (RMS) arm for deploying and retrieving payloads, were unique in a manned spacecraft, it was the orbiter's three main engines, which operated for over eight minutes at launch (and initially in parallel with the SRBs), and also its thermal protection system, that enabled it to survive re-entry through the atmosphere, which proved to be the most demanding and time-consuming challenge during the development years.

Engine Complexities

The contract for the development of the SSME (Space Shuttle Main Engine) was won by the Rocketdyne Division of Rockwell in July 1971. The contract was then valued at about $500m, but increased costs and inflation, together with additional orders from NASA, had increased its value to almost $1,300m by 1979.

The main engine was described as being the most sophisticated liquid propellant rocket engine ever attempted, and it represented a major advance in propulsion technology. As a reusable powerplant it was intended to go for over fifty flights without overhaul and require minimum maintenance between missions. About 14ft (4.3m) in length and 8ft (2.4m) at its maximum

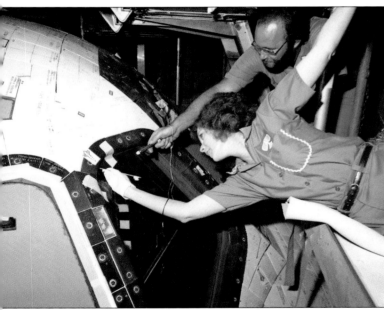

5 *The suit worn by shuttle astronauts during launch and return from orbit is a modified version of a USAF high-altitude pressure suit. It does not have the versatility (or bulk) of an EVA spacesuit.*

6 *This photo emphasizes the size and complexity of the shuttle's "robot arm" – remote manipulator system to give it the NASA name. The system was developed and built by SPAR of Canada.*

5

6

diameter, each engine had a design thrust of 375,000lb (170,000kg) at sea level and 470,000lb (213,200kg) in vacuum. The amount of thrust was digitally controlled between 65 percent and 109 percent of the rated power level to provide optimum launch characteristics, and the same control unit, when commanded from the orbiter, started and stopped the engine besides monitoring its condition. Inside the engine, a high pressure fuel turbine operated at over 35,000rpm to deliver propellants to the combustion chamber. For its power, the engine weight at approximately 7,000lb (3,175kg) was light.

Michael Collins wrote that the three engines, operating under great pressure and at high speed close to one another, scared him: in his view their form of combustion was too close to an explosion. It was perhaps inevitable that the development of such advanced engines should suffer from a variety of problems. In 1977 and 1978 independent reviews of progress were conducted by a National Research Council team, and in the latter year two fires in test engines further delayed the test program.

Thermal Protection

Equally critical to the successful development of a reusable vehicle – with a planned lifetime of one hundred missions – was the system chosen to protect the mainly aluminum and graphite epoxy outer structural skin of the orbiter from reentry temperatures that would reach nearly 3,000°F (1,650°C), but also from the -250°F (-157°C) encountered during orbital night. Between 1970 and 1973 NASA spent over $11m in studying various approaches and eventually opted for reusable surface insulation based mainly on silica tiles. Rockwell as system integrator awarded a contract worth around $50m in May 1973 to the Lockheed Missiles & Space Company for this work.

Different materials were chosen for different areas of the orbiter's outer skin according to the heat encountered during re-entry. Reinforced carbon-carbon (RCC) was used, for example, on the wing leading edges and the nose cap, where temperatures were to exceed 2,300°F (1,260°C). Black coated, high-temperature reusable surface insulation (HRSI) tiles of silica fiber were used where temperatures would be below 2,300°F, such as the entire underside of the orbiter where RCC was not applied, areas of the upper forward fuselage, parts of the aft OMS pods and the leading as well as trailing edges of the vertical stabilizer.

Where temperatures would be below 1,200°F (650°C), on parts of the forward, mid and aft fuselage and the upper wing, low-temperature reusable surface insulation (LRSI) silica fiber tiles were used. These had a white surface coating (the only difference from the HRSI tiles)

1

to provide better thermal characteristics in orbit. Finally, in areas where temperatures would not exceed 700°F (370°C), for example, on sections of the fuselage sides and on the upper payload bay doors, a flexible reusable surface insulation (FRSI) was used, composed of white blankets of coated Nomex felt.

Initially it was calculated that around 32,000 tiles would be required to cover each orbiter, but early on improvements and modifications were introduced which eventually reduced the numbers of tiles and blankets to under 27,500. Thus, after the first space-going shuttle Columbia was delivered, an advanced, flexible, reusable surface insulation (AFRSI) was developed. This basically took the form of a quilted white blanket which replaced most of the LRSI tiles, with gains in durability, lower fabrication and installation costs and also weight reduction. Black tiles called fibrous refractory composite insulation (FRCI) were later developed and replaced some of the HRSI tiles in various parts of the orbiters.

2

The new era of the shuttle demanded numerous modifications to facilities at the Kennedy Space Center, perhaps the most dramatic of which were the new launch control systems operated from new firing rooms in the Launch Control Center.

1 The mobile launch platform exits the VAB on the way to the pad.

2 Orbiter, SRBs and ET are in position at the pad and the crawler (left) makes its way back.

However, during the early manufacturing process it was found that, while many tiles could be produced by milling machines, others had to be individually contoured and therefore worked by hand. This was time consuming and costly, characteristics of much of the installation process, too. A high rejection rate for some tiles, the HRSI in particular, had a knock-on effect because the tiles were usually installed sequentially, and rejection affected installation work even when other types of tiles were available.

When the orbiter Columbia arrived at the Kennedy Space Center from Rockwell early in 1979, 10,000 tiles remained to be installed, and thereafter NASA managed

only half of its planned installation rate of 650 tiles per week. Subsequently the situation worsened even more, because tests in 1979 showed that the bonding of some tiles was not as strong as NASA believed necessary to protect the skin of the orbiter. "Pull testing" of critical tiles to verify their structural integrity began, and those that failed were either replaced entirely or rebonded. In addition, a densifying coat which doubled a tile's bonding strength was developed and applied to tiles that were rebonded. NASA initially calculated that about 2,000 tiles would need to be treated in this manner. However, by the beginning of 1980, while 14,000 tiles had been installed at Kennedy, over 10,000 had been removed and over 10,000 remained to be reinstalled.

Changes at Launch Center

The creation of new or the modification of old facilities to meet the requirements of the shuttle were called for at a number of NASA locations, particularly at the Kennedy Space Center, where by 1980 the likely total cost was being estimated at up to $800m. One major new addition was the two-runway, 15,000ft (4,572m) long, 300ft (91.4m) wide Shuttle Landing Facility with an associated mate/demate structure that was to be used for attaching the orbiter to, or lifting it from, the Boeing Shuttle Carrier Aircraft during ferry operations. A second major new feature was the Orbiter Processing Facility, which was likened to a sophisticated aircraft hangar, where postflight servicing and checkout of the shuttle was to be conducted, together with many of the modifications needed for forthcoming flights. The high bays, work platforms and other internal structures of the Vehicle Assembly Building, which achieved such fame during the Apollo-Saturn era, had to be modified for the shuttle, but it was there in due course that SRB segments were joined to form two finished boosters. The external tank was then attached to them and in its turn the orbiter was towed from the OPF, raised to a vertical position by overhead cranes and mated to the tank.

The erection of the shuttle components was to take place on one of the Apollo-Saturn mobile launcher platforms which also had to be modified for the shuttle. (The unfueled shuttle stack and platform together weigh 11m lb (5m kg.) The most obvious change was the removal of each launcher's 380ft (116m) high umbilical tower, with its nine swing arms and large crane that serviced the Saturn V, and the installation of the upper sections out at the launch pad as a fixed service structure. A new rotating service structure was constructed at each pad. These provided environmentally controlled conditions for inserting vertically handled payloads into the orbiter cargo hold and, attached to a hinge on the fixed service structure, rotated through 120 degrees away from the vehicle prior to launch to prevent suffering damage from heat and blast.

The combined shuttle and launcher platform moved to the pad from the VAB on one of two massive crawler-transporters in an operation reminiscent of Apollo days. But very different from that time were the new firing rooms in the Launch Control Center. Here, each was equipped with the automated, computer-operated Launch Processing System, which monitored and controlled assembly and check-out of the shuttle as well as launch. It was largely because of the LPS that the countdown for a shuttle launch (excluding built-in holds) takes only around 43 hours – half the time required for an Apollo launch – and requires 90 persons in the firing room compared with the 450 previously on duty there.

Increasing sophistication has similarly been introduced in the two Flight Control Rooms of the legendary Mission Control Center at the Johnson Space Center. With the passage of years, the flight controller teams became more specialized, with one being responsible for ascent-to-orbit and return-from-orbit activities, two for orbital operations and a fourth for planning mission activities during the following day. The Center also had extensive responsibilities for training, planning and vehicle testing in preparation for missions. A US Congressional Research Service report published in 1981 forecast the likely cost of a dedicated DOD facility of a similar nature to the MCC at about $200m.

Astronaut Assignments

Shuttle hardware was one thing, but there was also the matter of the crews to fly and operate the new vehicle. Astronaut input into the new program for most of the development period, including the ALT flights, was provided by men recruited up to 1969. The first intake of the shuttle era (group 8) joined NASA in 1978, and six of the 35 new entrants were women. Additional groups were selected in 1980, 1984, 1985, 1987, 1990 (the 23 members of this group including the first woman pilot astronaut) and 1992, all of whom underwent a one year training period before becoming regular members of the corps. At the end of 1991 there was a total of just under 100 flight status astronauts in the program.

The new selections revealed the different character of shuttle missions. Career astronauts are now divided into two categories: pilots and mission specialists. The requirements demanded of the former, who as commanders or pilots fly the shuttle and are responsible for overall mission safety and success, is broadly similar to those of their predecessors. They must have had at least 1,000 hours flying time in jets, although test pilot experience is not essential. The early emphasis on an engineering background has been eased somewhat with the requirement for a minimum of a bachelor's degree in mathematics, the sciences or engineering. The physical requirements continue to be strict, although the increased space available in the shuttle is reflected in the greater maximum permitted height of 76in (193cm).

Mission specialists are required to have a detailed knowledge of shuttle systems and are responsible for conducting on-board experiments, spacewalks (EVAs) and for operating the remote manipulator system (RMS) arm – for example in the deployment or retrieval of satellites. They, too, must have a bachelor's degree in mathematics, the sciences or engineering, with related professional experience, although a doctorate is regarded as fulfilling this latter requirement. An additional category of astronaut (normally referred to as being of non-career status) is the payload specialist: a scientist or technologist who conducts experiments or operates equipment on a specific shuttle mission, having been nominated by a payload sponsor or customer. Extensive knowledge of shuttle systems and procedures is still required and training for a mission might begin two years before the flight. (Following the Challenger tragedy of 1986, the role of the payload specialist was examined critically by NASA, but such specialists returned to space during the ASTRO-1 mission in December 1990 and more flew on scientific and applications missions in 1991 and 1992.)

All new entrants undergo a basic training course lasting one year, during which they are regarded as candidate astronauts. Extensive background learning in academic disciplines continues, to which is added tuition in shuttle systems, first of all by means of lectures and by reading operations manuals but subsequently in computerized, interactive trainers devoted to specific

shuttle operations and systems. The candidates all gain experience of weightlessness, both by making flights aboard a KC-135 jet flying parabolic maneuvers and also in the Johnson Space Center neutral buoyancy tank, known in NASA jargon as the Weightless Environment Training Facility.

Pilots and mission specialists alike are trained to fly NASA's T-38 jets, the former to maintain their skills and the latter to become familiar with high performance jet aircraft. Because of the unique landing characteristics of the shuttle, the pilots, during training, fly the KC-135 to gain experience of handling a large, heavy aircraft and they also fly the Grumman Gulfstream II, known as the Shuttle Training Aircraft (STA), which has been specially modified to reproduce the landing characteristics of the shuttle orbiter. Once assigned to a mission, pilots receive about 50 hours training in the STA, which is equivalent to around 300 shuttle approaches.

3 *May 1984: astronaut Judith Resnik prepares for a flight in a T-38 aircraft during training. Later that year she became the second American woman to fly in space.*

3

4

The first year's basic training successfully accomplished, astronauts pass to advanced work. What is termed system-related training largely takes the form of gaining detailed knowledge of shuttle systems in trainers which permit self-paced, interactive instruction. Phase-related training concentrates on the specific skills needed for an astronaut to perform successfully in space, and much of this is conducted in the $100m Shuttle Mission Simulator (SMS) at Johnson: "the only high fidelity simulator capable of training crews for all phases of a mission beginning at T-minus 30 minutes, including such simulated events as launch, ascent, abort, orbit, rendezvous, docking, payload handling, undocking, de-orbit, entry, approach, landing and rollout."

Commander and pilot training in the SMS takes place primarily in a station similar to those used in training and rating commercial airline pilots, complete with motion, visual displays and sound simulations. A second station used by the whole crew does not simulate motion but is devoted mainly to simulating specific payload activities planned for future missions. Crews are assigned to a particular flight about ten months before

5

4 *Mission specialist Norman Thagard practices EVA procedures in the 25-ft deep pool of the Weightless Environment Training Facility at the Johnson Space Center.*

5 *Anna Fisher, a member of the first group of women astronauts selected by NASA, during water survival training.*

1 *During 1977 the American public witnessed the first tangible signs of America's return to manned spaceflight when the ALT (Approach and Landing Test) missions were flown in the atmosphere. In the early flights the shuttle Enterprise, which was destined never to fly in space, was unoccupied and remained attached to the Boeing carrier aircraft. But later it was crewed by two pairs of astronauts who subsequently piloted the vehicle to landings after separation from the Boeing. This imaginative Rockwell International picture shows Enterprise being mated to the carrier aircraft.*

the scheduled lift off. Their training then increases in intensity and is directed by an assigned management team. During the final three months, the crew members participate in integrated simulations, where they work at their designated stations and interact with flight controllers in much the same way as they will during the actual mission. After being assigned to a mission each astronaut can expect to spend at least 300 hours in the SMS, and instructors have at their disposal almost 7,000 malfunction simulations with which to test a crew's reactions.

The essence of the training is to simulate the reality of missions and to anticipate problems both major and minor. Perhaps the best evaluation of the training is the observation sometimes made by returning crew members that the actual flight experience was "just like in training" – or, indeed, easier.

characteristics as well as the performance of its many systems. Since the tests were not to take place in space, it only carried simulated SSME, OMS and RCS units and just a few prototype insulation tiles.

The ALT program lasted from February until October 1977, during which *Enterprise* left the ground on top of the modified Boeing Shuttle Carrier Aircraft (SCA) for a series of 13 tests. Two crews of two astronauts were in the cockpit for the eight manned flights: Apollo 13 lunar module pilot Fred Haise and 1969 entrant Gordon Fullerton flew five of the missions, with the remainder bèing flown by X-15 pilot and 1966 entrant Joe Engle, with Dick Truly, who, like Fullerton, had transferred to NASA when the USAF's Manned Orbiting Laboratory project had been cancelled in 1969, in the second seat.

Between February 18 and March 2, the mated aircraft

1

2

2 *Stars of the "Star Trek" TV show join NASA Administrator Fletcher (extreme left talking to DeForest Kelley – Dr. McCoy) for the roll out of the test orbiter Enterprise at the Palmdale, California, plant of Rockwell International.*

First Flights

All this, however, was in the future. The astronauts who joined from 1978 onwards would not fly aboard the early flights: these were to be the responsibility of some of the older hands. Even then, before the shuttle went into space it needed to be tested during flights in the atmosphere.

Orbiter Vehicle 101 (OV-1 – the official designation during its construction) was rolled out from Rockwell International's facility at Palmdale, California in September 1976. Following a campaign by fans of the *Star Trek* television show, it was named *Enterprise* and prepared for the Approach and Landing Test (ALT) program taking place the following year at NASA's Dryden Flight Research Center, located at the Edwards Air Force Base. The purpose of the ALT flights was basically to test the subsonic airworthiness of the vehicle – to verify its aerodynamic and flight control

flew with *Enterprise* unmanned and with its systems inert. The tests verified the flight characteristics of the two vehicles and the final flight reached a height of 30,000ft (9,144m). There followed three manned but captive flights during June and July 1977, when the astronauts tested onboard systems and refined plans for the separation procedures to be adopted in the following missions.

From August 12 to October 26 five free flights of *Enterprise* took place. During the first three a fairing covered the three dummy main engines to reduce drag and to alleviate buffeting of the mated aircraft before separation. With the astronauts having gained valuable experience with these flights, the fairing was removed for the fourth flight in order to study the turbulence that would be created around the engines when the shuttle flew back from space. The first four missions landed at the Rogers Dry Lakebed runway at Edwards, but the fifth landed on the 15,000ft (4,572m) concrete runway as a test of its stopping characteristics – of critical importance given the intention of landing shuttles on a runway of similar length at the Kennedy Space Center.

Although more flights had been planned, NASA announced that the test objectives had been met and that further flights were therefore unnecessary. Although ALT had been part of a rigorous proving program for the shuttle so far as NASA was concerned, it also provided valuable publicity for the project in that (with flights into space some years away) it enabled members of the public to see in flight and in some detail the new vehicle which was scheduled to play such a major role in America's future in space.

At first it was intended that, after being used for various ground tests, *Enterprise* would be returned to Palmdale, where it would be modified for spaceflight. Ultimately NASA decided that this would be too costly and, after being exhibited both at home and abroad, *Enterprise* was finally ferried in November 1985 to Dulles Airport, Washington, D.C., where it became the property of the Smithsonian Institution.

Mission Profile

The ALT flights provided valuable information about the final stages of a mission: what would be the profile of the various stages of a mission actually into space? While no one mission is "typical," an outline account of the main highlights of such a flight from launch to landing can be given. At T-6.6 seconds ("T" is launch time) the main shuttle engines are ignited and throttled up to 90 percent thrust in three seconds. If the onboard computers verify this fact, the ignition sequence of the solid rocket boosters starts at T-3 seconds. At T minus zero, the hold-down explosive bolts are blown and the SRBs ignite, the shuttle now being committed to launch. The vehicle clears the launch tower at about T plus seven seconds, and control is then handed over from launch control to the Johnson Space Center.

Almost immediately a roll program is begun so that the shuttle heads due east to take advantage of the 1,000 mph (1,609kph) eastward rotation of the Earth. KSC is at latitude 28.5°N, so launch in a due easterly direction gives the vehicle an eventual orbital inclination of 28.5 degrees north and south of the Equator. If a mission requires it, safety considerations permit a launch in a northeasterly direction sufficient to achieve an inclination to the Equator of 57 degrees, which can be further increased to 62 degrees if required. At this time the orbiter is suspended beneath the external tank, with the crew in a "heads down" position. In due course acceleration will subject them to a force of 3Gs.

About one minute after lift-off the point of maximum dynamic pressure on the shuttle – "max Q" – is reached at an altitude of 33,600ft (10,240m). The SSMEs are throttled back to about 65 percent to keep the dynamic

The ALT flights were designed to "verify orbiter subsonic airworthiness, integrated systems operations and pilot guided approach and landing capability – and satisfy prerequisites to automatic flight control and navigation ...."

3 The tail cone was removed for the last two flights and the orbiter looked as it would returning from space.

4, 5 September 13, 1977: Enterprise separates and then turns and banks for landing.

6 A display panel on the forward flight deck of the orbiter Challenger shows the abort mode switch in the ATO mode. The first such abort occurred on July 29, 1985, at the launch of mission 51-F.

3

In the event that engine or other major malfunctions occur during a launch, procedures for various forms of abort modes have been devised. *Intact aborts* permit the safe return of the orbiter and its crew to a pre-planned landing site. The most benign of these is *Abort to Orbit* (ATO), when a limited SSME failure enables the orbiter to achieve a temporary orbit at a lower than planned altitude. The situation can then be evaluated by controllers and crew, following which the orbiter can de-orbit early to land or the OMS thrusters be used to raise the orbit and allow the mission to continue. The latter happened in the only ATO so far declared in the shuttle program during the STS51-F Challenger mission in July, 1985.

Abort Once Around (AOA) is somewhat more severe, in that a shuttle main engine failure does not enable orbit to be even temporarily achieved and the orbiter is flown once around the Earth to land either at the Edwards AFB in California or at White Sands in New Mexico. A *Trans-Atlantic Abort Landing* (TAL) occurs when the shuttle flies a ballistic trajectory across the Atlantic Ocean to land at designated sites in Africa or Spain. The weather at these sites is a factor in deciding if a launch may proceed from the Kennedy Space Center. The most severe of the intact aborts is *Return to Launch Site* (RTLS). This is a critical and complex procedure which follows the jettisoning of the solid rocket boosters. Any operational main

6

4

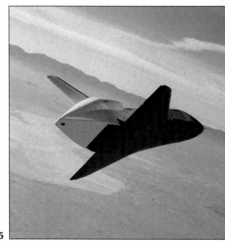

5

engines remaining, OMS and aft RCS thrusters continue to power the orbiter and fuel tank down-range until a pitch around maneuver is commanded to point the vehicle back to the launch site. MECO takes place at a time dependent on many factors and the orbiter separates from the tank, subsequently landing at Kennedy if all has gone well.

A *contingency abort* envisages the possible loss of the shuttle but the survival of the crew, for example by parachuting from an orbiter in gliding flight using the equipment introduced after the Challenger disaster of 1986. An SRB malfunction, as in that accident, is not regarded as survivable by the flight crew.

4

3

pressures on the vehicle within limits but, once the atmosphere has thinned sufficiently and the shuttle has passed through max Q, the orbiter engines are throttled up once again to full power (104 percent).

At a little after two minutes into the flight the SRBs have expended their fuel and are jettisoned: the shuttle at this point is around 30 miles (48km) high and moving at 2,890mph (4,650kph). The SRB casings continue climbing briefly but then drop back to an eventual parachute landing in the ocean. Meantime, the orbiter's main engines are taking it higher in its ascent trajectory until, at about eight minutes into the flight and at an altitude of 60 miles (96km), the computers command MECO: main engine cut-off. The vehicle's speed at this point is just under 16,700mph (26,870kph), with the orbiter flying upside down relative to the Earth. The large fuel tank is now jettisoned: the orbiter moves away from it using RCS thrusters and then the OMS thrusters are fired to accelerate the vehicle still further, to a speed

of about 17,400mph (almost 28,000kph), when it achieves orbit at a minimum altitude of 115 miles (185km). The fuel tank continues on a ballistic trajectory and re-enters the atmosphere to break up over the Indian or Pacific Oceans.

Orbit is achieved in two ways. The standard method involves two OMS burns: the first places the vehicle in an elliptical orbit and the second (at apogee or high point in the orbit) results in the shuttle achieving a near circular orbit. Further firings of the OMS during the mission can raise or lower the orbit depending upon whether the shuttle is facing toward or away from its direction of flight. The second method of achieving orbit is termed direct: the high point of the elliptical orbit is obtained by a continued firing of the SSMEs, with just one OMS firing to circularize the orbit, thus saving OMS fuel for on-orbit activities. The first direct insertion orbit of the shuttle program occurred during the Challenger STS41-C mission in 1984 when, with the orbiter in a circular orbit of just under 290 miles (467km), the Sol Max satellite was captured, repaired and redeployed.

Conditions in Orbit

In orbit the crew enjoy a shirt-sleeve environment (similar to that on Earth) provided by the Environmental Control and Life Support System (ECLSS), which also cools or heats orbiter components and systems. One result of the normal pressure in the crew compartment is that astronauts who are to conduct extravehicular activity must enter an airlock situated in the middeck, where they don their spacesuits and pre-breathe pure oxygen for several hours to prevent suffering from the "bends." The atmosphere within the airlock is then vented to space, the outer hatch is opened and a safety tether immediately attached to each suit before the EVA begins. The airlock system removes the need for depressurising the crew cabin in the orbiter at the beginning of extravehicular activity.

Activity in orbit takes place in all areas of the crew compartment – forward flight deck (cockpit), aft flight deck and in the middeck – with a total volume of 2,525 cubic feet (71.5 cubic metres) available. Control of orbiter attitude and orbital maneuvers is conducted by commander and/or pilot from either part of the flight deck, while use of the manipulator arm, any experiments in the cargo hold and EVAs are controlled or monitored from the aft flight deck only. The RMS arm is normally used to deploy or retrieve satellites, but some spacecraft are deployed by being "spun-up" on a bearing table and ejected from the cargo hold direct.

Many experiments are performed in the middeck, which also contains storage compartments and is the general living area. It is here that the crew prepares the now much more varied food (with the availability of a hot-plate oven); sleeps (in bunks or sleeping bags or merely restrained by straps); and enjoys the privacy of a toilet much like one on Earth, save that body wastes separate not because of gravity but because of a strong flow of air which pulls the liquids and solids into a sealed disposal chamber. With a typical mission length of about seven days a shower facility is not provided, but hands can be washed in a small domed device and dampened washcloths are extensively used. Clothing broadly follows the pattern set by Skylab, and flight suits, trousers, shirts, shorts, soft slippers and other items are available in attractive colors. Whether as part of an experiment or as a favorite off-duty pastime, the six large windshield windows in the forward flight deck and the two overhead and two rear-looking windows behind provide crew members with an excellent means of observing and photographing the Earth's surface.

The Final Stages

With mission events concluded, the payload bay doors are closed and Houston gives clearance for the de-orbit burn of the OMS engines. The orbiter is first maneuvered so that its tail faces the direction of travel and then the burn takes place for about 2.5 minutes – the precise length of time depending on the vehicle's mass and orbit. This slows the orbiter by some 200 mph (322kph) and puts it on a new trajectory for a landing approximately 5,000 miles (8,045km) away around the world. RCS thrusters then fire to turn the vehicle back to a nose-first attitude at an "attack angle" of 40 degrees to the direction of flight, so that the undersurfaces of the wings and body encounter the most extreme heat on re-entry. Subsequently, the thrusters are in frequent use to effect pitch, roll and yaw changes in the orbiter's attitude, until its aircraft-like control surfaces become operable lower

down in the atmosphere. For most of its return journey the shuttle is essentially a high speed missile under the control of onboard automatic systems which are being closely monitored by the flight crew.

At an altitude of approximately 75 miles (121km) the orbiter is still traveling at more than 17,000mph (27,353kph), but is only 30 minutes from touchdown. The normal communication blackout occurs at this time and continues until 12 minutes before touchdown. The speed of the vehicle is slowed by a series of banks and roll reversals (astronaut Vance Brand has likened the orbiter's return to descending a steep spiral staircase) and, by the time it has descended to an altitude of just under 10 miles (16km) and a distance of 22 miles (35km) from the runway, it has gone sub-sonic, announcing its return to the states of Florida or California by two resonant bangs. As it nears landing the shuttle can either make a straight-in approach to the landing strip or (more usually) perform an S-turn to further dissipate energy if this is required.

While provision has been made for totally automated landings, the commander and pilot normally take over control at this stage for the final approach and landing. The initial glide slope of the approach is 19 degrees (six times steeper than that of a commercial airliner), but just before touchdown this is modified to 1.5 degrees. Touchdown occurs about half a mile beyond the runway threshold at a speed of about 220mph (354kph). The flight crew, which just one hour before was still enjoying zero gravity in orbit, experiences a maximum 1.7Gs under deceleration, and some inflate

5 Former X-15 pilot and ALT commander Joe Engle went into space with ALT colleague Dick Truly on the second mission, launched on November 12, 1981. The established need for exercise was met by using a specially designed treadmill, with straps to keep the astronaut in position.

5

6

bladders in the legs of their altitude protection suits to prevent blood flowing from their brains and pooling in their legs, which could cause temporary black outs. It can take perhaps 30 minutes for the crew members to find their "1G-legs," during which time recovery convoy crews outside have been testing for (and if necessary dealing with) any explosive or toxic fumes remaining around the engines and thrusters or in the payload bay. When all is well, the orbiter hatch is opened and after a brief preliminary medical examination the crew can leave.

There is enormous detail and complexity in each mission, and yet there is also one simple and stark fact: when it returns from orbit, the shuttle is a high speed glider which has just one chance of landing. Since it is without power, there can be no going around again.

And this was the challenge that, after many delays, resulting from both political and technical difficulties, it was finally time to confront.

1 A launch tower view of an orbiter as it begins the climb into space. The initial lift-off is relatively slow and it takes about seven seconds for shuttle and boosters to clear the tower.

2 A few more seconds and ground-based cameras begin to record dramatic rear-end views of both SRBs and SSMEs firing.

3 Within a very short time the orbiter is miles away from the launch pad, though clearly visible as a bright point of light as high altitude winds disturb the exhaust trail.

4, 6 Night launch (STS-8) and dawn return (STS-5).

10

THE SHUTTLE IN SPACE AND SALYUT 7

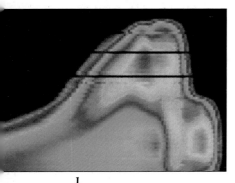

1 *NASA's flying telescope – the Kuiper Airborne Observatory – took infrared images of Columbia as it re-entered the atmosphere during the test missions to obtain external evidence of the heat to which the orbiter's surfaces were being subjected. In this color-coded view, red is the highest and violet the lowest temperature. The relative coolness of the wing's leading edge was not expected.*

2 *Gemini and Apollo veteran John Young (right) was chosen to command the first flight of the shuttle orbiter into space, with Robert Crippen (who had transferred to NASA from the USAF's Manned Orbiting Laboratory) as pilot. They are shown here on the forward deck of Columbia while it was still in the orbiter processing facility (OPF) at the Kennedy Space Center.*

By April 1981, after well over a year and a half at the Orbiter Processing Facility, resulting primarily from problems with the tiles of the thermal protection system, the space shuttle Columbia was at last ready for the first proving flight. The purpose of STS-1 (which marked the first use of solid rockets on a manned vehicle and the first time that NASA had put astronauts aboard the first flight into space of a new spacecraft) was to demonstrate that the system could launch, go into orbit and return safely, with on-board instrumentation and the astronauts assessing the orbiter's performance. Commanding the mission was John Young, an astronaut for almost twenty years, and Robert Crippen, who had joined NASA in 1969 when the USAF Manned Orbiting Laboratory project had been cancelled. A computer problem caused the cancellation of the first launch attempt on April 10 but all went well two days later, and Young and Crippen spent over 50 hours in space before bringing Columbia down to a landing at Edwards Air Force Base on April 14, with Crippen exclaiming excitedly over the radio "What a way to come to California!" Columbia suffered lost and damaged tiles at lift off because of an overpressure wave created when the SRBs ignited, but the problem was solved for later launches by a modification to the water sound suppression system at the 39A launch pad.

Columbia had more problems at the start of the next mission. A scheduled launch on October 9, 1981, was postponed when a spill took place as the orbiter's forward RCS thrusters were being fueled. Instrumentation and auxiliary power unit (APU) problems caused a further postponement on November 4, but launch finally took place eight days later, although even then there was a further hold of some hours. Aboard STS-2 were ALT pilots Joe Engle and Dick Truly.

This was the first time a manned spacecraft had been reflown with a different crew, but a fuel cell problem caused the duration of the mission to be cut from the planned five days to just over two. Nonetheless most of the mission objectives were accomplished. The performance of spacecraft systems was again tested and the RMS arm was operated for the first time in all its modes. In the cargo bay, and mounted on the European Space Agency-built, unpressurised Spacelab pallet, was a payload (called OSTA-1 for NASA's Office

of Space and Terrestrial Applications) largely devoted to the remote sensing of the Earth's surface and atmosphere. Much publicity resulted when an imaging radar system in the package "saw through" several feet of dry sand in Egypt to reveal the original surface, including dried river beds. Engle and Truly landed at Edwards AFB in the afternoon of November 14, and this time the tile damage on Columbia was found to be only minor.

With the second orbiter nearing completion at Rockwell International's plant in California, Columbia lifted off again on March 22, 1982, after a delay of just

one hour. Over 1,000lb (543kg) in weight was reportedly saved by using a thinner and unpainted insulation on the fuel tank, which appeared brown as opposed to the white of the first two missions. Commanding the STS-3 mission was Skylab veteran Jack Lousma, with ALT pilot Gordon Fullerton alongside.

The astronauts conducted further tests of the manipulator arm and examined the effect of solar heating on the orbiter itself by exposing nose, tail and top sections to the Sun for various periods of time. Another package of experiments was carried in the cargo bay (this time called OSS-1 for Office of Space Science and Applications), and one of these obtained data on the contamination which Columbia introduced into its near environment – an issue of critical concern for future scientific experiments. Among the experiments inside the orbiter was a continuous flow electrophoresis system (to investigate the separation of biological components): an electrophoresis experiment had been flown during the 1975 ASTP mission and was to be flown again on a considerable number of forthcoming shuttle missions.

Problems encountered during the mission included initial sickness, a malfunctioning toilet and an unreliable auxiliary power unit. The planned seven day duration was extended to eight because the winds above the back-up landing site at the Northrup Strip in New Mexico were too high – the runways at the Edwards AFB prime site having been rendered unusable by heavy rain. Lousma and Fullerton finally touched down on the morning of March 30. There was some brake damage and Columbia, with over fifty tiles lost or damaged, also suffered extensive contamination from a dust storm.

STS-4 was the last of the research and development flights, and the orbiter was launched with no schedule delays and completely on time in the late morning of June 27, 1982. (A parachute failure, however, resulted in the loss of both solid rocket boosters.) The purpose of the seven day mission was essentially to continue testing Columbia and its systems and to elaborate a series of investigations and experiments, most of which had been flown on the earlier missions. However, new additions were a classified USAF payload and the first "Get Away Special," an opportunity provided by NASA for investigators to fly experiments at low cost on condition that the hardware be automated or require the minimum of astronaut attention. Many such "specials" were to be flown thereafter.

When Apollo veteran Thomas Mattingly and Henry Hartsfield (who had transferred to NASA from the Manned Orbiting Laboratory project in 1969) brought Columbia in for the first shuttle landing on the 15,000ft (4,572m) concrete runway at Edwards, it was appropriately July 4. The President and First Lady were present to welcome them, and overhead the second orbiter, Challenger, on top of the Boeing 747 carrier aircraft, flew in salute. The test flights were over.

Like the previous flight, STS-5 lifted off on time and with no schedule delays. Aboard Columbia in the early morning of November 11, 1982, were two commercial communications satellites and four

astronauts: ASTP command module pilot Vance Brand, Manned Orbiting Laboratory transferee Robert Overmyer, and two scientist-astronauts, Joseph Allen and William Lenoir, who joined NASA in 1967 and were now flying as the first mission specialists. There were numerous experiments on board including Get Away Specials and investigations forming part of a student involvement program, but the major tasks of the astronauts were to launch the communications satellites and to conduct the planned EVA (spacewalking) activity.

The satellites were each successfully ejected from the cargo bay and after the orbiter had moved to a safe distance a Propulsion Assist Module (PAM) – a

1 *A night view of Columbia as, towards the end of 1983, it is prepared for the longest and most complex flight to date: that of the ten-day Spacelab mission.*

2 *The Spacelab pressurized laboratory located in the orbiter's cargo bay as seen through an aft compartment window. In the foreground is the connecting tunnel.*

3 *In a crew self-portrait that was to become a tradition are Spacelab 1 (STS-9) astronauts (clockwise from top): mission specialist Owen Garriott; pilot Brewster Shaw; US payload specialist Byron Lichtenberg; mission specialist Robert Parker; commander John Young; and ESA payload specialist Ulf Merbold.*

4 *The first American woman to go into space, Dr. Sally Ride, who flew on the STS-7 mission of Challenger in June 1983.*

solid rocket motor – attached to each fired automatically to place them into a highly elliptical orbit: subsequently they were maneuvered into the desired geostationary orbits by controllers employed by the owners. The second major activity, however, was not successful: different components in the two specially designed shuttle spacesuits that should have been worn by Allen and Lenoir malfunctioned, and the privilege of performing the first EVA from a shuttle orbiter fell to the next crew to fly. Columbia landed at Edwards on the morning of November 16 after a mission lasting a little over five days.

Challenger had arrived at the Kennedy Space Center in July 1982 and its first flight was scheduled for January 20, 1983. However, the launch was postponed twice: the first time for a hydrogen leak that required the removal and repair of two of the orbiter main engines and the replacement of the third, and the second time because the satellite in the payload bay was contaminated at the pad during a severe storm. Launch eventually took place on April 4, with the orbiter crewed by four astronauts who had all joined

NASA ⊓ 1969) and Story Musgrave (a scientist-astronaut from the 1967 group) performed various tethered tests in the cargo bay for more than four hours. A variety of other experiments were performed and Challenger landed at Edwards on April 9, 1983, after five days in space.

The return of Challenger to space in the STS-7 mission went much more smoothly. The launch took place on schedule and with no delays early on the morning of June 18, 1983. For the first time a crew of five astronauts flew on the shuttle, one of whom was Sally Ride, the first American woman in space. Robert Crippen was commander, but all other members of the crew were group 8 astronauts who had joined NASA in January 1978 and were the first of that group to fly in space.

Two PAM-equipped commercial communication satellites were launched successfully and a large program of experiments, both in the cargo hold and in the crew quarters, was carried out. Norman Thagard, one of the mission specialists and a medical doctor, conducted a series of tests on himself and fellow crew members to gain more information on the nausea and sometimes vomiting not infrequently occurring during the early stages of space missions, which by this time was coming to be referred to as Space Adaptation Syndrome.

Another first aboard the shuttle resulted in considerable publicity. A satellite developed and built in West Germany (SPAS-01 – Shuttle Pallet Satellite) was released into space, and subsequently retrieved, by the manipulator arm. The satellite was capable of independent powered flight, and, during the hours it flew alongside, a NASA-supplied Hasselblad camera was triggered from Challenger to take impressive color images of the orbiter against a backdrop of Earth and sky, the first time either of the first two spacegoing shuttles had been pictured in this way. Challenger was scheduled to land at the Kennedy Space Center but bad weather there forced a diversion to Edwards, where it touched down on the morning of June 24 after a six day mission.

The STS-8 mission, with a crew of five, which included Guion Bluford, the first American black astronaut to go into space, was commanded by Richard Truly who, like Crippen, was making his second trip aboard a shuttle. Challenger made the first night launch of the shuttle era in the early hours of August 30, 1983, with its primary payload the INSAT 1B communications and applications satellite that was being launched for India. The satellite was deployed successfully on the second day of the mission.

Several experiments involved the shuttle itself: in one Challenger's nose was pointed away from the Sun for 14 hours to test the effect of the extreme, external cold on the flight deck, and in another the orbital height was reduced to just under 140 miles (225km) as part of an experiment attempting to explain the cause of the glow which could surround parts of the orbiter during darkness. Getaway Special, electrophoresis, animal and space adaptation syndrome investigations were conducted before Truly, with group 8 astronaut Daniel Brandenstein alongside him as pilot, brought Challenger in to make the first shuttle night landing at Edwards AFB at just after midnight on September 5, at the end of a six day mission.

1

2

NASA during the Apollo era but only one of whom, the commander, Paul Weitz, had actually flown in space.

A major purpose of the STS-6 mission was to launch the first of three TDRS (Tracking and Data Relay Satellite) spacecraft which, when in geostationary orbit, would keep low Earth orbit vehicles such as the shuttle in almost continual contact with a central control center, making it possible eventually to dispense with a large part of the network of ground tracking stations operating to that time. A two-stage IUS (Inertial Upper Stage) was to inject the TDRS into its final orbit, but it malfunctioned. Fortunately, the loading of extra fuel aboard the satellite itself enabled controllers over the ensuing few months to nudge TDRS gradually into geosynchronous orbit, where it could begin normal operations. The first EVAs from a shuttle orbiter went far more smoothly, and Donald Peterson (who joined

First Spacelab Flight

After a lengthy schedule delay caused by a faulty solid rocket booster, Columbia returned to space on November 28, 1983, for the longest (ten day) and most

3

space; Ulf Merbold, one of the payload specialists, was the first European to fly on a NASA mission; the TDRS was used for the first time to transmit a large volume of data back to Earth; and it was the first time that a shuttle had gone into an orbit inclined at 57 degrees to the Equator, which took it over most of Europe.

The astronauts split into two teams to derive the most benefit from the crowded program of experiments, and the success of the work stemmed in no small measure from the existence in Houston of a recently established Payload Operations Control Center which enabled the investigators to talk directly with the crew members conducting their experiments by proxy aboard Columbia.

The mission was commanded by John Young, while Skylab scientist-astronaut Owen Garriott and Robert Parker (who joined NASA in 1967 as a scientist-astronaut) were the mission specialists. Subsequently there was criticism in some quarters that ESA had entered into a project which cost the Agency around $1,000m and for which it received participation in half a mission but, whatever the justification of that view, the basic success of the hardware was a demonstration of Europe's capability in space.

Computer malfunctions caused a delay of some hours in Columbia's return to California on December 8, 1983, and when the orbiter landed several small fires were discovered in the tail section, caused by leaks of hydrazine from the auxiliary power units.

5 Astronauts Story Musgrave (left) and Donald Peterson conduct the first EVA from the shuttle in April 1983. They remained tethered to Challenger as they performed various tests during a four-hour EVA.

6 The first view of an orbiter in space taken from outside: Challenger shown during the STS-7 mission.

7 An experiment rack is lifted prior to installation in the Spacelab module for the STS-9 mission.

4

5   6

complex shuttle mission so far: the flight of Spacelab 1. Spacelab was the result of a joint ESA-NASA project initiated ten years earlier: it was aimed at conducting major scientific and applications investigations in space by means of a pressurized laboratory module and unpressurized U-shaped pallets located in the orbiter's cargo hold. Spacelab 1 hardware was built and funded by ESA and, although this first flight was meant primarily to demonstrate the satisfactory functioning of the system, over seventy investigations from Europe and the US were conducted in such disciplines as astronomy, atmospheric physics, earth observations, materials sciences, life sciences and space plasma physics.

A number of firsts were recorded on this STS-9 mission. It was the first time that six astronauts had been launched into space aboard a single vehicle; it was the first time that payload specialists went into

Discovery, the third orbiter, was delivered to the Kennedy Space Center in November 1983, and when Columbia returned from the Spacelab mission it was withdrawn from operations for a period that ultimately extended to two years. A refurbishment of the first operational orbiter had been planned for some time, but the availability of Challenger and Discovery, and the need for Rockwell International to work on the fourth orbiter Atlantis (as well as its commitments to the USAF B-1 bomber program), led both to a postponement of the work to bring Columbia up to the standard of the later orbiters and, more importantly, to the spacecraft being used as a source of spares for the other vehicles – an adequate supply of spares being a constant source of difficulty for NASA and of contention between it and the legislators in Congress.

As 1984 began it was possible to see a pattern emerging in shuttle missions: one of work for com-

7

1 *February 1984: Challenger photographed in orbit by astronaut Bruce McCandless, who tested a new EVA device allowing crewmen to dispense with tethers and umbilicals.*

the original order or sequence of the mission in the year. Increasing numbers of astronauts from group 8 who joined NASA in 1978 began to fly on the orbiters, but the established practice of having at least one crew member who had been in space before (usually the commander) continued to be observed.

Two of the missions launched on schedule and two after delays of one day and five days respectively. Discovery's first flight, however, went through a number of postponements between June 25 and August 30, the most serious of which was on June 26, when the first launch abort after main engine ignition (but before the SRBs ignited) in the shuttle program was commanded by the onboard computers when an anomaly in the orbiter's number three engine was detected. The engine had to be replaced. The 41-B mission made the first landing of the shuttle era at the

mercial customers (the launch of communication satellites in particular, but individual experiments aboard the orbiter as well); the launch of scientific satellites and conducting scientific and applications experiments for NASA itself and other institutions; and carrying out a large and highly varied range of investigations such as those submitted under the Shuttle Student Involvement Project (SSIP). In addition, the demands of launching, operating and landing a complex system such as STS resulted in problems and delays: there were successes but also many frustrations. The year 1984 was to be no different.

Untethered Space Walks

Five launches took place in the year – only half of those originally planned. Challenger flew three times and Discovery twice – the latter's first flight into space taking place from August 30 to September 5. The new year also saw NASA adopting a new flight designation: thus the first mission was 41-B, the first numeral indicating the year on which the mission was scheduled to fly, the second the fact that it would be launched from the Kennedy Space Center (1 – as distinct from the USAF shuttle launch complex being prepared in California, which would be 2); and the letter indicating

Kennedy Space Center and two other flights during 1984 terminated there. Mission 41-C should have landed in Florida, but because of rapidly deteriorating weather in the area was diverted at a late stage to Edwards. Discovery, at the end of its first mission, made a scheduled landing at Edwards since the greater flexibility offered by the number of runways and more reliable weather at the base was considered appropriate for the first landing of a new vehicle.

Vance Brand commanded the five man crew aboard Challenger during the 41-B mission which lasted from February 3-11, 1984. Two communications satellites (WESTAR-VI and PALAPA B-2) were launched, but the failure of both PAM motors meant that they could not obtain geo-synchronous orbit. A major feature of the mission was the first untethered flights in space by US astronauts wearing the Manned Maneuvering Unit (MMU), which was powered by inert-gas jets and which was regarded as an essential step in giving astronauts greater flexibility and capability in conducting EVAs. One of the MMU astronauts was Bruce McCandless, who was making his first spaceflight after joining NASA no less than eighteen years before in 1966. The work of McCandless and group 8 astronaut Robert Stewart with the MMU, and

also in testing the operation of a foot restraint fixed to the remote manipulator arm, looked forward to the next mission when the attempt would be made to repair the Solar Maximum ("Solar Max") scientific satellite in orbit.

Repair Mision

That next mission of Challenger was 41-C, which lasted from April 6-13. Robert Crippen returned to space once again as commander in a five man crew. The rendezvous with Solar Max required the raising of the orbiter's altitude to some 300 miles (483km). Over two days, and with close support from a control center at NASA's Goddard Space Flight Center, group 8 mission specialists George Nelson and James van Hoften retrieved Solar Max, locked it into the cargo bay and then repaired the satellite by replacing two

2 *During the 41-C mission in April 1984, astronauts George Nelson (right) and James van Hoften performed an EVA to repair the malfunctioning Solar Maximum Mission Satellite.*

3 *Success! Challenger separates from the repaired satellite, which can continue its study of the Sun.*

4 *This dramatic view, of the reddish glow created by the heat of re-entry as seen from the orbiter's forward deck, eventually became almost commonplace.*

5 *McCandless "flies" the nitrogen-propelled Manned Maneuvering Unit (MMU) during the 41-B mission and is photographed from the orbiter.*

4

5

6

7

6 *McCandless appears again in this picture, but at a far greater distance from the orbiter. His relative size seems to underline man's vulnerability in space, the dividing line between safety and tragedy being very thin and demanding constant care and attention. Fellow astronaut Robert Stewart also tested the MMU during the 41-B mission.*

7 *A later mission in 1984, the first flight of Discovery in August/ September, looked to the future with tests of an array (shown here) containing different types of solar cells that might be used to provide power for longer stays in space, such as those of the proposed space station.*

1 *Drama on the launch pad on June 26, 1984: scheduled for its first flight, the orbiter Discovery's main engines were ignited, but onboard computers detected an anomaly in one of them and commanded shutdown before the SRBs ignited.*

2 *A TV monitor records the moment when the main engines were being ignited. The 41-D mission eventually lifted off on August 30.*

1

2

3 *A major task of the 51-A crew during the last mission in 1984 was to retrieve two communications satellites which had not reached the correct orbit. Ultimate success in both cases was eventually achieved but this, and other "rescue" missions, showed that equipment designed to facilitate the rescue work rarely functioned as planned. In this view astronaut Dale Gardner, wearing the MMU on his back and a device on his chest to stabilize the satellite for capture, is dwarfed by the enormous size of the target.*

4 *Yet another EVA intended to correct a fault: 51-D astronaut Jeffrey Hoffman, with colleague David Griggs partly in view at left, works at attaching two hastily constructed "fly-swatter" like devices which it was hoped would correct a sequence start lever malfunction on a communications satellite. The attempt failed.*

5 *Crewmates Sally Ride (right) and Kathryn Sullivan synchronize their watches before the launch of their 41-G mission .*

malfunctioning units. This was the first ever planned repair of an orbiting spacecraft. Another major objective of the mission was to place in orbit the Long Duration Exposure Facility (LDEF) – a cylinder over 21,000lb (9,524kg) in weight carrying 57, mostly passive, experiments intended to yield information on the long term effect of the space environment on materials, seeds and other items. The LDEF was destined to spend far longer in space than was planned originally.

The much delayed first flight of Discovery for the 41-D mission lasted from August 30 to September 5, 1984. The crew of six astronauts was commanded by Henry Hartsfield and the three mission specialists included Judith Resnik, who thus became the second American woman to fly in space. The single payload specialist was Charles Walker, a member of the staff at McDonnell Douglas, who was the first corporate employee to operate his company's equipment – the Continuous Flow Electrophoresis System – aboard the shuttle.

Three commercial communications satellites were launched from Discovery and extensive tests were conducted of several types of solar cells which formed part of a wing that was 13ft wide by 102ft long (4m x31m) when unpacked from its containers and extended into space – the main purpose of the experiment being to demonstrate the feasibility of using large, lightweight solar arrays for supplying power to future large facilities such as the space station. When the array was extended, Discovery's vernier engines were fired to establish the amount of movement and vibration it exhibited when under stress.

The two missions scheduled to follow 41-D (E and F) were cancelled, and incorporating cargo from each aboard the orbiter resulted in a total payload weight of 47,000lb (21,320kg), the heaviest to date. Included in the payload was an IMAX camera which was being flown on a number of missions as part of a project to portray life aboard the shuttle in a large screen movie format shown in specially-built cinemas.

Challenger returned to space for the 41-G mission which lasted from October 5-13. Robert Crippen flew for the fourth time and commanded a crew which for the first time totaled seven astronauts, two of whom were women, with Sally Ride returning to orbit as a mission specialist, where she was joined by Kathryn Sullivan, another group 8 astronaut.

An important task during the mission was the deployment of the Earth Radiation Budget Satellite (ERBS), which formed part of a multi-satellite study to measure the amount of energy received from the Sun and reradiated into space, as well as the seasonal movement of energy from the tropics to the poles. A remote sensing package in the cargo bay included the second imaging radar system to fly on the shuttle (SIR-B) and a high-resolution, large-format photographic camera. A wide range of experiments devised in Canada was conducted by payload specialist Marc Garneau (the first Canadian to fly in space), and the presence of second payload specialist Paul Scully-Power from the US Naval Research Laboratory demonstrated the considerable contribution that a specialist could make to a chosen area of research, in Scully-Power's case oceanography. Sullivan became the first American woman to perform a spacewalk and, with group 9 (January 1980) mission specialist David Leestma, demonstrated by using equipment in the cargo bay that it was possible to refuel satellites in orbit, including some not originally designed to permit such activity.

Mission specialist Joseph Allen was the longest serving astronaut aboard Discovery during the 51-A

mission, which lasted from November 8-16 and was the last of 1984. The other four crew members were all group 8 astronauts, although commander Frederick Hauck and mission specialist Dale Gardner had each flown once before.

The first major task was to launch two communications satellites and this was accomplished successfully on the second and third days of the mission. The second task was to retrieve the two satellites – PALAPA B-2 and WESTAR VI – which malfunctioning motors had placed in incorrect orbits following their launch during the 41-B mission earlier in the year. The two satellites had been brought down by controllers from an altitude of over 600 miles (966km) to about 210 miles (338km) to facilitate capture. A

3

number of "burns" by Discovery brought the orbiter close to PALAPA, which was successfully recovered and fastened into the cargo bay on the fifth day of the mission. The second satellite was about 700 miles (1,126km) ahead and Discovery maneuvered successfully to retrieve it one day later. The EVAs involving the use of the MMU were performed by Allen and Gardner and, while they had practised the retrieval techniques extensively in training and special equipment had been devised for the purpose, problems during the mission resulted in much improvisation and hard work being required before success was achieved. It was a good note on which to end 1984, even if far fewer missions had been flown than originally planned.

An Abort to Orbit

Shuttle orbiters went into space nine times during 1985. Discovery flew four missions, Challenger three and Atlantis joined the four-orbiter fleet with two missions.

The first flight of the year by Discovery from January 24-27 was a dedicated Department of Defense mission, the first of the shuttle era, and when Atlantis flew for the first time from October 3-7 that, too, was a classified DOD mission.

The launch record from Pad 39A at Kennedy (the refurbished Pad 39B was intended to be ready to make a significant contribution to easing scheduling difficulties in 1986) was of six lift-offs taking place with no delays, or only short ones of up to one day. Mission 51-I of Discovery slipped three days from August 24 to 27 because of a mixture of weather and technical problems, and the 51-D mission of the same orbiter slipped almost one month, from March 19 to April 12, partly because of the transfer of payload from the cancelled 51-E mission and partly because of damage to one of the orbiter's payload bay doors sustained during preparation for the mission.

The most eventful launch, however, was that of Challenger, commanded by Gordon Fullerton at the head of a seven man crew, flying the 51-F mission from July 29 to August 6. Launch was initially scheduled for July 12, but the malfunction of a coolant valve in one of the shuttle main engines caused a computer shut down of all three SSMEs three seconds before the solid rocket boosters were ignited. Launch on July 29 took place after a short delay, but a premature shutdown of one of the main engines resulted in the first ever abort-to-orbit (ATO). The initial orbital altitude of 124x165 miles (200x265km) was later corrected to 196x197 miles (315x317km) by a series of OMS burns. Although much replanning was necessary the mission proceeded well, and to make up lost ground Challenger flew for an extra day before landing.

The DOD 51-C mission of Discovery commanded by Thomas Mattingly touched down at the Kennedy Space Center without trouble at the end of its three day mission on January 27. However, when Discovery, commanded by Karol Bobko, landed there on April 19 at the end of the 51-D mission, the demands of correcting movement by the spacecraft away from the center line on the runway caused extensive brake damage and a tire blowout. As a result, while the capability for nose-wheel steering on the orbiters was thoroughly tested, all later flights in 1985 landed on the less-restricted runways of Edwards AFB.

The seven civilian missions in 1985 demonstrated the versatility of the shuttle orbiter system. The main purpose of the 51-D mission (April 12-19) was to launch two communications satellites. There were no problems with Anik C-1 but the booster stage of LEASAT-3 did not fire. Observation of the satellite when Discovery closed on it again revealed that a sequence start lever, which should have opened automatically, had not done so. With guidance from Mission Control, astronauts made two "flyswatter" devices that could be used to tug at the lever and these were fixed to the end of the manipulator arm by mission specialists David Griggs and Jeffrey Hoffman during an EVA. Rhea Seddon on her first shuttle mission successfully hooked the lever and used the arm to tug hard at it, but there was no response from the satellite. Otherwise the crew was busy with the normal variety of experiments, but certainly one of the most unusual ever flown was a study of the behavior of mechanical toys in microgravity – the videotaped results being intended as part of a study package for schools.

The 51-B Challenger mission lasting from April 29 to May 6 and commanded by Robert Overmyer saw the return of the Spacelab pressurised module to space on its first operational flight. Spacelab 3 (the mission sequence had been altered) covered five basic discipline areas – materials sciences, life sciences, fluid mechanics, atmospheric physics and astronomy – and a major objective was to demonstrate that the orbiter-borne module could provide a stable enough environment for the delicate materials processing and fluid experiments. Two monkeys and 24 rodents were flown in

special cages, and this was the first time that US astronauts had flown with live mammals aboard. On a personal note, this was the first flight for mission specialist Don Lind, who had waited since April 1966 to get into space.

Daniel Brandenstein commanded the 51-G mission of Discovery from June 17-24. This was the first to be crewed by the newer (1978) class of astronauts without any representation on the part of earlier entrants to the astronaut corps. Three communications satellites were launched; the experiments included one flown on behalf of the Strategic Defense Initiative; and SPARTAN 1, a "free flyer" astronomical carrier weighing over 2,200lb (998kg), was deployed and then retrieved for return to Earth.

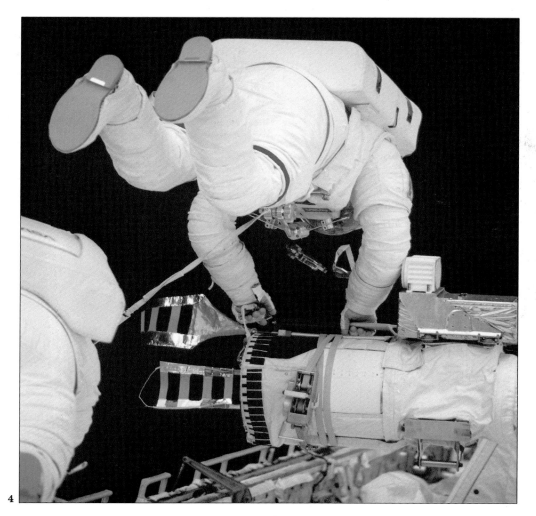

4

The eventful flight of Challenger during the 51-F mission (Spacelab 2) from July 29 to August 6, 1985, saw the successful verification of ESA's Spacelab in its pallet only form. Three pallets open to space in the cargo bay carried a battery of instruments and experiments which were operated by the astronauts from inside the orbiter cabin or by ground control. Support subsystems for the experiments required a pressurised environment, and this was provided in a unit located at the forward end of the cargo hold and known as the Igloo. The fine pointing accuracy of the solar and astronomy instruments on the pallets was controlled by the ESA-designed and built Instrument Pointing System (IPS). There were some problems with this system during the flight but trouble shooting and cooperation between crew and experts on the ground overcame them. Two of the mission specialists on the flight, Karl Henize and Anthony England, had been waiting since August 1967 to fly in space.

5

Joe Engle flew his last space mission (51-I) aboard Discovery from August 27 to September 3. Three communications satellites were launched successfully,

1

1 *Twelve years on from Skylab: a 12x8x12mm crystal of mercuric oxide grown aboard Spacelab 3 in the spring of 1985.*

2 *Astronauts on the 61-B mission conducted EVAs aimed at gaining experience of procedures required for building the space station.*

3 *Astronaut Jeffrey Hoffman in Houston practices for a 51-D mission experiment on the behavior of toys in space – part of an educational program.*

2

3

although one failed to function after reaching geosynchronous orbit. On the fifth and sixth days of the flight, mission specialists William Fisher and James van Hoften performed EVAs to repair the LEASAT-3 satellite which had failed to boost itself into the correct orbit when launched the previous April. The task was rendered more difficult by a fault with the remote manipulator arm, but a bypass system to provide ground control of the satellite was installed successfully: it later achieved the correct orbit and entered normal service.

German Specialists
Following the Atlantis DOD mission (commanded by Karol Bobko) in October 1985, the first ever flight by eight astronauts aboard an orbiter took place from October 30 to November 6, when a dedicated German Spacelab mission (61-A) was flown by Challenger. Henry Hartsfield commanded the crew which included two West German payload specialists as well as Wubbo Ockels of ESA, who had been back-up to Ulf Merbold for Spacelab 1. The scientific operations during the seven day mission – which concentrated on materials science, although a sled that could be moved backward and forward along rails with a subject strapped into the seat gave valuable insights into the human vestibular and orientation systems when subjected to microgravity – were controlled by a center operating from near Munich in Germany.

The final mission of 1985 (61-B from November 26 to December 3) was flown aboard Atlantis and commanded by Brewster Shaw. This was the second night launch of the shuttle era. Three communications satellites were successfully launched, but perhaps the main highlight and one looking forward very much to the concept of a space station, were EVAs carried out by mission specialists Jerry Ross and Sherwood Spring in which they assembled two different structures, one composed of small struts and the other of larger beams. In addition, there was the usual large complement of experiments and activity, with McDonnell Douglas payload specialist Charles Walker operating the electrophoresis system for the third time and the IMAX camera recording the EVA activity from the cargo hold.

Successes – and Difficulties
1985 had been the busiest year to date for the shuttle orbiters, with some considerable and varied achievements. Atlantis had joined the fleet in good time and part way through the year NASA management was looking forward cautiously to a possible launch rate of 24 missions a year (i.e. twice a month) within five years. There was talk of an "evolving maturity" in the program: foreign nationals from France, Saudi Arabia and Mexico had flown as payload specialists, as had the American politician Senator Jake Garn, with Representative Bill Nelson to follow. Further relaxation was taking place, with a teacher scheduled to fly early in 1986 and the selection process underway leading to a journalist flying by the end of that year.

But there was another side to the performance. The need to replace several thousand of the thermal protection tiles on Challenger at the turn of 1984-85 meant that Discovery had to replace it for the 51-C mission in January. Too much overhaul, repair and replacement work was having to be done on the SSMEs – the shuttle main engines – with the target of up to 55 missions between major overhauls of each engine being far from achieved. There were accidents, like the fall of a work platform from a crane that damaged Discovery in April and delayed launch by 15 days. The decision to switch all landings for a time to Edwards AFB after the 51-D landing damage at Kennedy added to turn-round times between missions. And there were the matters over which NASA had no control, such as the weather, or limited control, as when ships strayed into the SRB recovery area at sea or aircraft into the airspace around the launch site at Kennedy during critical periods.

The result was that, despite the very definite achievements in 1985, nine launches represented a shortfall of four on the scheduled program. Moreover, NASA's earlier plans had envisaged well over one hundred launches in the period 1981-85, whereas only 23 took place. The pressure of operating such a complex and sophisticated system to meet all of the many and varying demands placed on it – from acting as a commercial satellite transporter, where it faced intense competition from overseas, to a precision science platform – was enormous. The first launch of 1986, when Columbia was returned to duty for the 61-C mission, was seemingly to present a composite of all the difficulties.

MISSIONS TO SALYUT 7
While the program of U.S. space shuttle missions developed, Soviet space station flights faded into the background. Although the duration record was extended, no new ground was broken. However, Salyut 7 was launched on April 19, 1982, and Soyuz T-5 followed on May 13. The crew of Anatoli Berezovoi (commander) and Valentin Lebedev (flight engineer) spent 211 days in orbit, not coming down until December 10, 1982.

During that time they played host to two visiting crews. First, on June 24, was the three-man Soyuz T-6 crew which included the first French *spationaute*, Jean-Loup Chretien. One of the three crew members of Soyuz T-7 that followed on August 19 was Svetlana Savitskaya, the second Soviet woman cosmonaut. (Sally Ride was scheduled to fly aboard the seventh shuttle

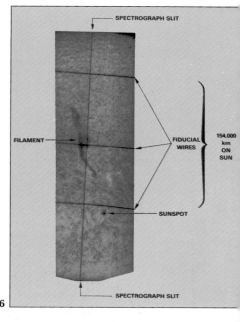

5

4

6

7

4 *61-B astronaut Sherwood Spring on the end of the remote manipulator arm checks joints on the ACCESS (Assembly Concept for Construction of Erectable Space Structures) tower extending from the cargo bay of the orbiter Atlantis: November/December 1985.*

6 *In the summer of 1985 scientist astronauts aboard Spacelab 2 operated a High Resolution Telescope and Spectrograph (HRTS) designed by the US Naval Research Laboratory to study the Sun. This color image, taken in the Hydrogen Alpha waveband, shows a sunspot and filament on the disc of the Sun during a period of relatively low activity. Picture 5 shows the location of the imaged area on the face of the Sun.*

7 *Auroral phenomena have long been a favorite photographic subject for orbiting astronauts. There are both advantages and disadvantages: the altitude enables large areas of aurora to be imaged (and studied), but the high orbital speed of spacecraft and often dimness of the feature leads to some "smear," despite the use of high speed films.*

mission the following summer: this was the Soviets' way of upstaging her.) In both cases the flights lasted eight days. The Soyuz T-7 crew came down in Soyuz T-5, leaving a "fresh" spacecraft docked to the space station. Four Progress flights provided supplies.

After these successful flights, 1983 represented a major setback. The first event was the launch of the Cosmos 1443 space station module on March 2, 1983. It docked with Salyut 7 on March 10. Soyuz T-8 was launched on April 20 with a three-man crew of Vladimir Titov (commander and no relation to the Vostok 2 pilot), Gennadi Strekalov and Alexander Serebrov (flight engineers). After reaching orbit, it was discovered the radar did not work. It was later determined that the antenna had been ripped off when the shroud separated during launch. Titov tried to make a manual rendezvous but the approach speed was too high. Soyuz T-8 went sailing past Salyut 7 and

narrowly avoided a collision. Return to Earth came after two days in orbit.

The situation initially improved with the launch of Soyuz T-9 on June 27, 1983, when Vladimir Lyakhov (commander) and Alexander Alexandrov (flight engineer) began working in the Cosmos 1443/Salyut 7 space station. The module extended the capabilities of the station but also imposed limitations. It doubled the volume, contained more supplies than a Progress, and its fuel supply could be used for orbital maneuvers. However, with Cosmos 1443 at the forward docking port and Soyuz T-9 at the rear port, the station could not receive additional Soyuz or Progress missions, so until Salyut could be fitted with multiple docking ports the Soviets would not be able to take full advantage of the modules. Cosmos 1443 undocked on August 14 and later de-orbited a return capsule. Soyuz T-9 was then shifted to the forward port.

1 *July 1984: Svetlana Savitskaya walks in space; the first woman cosmonaut to do so as well as the first to go into space twice. Her mission was (presumably deliberately) timed to precede NASA's 41-G mission the following October, in which Kathryn Sullivan and Sally Ride established the same records on the US side.*

2 *Soyuz T-10 commander Leonid Kizim during repair work on Salyut 7's propulsion system. He and colleague Vladimir Solovyov conducted no less than six EVAs.*

Danger at Launch

Events now turned against the Soviets. Three days after Cosmos 1443 separated, Progress 17 was launched. During the refueling operation a fuel line aboard Salyut 7 began leaking, and this rendered its propulsion system useless. Despite the setback, preparations went ahead for the Soyuz T-10 flight: Titov and Strekalov (from Soyuz T-8) would take over the station from Lyakov and Alexandrov and would add solar panels to the station's arrays.

The launch was scheduled for September 26. Ninety seconds before lift-off, a fire broke out at the base of the rocket. The crew noticed a strange vibration

1

2

and flames outside the spacecraft. Ground control sent the command to trigger the escape rocket but the fire had already cut the cable. The radio abort command was ordered to be sent but it took several seconds to arm the console and send the dual signals. Six seconds before the booster toppled over, the abort rocket fired. In three seconds, the capsule had reached Mach 1 going straight up. The crew was battered by 17 Gs but the emergency parachute opened and they made a rough landing two miles (3.2km) from the pad. The booster debris burned for 20 hours: although details of the abort were published within days in the West, the Soviet press said nothing for some three years.

Lyakhov and Alexandrov continued their mission, making two spacewalks in early November to install the new solar panels. Progress 18 also brought up supplies. During this time, the British press was filled with stories that the Soyuz T-9 had exceeded its safe lifetime, and that the crew was "lost in space." In fact,

the Soyuz T spacecraft could operate for six months, and when the crew returned safely on November 23 after 149 days the British press hardly noticed.

First EVA by a Woman

After the setbacks of 1983, the following year was much more satisfactory. As in 1982, the flight activities involved a long-duration flight and two guest missions. Soyuz T-10 was launched on February 8, 1984. Its crew was Leonid Kizim (commander), Vladimir Solovyov (flight engineer) and Dr. Oleg Atkov (cosmonaut researcher). Atkov was a medical doctor and conducted tests on the crew. Although their time was taken up with astronomical, biological and Earth resources studies, repair of Salyut's propulsion system had a high priority. Kizim and Solovyov made no less than six spacewalks. The work was as difficult as the Solar Max repair carried out by US astronauts, but after over 22 hours of labor Salyut 7 was fully repaired. Once more the Soviets had shown the ability to improvise and overcome setbacks. Their most difficult test was yet to come, however.

The two guest missions followed the earlier pattern. Soyuz T-11 (April 3) carried Rakesh Sharma, the first Indian cosmonaut, as part of its three-man crew. He photographed the Indian sub-continent and tested yoga as a means of adapting to weightlessness. The mission ended after the standard eight-day flight time.

Soyuz T-12 followed on July 17, 1984, with Svetlana Savitskaya aboard. On July 25, Savitskaya and mission commander Vladimir Dzhanibekov made a spacewalk. She thus became not only the first woman to fly twice in space but the first to make a spacewalk. This upstaged similar achievements scheduled aboard the space shuttle later that year. The third crew member, Igor Volk, was overshadowed, but in fact had a much more significant role. Volk was an experienced test pilot and his spaceflight was in preparation for work as the chief test pilot for the Soviet shuttle. Unusually for a guest flight, Soyuz T-12 returned to Earth after 11 days. Kizim, Solovyov and Atkov made their own return on October 1 after 236 days in space. The crew looked frail, tired and pale but had suffered no long-term harm.

An Electrical Fault

The schedule for 1985 followed that of the previous year. There would be a single, nine-month flight supported by two guest flights (one with an all-woman crew) and a number of Progress missions. (Five of the latter had been flown in support of Soyuz T-10.) These plans were cancelled in February 5 when all contact with Salyut 7 was lost: not even the radio beacon was transmitting. The Soviets issued a statement that the station's mission had been fulfilled and that "the station has been mothballed and continues its flight in an automatic mode."

Some were not willing to write off Salyut so quickly, however. If cosmonauts could board the station, they might be able to repair the failure. It would be difficult because Salyut would be tumbling slowly and the Soyuz would have to dock unaided. To command the flight, Vladimir Dzhanibekov was selected: this was his fifth spaceflight and on Soyuz T-6 he had needed to make a manual docking with Salyut 7. Viktor Savinykh was flight engineer. Soyuz T-13 was launched on June 6 1985. It made a slow, two-day rendezvous (one day is normal) and the final maneuvers were conducted using a laser rangefinder. Once docked, they checked for any toxic gas – as from a fire: there was none and they opened the hatch.

Salyut 7 was dark and freezing, with the walls and windows covered in frost. Even with fur-lined suits, gloves, hats, extra socks and boots, they could only work for an hour at a time. They found that the failure had been triggered by a faulty charge-measuring device on one of the eight batteries attached to the solar panels. The device was intended to prevent power from being drawn from a battery if it fell below a certain level: when the device failed, it both drained the battery and prevented the others from being charged. Once they had been drained, Salyut was dead.

The crew connected the six still-functional batteries directly to the solar panels. After long, cold hours of

3 *Alexander Volkov (left) Vladimir Vasyutin (center) and Viktor Savinykh suddenly ended a scheduled long-term stay on Salyut 7 on November 21, 1985, and returned to Earth. Vasyutin, commander of both Soyuz T-14 and the mission to Salyut 7, had been taken ill and could not be treated by drugs on board the space station. Flight engineer Savinykh had gone up to Salyut earlier in Soyuz T-13 to help Vladimir Dzhanibekov repair a major electrical problem which at one time made the demise of the space station seem certain.*

4 *Cosmonauts at work in Salyut. On the wall at right is a map which shows the station's ground track.*

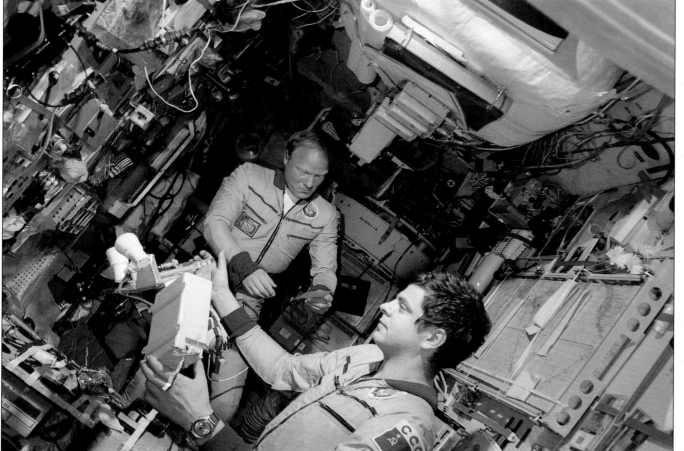

work, the voltage indicator trembled, then began to rise. Several hours later, the lights started to glow: Salyut 7 was beginning to live again. Over a 10-day period, Dzhanibekov and Savinykh checked every system aboard the station and replacement parts were brought up during two Progress missions.

It was now time for the long-duration crew to take over. Soyuz T-14 was launched on September 17, carrying Vladimir Vasyutin (commander), Georgi Grechko (flight engineer) and Alexander Volkov (researcher). This mission involved a crew switch. After eight days Dzhanibekov and Grechko returned to Earth on Soyuz T-13, leaving Vasyutin, Volkov and Savinykh aboard Salyut. The station was again expanded when a module – Cosmos 1686 – docked on October 2.

On October 25, 1985, the Soviets said the "cosmonauts are in good health and feeling well." The statement was not repeated. In late October, listeners in New Jersey noted the crew was using a voice scrambler. Then, on November 21, Soyuz T-14 returned to Earth and it was revealed that Vasyutin had become so ill that an emergency medical evacuation was required. He was, in fact, suffering from prostatitis: he had it before the flight but the pre-launch medical tests would not have detected it. Treatment required

powerful antibiotics which were not available on the station. If left untreated, blood poisoning and death could have occurred. So, with Vasyutin too ill to continue, Savinykh had taken over command. Back on Earth, Vasyutin spent a month in hospital, while the other two crewmen were placed in medical isolation. So ended Salyut 7's mission.

For more than a decade, the Russians had been slowly rebuilding their space effort. Each flight built on the last; the duration of each mission being extended by about a month. With this step-by-step program came the experience and maturity that allowed the Soviets to overcome problems. The repair of Salyut 7, for example, was certainly the equivalent of the Skylab rescue. At the same time, the urge for cheap "firsts," which caused so many problems in the 1960s and early 1970s, still existed. The two flights of Savitskaya were a prime example, and after the all-woman flight was cancelled, the female cosmonauts left the program. The difference with these later "firsts," however, was that they were relatively harmless and were not allowed to affect goals or policy in any significant manner.

As 1986 dawned, both the US and Soviets had ambitious space plans. When the year ended, the US program lay in ruins and the Soviets were heading for the permanent habitation of space.

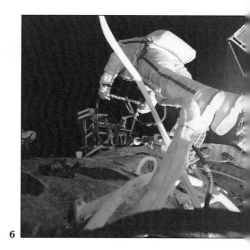

5 *Soyuz T-11 in April 1985 was a "guest" mission. Shown here with cosmonauts Yuri Malyshev and Gennadi Strekalov is (bottom) Indian Rakesh Sharma.*

6 *Another view showing Savitskaya during her EVA. During the T-12 mission she spent eleven days in space.*

11

THE SHUTTLE:
TRAGEDY AND RECOVERY

Early in 1986 NASA faced a critical scheduling period in which payloads of major scientific importance had to launch on time to catch specific "windows" – for example, the Galileo probe to Jupiter and Ulysses via Jupiter to the Sun – or to study a specific event as in the case of the launch of the Astro mission aboard Columbia on March 6 so that the telescopes could be directed at Halley's Comet at the same time as the ESA, Soviet and Japanese probes were closing with the most famous of all comets.

1

1 *Group portrait of the 51-L crew:*
(front row – left to right) Mike
Smith, "Dick" Scobee and Ronald
McNair; (standing – left to right)
Ellison Onizuka, Christa
McCauliffe, Gregory Jarvis and
Judith Resnik.

2, 3, 4, 5 *and* 6 *A tragedy unfolds.*
In 2, *a bright point of light between*
SRB and fuel tank reveals where
hot gases act like a blow torch on the
skin of the tank; 3 *and* 5 *show the*
aftermath of the explosion, while in
4 *flight directors in Houston register*
dismay. In 6, *the SRBs continue*
their burn until destroyed.

A NASA account of the launch of the 61-C mission of Columbia, which was commanded by Robert Gibson and which placed a communications satellite successfully into orbit, read as follows:

"Jan 12, 1986: 6:55 am EST. Launch was originally scheduled for December 18, 1985, but had to be delayed until Dec 19 when it was decided that additional time was needed to close out the orbiter aft compartment. Countdown on December 19 was halted at T-14 secs due to an indication that the right SRB hydraulic power unit (HPU) was exceeding RPM redline speed limits. (This was later determined to be a false reading.) Launch was delayed 18 days and rescheduled for Jan 6, 1986, but was scrubbed at T-31 secs due to an accidental draining of approximately 14,000 pounds of liquid oxygen

from the external tank. Launch was rescheduled for the following day Jan 7 but bad weather at both trans-Atlantic abort landing (TAL) sites … caused another scrub at T-9 minutes. Launch was delayed 48 hours and rescheduled for Jan 9 but was postponed 24 hours due to a launch pad liquid oxygen sensor that had broken off and become lodged in a main engine number 2 prevalve. Launch was rescheduled for Jan 10 but heavy rains prompted another 48 hour delay. The final launch countdown sequence on Jan 12 proceeded without problem."

One journal described the events as being more reminiscent of the early days of Project Mercury than of the operational shuttle program. Unfortunately, events were to conspire against NASA even more: the landing of 61-C, which, because of the modifications to Columbia's nose wheel, was due to take place at the Kennedy Space Center with obvious advantage to the turnaround time for the following Astro mission, in fact took place at Edwards AFB early on the morning of January 18 (and even then some brake damage was sustained). A NASA account read:

"Landing at KSC was moved up a day to Jan 16 to save orbiter turnaround time. This early landing attempt was abandoned due to unacceptable weather at KSC. Bad weather at KSC delayed by 24 hours a second landing attempt on the originally scheduled landing date Jan 17. Landing was rescheduled for Jan 18 at KSC but persisting bad weather forced a one revolution extension of the mission and a landing at Edwards. Orbiter returned to KSC Jan 23, 1986." [i.e. a delay of five days].

In the meantime Challenger was about to inaugurate the refurbished Pad 39B, used for the first time since ASTP over 10 years before. Mission 51-L was to launch the second Tracking and Data Relay Satellite (TDRS-B), but the major innovation would be the flight into space of Christa McCauliffe, the successful entrant from an initial 11,000 teachers who applied to participate in the Teacher in Space Project – one element in the NASA scheme to provide a limited number of private citizens with the opportunity to join shuttle missions. McCauliffe would conduct "live" televised lessons for schools back on Earth and also prepare material for later use.

Commander of 51-L was "Dick" Scobee who, like mission specialists Judith Resnik, Ronald McNair and

Ellison Onizuka, had joined NASA in 1978 and had already flown one mission. It was US Navy pilot Michael Smith's first mission in the right seat of the cockpit, and it was a first flight, too, for Hughes Aircraft Co. aerospace engineer Gregory Jarvis.

After a delay of almost a week caused by technical and weather problems, and a hold of two hours immediately prior to launch, Challenger lifted off at 11:38 EST on the morning of January 28, 1986. Some 73 seconds into the flight and at a height of around 47,000 feet (14,290m) there was an explosive burn of hydrogen and oxygen propellants from the large fuel tank and within moments the orbiter broke up. Range safety officers destroyed the SRBs, which were still flying, at 110 seconds. The crew compartment, which for a short time had continued upwards to an altitude of 65,000 feet (19,760m) after break up, hit the surface of the Atlantic off the Florida coast some 165 seconds after launch. The terminal speed was estimated at over 200 mph (322kph): it was thought subsequently that some members of the crew might have survived the initial vehicle break up, but it was impossible for them to survive the final impact.

Commission of Enquiry

NASA had encountered tragedy in 1967 during the early days of Apollo. But the fire in the command module had claimed the lives of three career astronauts during the course of a technical test conducted away from the public gaze. The Challenger tragedy occurred in full view of onlookers, including loved ones of the crew, and of the TV cameras, which beamed the pictures

was the only member currently working for NASA.

The Presidential Commission worked incisively and quickly, presenting its report in a little over four months. The cause of the tragedy was clear cut: a failure in the joint between the two lower segments of the right SRB (each booster is made up of four segments assembled at Kennedy) which, during the propellant burn of the rocket motor, had allowed hot gases to leak and impinge on the fuel tank with disastrous consequences.

Critical to the sealing of SRB joints were rubber "O" rings which had to be seated correctly in grooves and their resilience was adversely affected by low temperatures. There had been a number of previous launches after which the rings in the recovered SRBs had shown abnormal and potentially dangerous deterioration but no decisive action had been taken by NASA or Thiokol managers. Commission members likened this to a "kind of Russian roulette," where those involved accepted increasing risk apparently because they "got away with it last time." In the case of 51-L, some Thiokol engineers opposed launching Challenger on January 28 because of the very cold temperatures predicted, but they were subsequently overruled by their own senior management reacting to pressure from NASA managers at the Marshall Space Flight Center, which had responsibility for the SRBs. The Commission found that the long term problems resulting from the "faulty design" of the joint had not been raised with senior NASA management, and that on the day the decision to launch Challenger was "flawed" because those making the ultimate decision were unaware both of the immediate concern which had been expressed as well as the longer term problems.

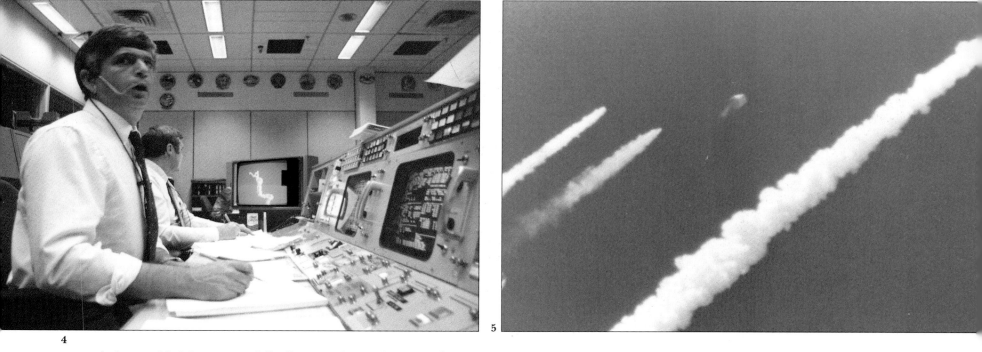

around the world. Moreover, while five members of the crew were career astronauts, Mrs. McCauliffe was a wife, mother and teacher who was representing all of those many "private citizens" who wished to fly, and Gregory Jarvis was a payload specialist who should have flown on 61-C, the previous mission. Despite some criticism, the investigation of the 1967 tragedy was held by NASA itself. The circumstances of 1986 made this impossible, and within a week President Reagan moved to establish a commission of enquiry. It was chaired by former Secretary of State William Rogers and its 13 other members from the worlds of science, engineering and law included former astronaut Neil Armstrong and the legendary "Chuck" Yeager, the first person to break the sound barrier. Shuttle astronaut Sally Ride

The Commission explored the background to the tragedy and identified the "unrelenting pressure" which NASA faced in meeting the demands of an accelerating flight schedule, although it somewhat unfairly laid blame solely on NASA (without including the legislators in Congress) for not providing "adequate resources for … [the schedule's] attainment." It referred to the launch rate going up as staff levels were frozen or effectively lowered at the Johnson Space Center as the space station began to demand attention. It reported on the heavy work load at Kennedy which resulted in high overtime rates, and which led to accidents. It commented on NASA's willingness at short notice to fly "can do" missions (e.g. repair of satellites) which were spectacular and exciting but which stretched limited resources further

1 *One year after the tragedy, all of the recovered debris from Challenger was placed in two abandoned missile silos at the Cape Canaveral Air Force Station. It was stored in such a way that it could readily be retrieved at a later date if necessary.*

2 *A picture taken on the morning of the launch shows the extent of ice at the pad.*

and were inconsistent with the long term planning needs of a "space truck" service. It was concerned about the critical shuttle spares situation which led to cannibalization of items between the orbiters. In general, its view was that the inadequate resources had led to strain in achieving the "modest" rate of nine launches in 1985 and that, with the capabilities of the system "stretched to the limit," the indications were "that NASA would not have been able to accomplish the 15 flights scheduled for 1986."

Wide Recommendations

The Commission's recommendation that the SRB joint and seal must be changed by redesign, or a new design eliminating the joint be introduced (with an independent body charged to oversee the change), came as no surprise. But it went considerably further. It identified the management problems that had contributed to the tragedy as stemming largely from the lack of a centralized program control, which should be firmly reestablished, and qualified astronauts should be encouraged to move into management positions. There should be a review of all "criticality" items and hazard analyses and an Office of Safety, Reliability and Quality Assurance established, with a chief reporting directly to the NASA Administrator. In stressing a general need for improved communications, the Commission was highly critical of the Marshall Space Flight Center for its tendency to "management

isolation," which it insisted must be eliminated "whether by changes of personnel, organization, indoctrination or all three." A policy also needed to be developed to govern the imposition and removal of launch constraints.

The Commission had reviewed all the shuttle systems and made recommendations on two further areas. On landing safety, it pronounced the orbiter's tires, brakes and nosewheel steering in need of improvement: until this was done and fully tested there should be no

landings at the Kennedy Space Center, and (once landings at KSC resumed) during periods of unpredictable weather NASA should plan on landing at Edwards, which "may necessitate a dual ferry capability." The Commission had looked closely at crew escape systems but realized that there were limitations on what was possible. It did, however, state that every effort should be made to provide a means of escape during an emergency when an orbiter was in controlled gliding flight.

In stressing again the "relentless pressure on NASA" resulting from the shuttle being the nation's principal space launch capability, William Rogers and his colleagues recommended firmly that "such reliance on a single launch [system] should be avoided in the future."

The Report was in general eminently fair and, while not hesitating to criticize where appropriate, urged that NASA continue to be supported as a national resource and "a symbol of national pride and technological leadership." NASA's spectacular achievements in the past were applauded and the Commission's findings and recommendations were presented as contributing to the future successes that were both expected and required of NASA as the 21st century approached.

The Challenge to NASA

Overcoming the trauma of the Challenger tragedy and returning the orbiter fleet to space was to be a long haul. Morale at NASA had plummeted – the more so because the loss of Challenger took place during what amounted to a temporary management vacuum at the top. The Administrator James Beggs had resigned in December 1985 to fight government charges (subsequently dropped) over the handling of defense contracts when he was at General Dynamics. The Deputy Administrator had only been in office for a few weeks and Houston, lead center for shuttle operations, was also without a director.

While there was some initial reluctance to release information about the accident (for which it was criticized by the Commission) NASA launched its own investigations into the tragedy in advance of the Commission's Report, and, before the operation was ended in August 1986, over 6,000 NASA, DOD and other personnel had worked on the task of retrieving Challenger wreckage from the Atlantic. President Reagan persuaded James Fletcher to return to NASA as Administrator, although his appointment was not

Numerous Changes

There was also to be more bad news before recovery began. With launch schedules (including those of such major science missions as Galileo and Ulysses) in disarray, the unsuccessful launches in the spring of the hitherto very reliable Delta, of an Atlas-Centaur and of a USAF Titan III, effectively barred the US from space for the time being. It was no consolation that failures of the European Ariane and Soviet Proton launchers also occurred at about the same time.

Nonetheless, NASA began to rally. One of Fletcher's first acts was to publish an almost immediate *Report to the President* on the manner and time-scale in which it was planned to implement the Presidential Commission's recommendations. There was naturally caution on the date of resumption of flights, but the initial program called for a certification review of the SRB, which was of fundamental importance for obvious reasons, by the end of 1987.

In scrutinizing and modifying shuttle systems where appropriate, NASA, over the remainder of 1986 and 1987, went much further than the changes recommended by the Rogers Commission. Redesign of the SRB was decided on and ultimately a total of 145 changes were introduced. There were eight changes to the external fuel tank, 30 to the main engines (SSMEs) and no less than 220 to the orbiter. The last included changes to the RCS, OMS and APU systems; to the fuel

3 The Presidential Commission was taken to the VAB on a visit to KSC on February 14, 1986. Here, Robert Crippen points out features of an orbiter's tile work to Commission member Neil Armstrong.

4 Top of the aft segment of the right solid booster retrieved from the sea. It shows the location of the "O" rings which figured prominently in the inquiry.

4

5

6

confirmed until May 1986. At the same time, there were some not unexpected resignations among both NASA and contractor personnel.

In those early months and subsequently there was no shortage of individuals and institutions wishing to heap scorn and criticism upon NASA. In addition, the feelings of space scientists who had spent years developing experiments that were due to be placed in Earth orbit or launched toward interplanetary space by the shuttle can well be imagined. While Fletcher had the advantage of six years experience as Administrator during the birth of the shuttle project (1971-77), he was not well suited to providing the relentless, no-nonsense drive – and yet cheerful encouragement – that NASA employees needed. In 1967, the recovery had been driven by the objective of still meeting the Kennedy deadline. In 1986 there was no such goal: NASA seemed indecisive and to go too far in seeking to remove any possible chance of shuttle malfunctions – an impossible objective.

cells; the thermal protection system on the underneath of the orbiter's nose and on its wing elevons; to the main landing gear – with the modified nose wheel steering system already installed on Columbia to be installed on Discovery and Atlantis; and also to the disconnect valve between the external tank and the orbiter, where a possible malfunction (identified in the Commission's Report) could have cut off the fuel supply to the SSMEs during launch. After extensive study, an escape system for use during gliding flight was adopted which comprised a side hatch that could be jettisoned by explosive bolts, followed by the extension of a telescopic pole down which crew members would slide (to clear the left wing of the orbiter) before opening their parachutes. An egress slide was also provided for rapid and safe exit from the orbiter should an emergency develop after landing.

NASA also took the opportunity during the period following the tragedy to introduce and plan numerous changes not called for in the Commission's Report.

5 It was April before the remains of the 51-L crew were released to family and friends. Here an unidentified coffin is being loaded aboard a Military Airlift Command C-141 at the Kennedy Space Center. A military honor guard and NASA representatives were present at the Shuttle Landing Facility for the departure.

6 During a memorial service held at the Johnson Space Center on January 31, 1986, attended by President Reagan and the First Lady, four T-38s fly overhead and one breaks away in the "missing man" formation, a traditional USAF farewell to a departed member.

1 *For obvious reasons, the solid rocket boosters were thoroughly tested after the extensive modifications recommended by the Rogers Commission were introduced. Here a static firing is taking place at a Thiokol facility on August 30, 1987.*

3 *A small reminder of the continuing dangers of spaceflight: at the end of 1988, after the resumption of flights, the orbiter Atlantis suffered some tile damage on its starboard side. This photograph was taken after its return to California at the end of the STS-27 mission.*

2 *A US Navy parachutist slides down a pole to exit a C-141 aircraft in a test of one of two proposed shuttle escape systems; the other being based on extraction of crew members by rocket. In the system shown here, a lanyard attachment enabled the crew member to slide to the end of the pole and thence away from the vehicle. This was required because a straightforward jump from a hatchway would almost certainly result in the crew member being struck by parts of the fuselage. This was the system finally selected.*

Thus, upgraded computers were to be introduced into the orbiters (they began to fly in space early in 1991) with over twice the memory and three times the processing speed, as well as being half the size and half the weight, of the early 1970s era computers installed in the spacecraft when built. A boost to morale was provided in August 1987 when, following the award of a $1,300m contract from NASA, Rockwell International began work on a new orbiter which in due course was given the name Endeavour. However, the changes and innovations did not only concern the orbiters and associated flightware. For example, although pad equipment was not a contributory factor in the Challenger accident, over one hundred modifications were made initially at Pad 39B at the Kennedy Space Center. These included extensive changes to the Crew Emergency Egress System and the addition of an SRB joint heater umbilical capable

of keeping the booster joints at around 75°F (24°C). The modifications were also subsequently made to Pad 39A. The smoothness of the landing strip at KSC was also further improved.

Astronaut Managers

Major changes were introduced into NASA's management structure as required by the Presidential Commission. The "lead center" concept was abandoned and headquarters authority over the shuttle program was reasserted. The movement of astronauts into management positions became marked. At an early stage ALT and shuttle astronaut Richard Truly was appointed to the post of Associate Administrator for Space Flight, where he had responsibility for both the shuttle as well as the space station. Some time after flights into space

resumed, James Fletcher stepped down as Administrator and Truly moved up to that top position.

The structure of the new management organization took some time to consolidate, but by the end of 1990 Truly had been joined by a number of his former astronaut colleagues in senior positions. His former position as Assistant Administrator at the (slightly) renamed Office of Space Flight was taken by William Lenoir, who had joined NASA in 1967 as a scientist-astronaut and who had gone into space on the fifth shuttle flight. In 1984 Lenoir took a job outside NASA but five years later agreed to return at Truly's request for a three year period. The highly experienced Robert Crippen was immediately involved in dealing with the aftermath of the Challenger accident, and by the end of 1990 was Director, Space Shuttle. (This office had previously borne the title of National Space Transportation System – NSTS – but with the ending of almost total reliance on the shuttle as a launcher the title obviously became inappropriate.)

Two assistants reported to Crippen. One – Deputy Director, Space Shuttle (Program) – was broadly in charge of policy and programming and, although an HQ employee, was based at the Johnson Space Center. The other – Deputy Director, Space Shuttle (Operations) – was also an HQ employee but located at the Kennedy Space Center. This post at the end of 1990 was held by Brewster Shaw, who had flown the shuttle three times. The post involves a number of responsibilities and the Deputy Director is charged with "Final vehicle preparation, mission execution and return of the orbiter for processing for its next flight." The lessons to be learned from the poor communications and obscure lines of responsibility which contributed so much to the events of January 1986 were accepted by NASA, and it is the Deputy Director (Operations), chairing a Mission

Management Team that is active as the final countdown to launch takes place, who is responsible for the final "go/no-go" decision being given to the Launch Director.

Shuttle veteran Henry Hartsfield also was appointed to a post in senior, technical management at NASA HQ (before moving at the end of 1990 to a space station management post at the Marshall SFC). But the policy of moving astronauts into management did not, however, only concern headquarters. Given its major role in manned spaceflight, it was appropriate perhaps that Skylab and shuttle astronaut Paul Weitz should be appointed Deputy Director at the Johnson Space Center during the period leading up to the return of the shuttle to space. Group 8 shuttle astronaut Steven Hawley in due course became Associate Director at NASA's Ames Research Center. The advancement of astronauts into top NASA management received further confirmation at the beginning of 1992 when Robert Crippen was appointed director of the Kennedy Space Center. In addition, senior management appointments by NASA's new Administrator Daniel Goldin in the spring of 1992 included three astronauts – Bryan O'Connor, Charles Bolden and Frederick Gregory.

Another major policy development following on Challenger arose from one of the Presidential Commission's most predictable findings: that "reliance on a single launch capability should be avoided in the future." The immediate NASA response was to state that it "strongly supports a mixed fleet [of launchers]" and other advice continued to press for the reappearance of a number of expendable launch vehicles (ELVs). Thus a report by a task force appointed by the NASA Advisory Council firmly recommended in March 1987 "a major evolution ... [of policy] ... from one that has maximized the use of the Shuttle to a policy that preserves the [shuttle] fleet for those critical missions that require its unique capabilities. These features include two-way crew transportation, manned on-orbit tasking and intervention, spacecraft servicing and reboost and the ability to return cargo from space." Broadly speaking, that policy has been pursued subsequently, together with a departure (as required by a January 1988 Presidential Directive) from the previous energetic policy of using the shuttle as a launch system for commercial satellites, save in certain circumstances such as those involving national security or foreign policy. This has been made possible, under the stimulus of demand from the USAF as well as commercial companies, by the development of modified versions of previously-successful expendable launchers such as the Atlas I, Delta II and Titan 4.

Loss of Centaur

Inevitably there were negative as well as positive results in the many changes introduced during the rebuilding work that led to the return of the shuttle to space. Although its needs had been strongly reflected in the initial concept of the shuttle, the USAF had always had a somewhat ambivalent attitude to the system and, with plans for ELVs gathering pace, it announced in 1988 that the Vandenberg shuttle complex in California – built to place the shuttle into polar orbits, which were essential for most military purposes – was being "mothballed." A more significant loss for the world of science was the decision to ban the Centaur high energy upper stage (which was due to boost, for example, the Galileo spacecraft to Jupiter) from the shuttle orbiter cargo hold. This was on the grounds that the highly volatile cryogenic propellants aboard the Centaur in certain circumstances could present an unacceptable safety risk. This in turn forced NASA to plan planetary "gravity assists" for those spacecraft due to be launched from

the shuttle using stages with less power than the Centaur, with an inevitable increase in the length of missions.

Tests and Delays

By the time NASA made its promised one year progress report to President Reagan in July 1987, Discovery had been chosen to take astronauts back into orbit with a planned launch date in June 1988, and managers were thinking in terms of a maximum flight rate of 14 missions per year. The lengthy check out of Discovery began in August 1987, but progress toward the launch was substantially delayed when a redesigned nozzle part on a solid rocket motor failed during a test firing by Thiokol at the end of 1987. Modifications were successful, but delays in making some of the mandatory changes to Discovery, more SSME problems and the relatively late

4

decision on which of two escape systems should be selected resulted in launch being put back to September 1988.

But at long last it was time to return to space, and NASA decided to revert to a single numeral designation for flights, all of which would lift off from Pad 39B at the Kennedy Space Center (until the modifications to 39A were completed) and would land at Edwards until further notice.

STS-26 was flown aboard Discovery from September 29 until October 3, 1988, and its primary payload was a Tracking and Data Relay Satellite. The five man, all veteran crew, commanded by Frederick Hauck, was delayed for 98 minutes before lift off but once in orbit promptly deployed the satellite and then turned to the mixture of scientific, technological and student experiments which had become so familiar in earlier shuttle

4, 5 The US returned to space when Discovery lifted off on September 29, 1988. All of the STS-26 crew had flown in space before and the main task of a relatively brief mission, which was obviously of major psychological as well as practical importance, was to carry a Tracking and Data Relay Satellite (TDRSS) into orbit. That had been 51-L's payload as well, and it was important for NASA to consolidate its new satellite communications system, which would greatly improve contact between crews and mission controllers in Houston.

5

1 *An ASTRO-1 image of the famous globular cluster Omega Centauri recorded in ultraviolet light, which facilitated study of hotter stars.*

1

2

2 *The battery of telescope and other instruments carried into space by the much-delayed ASTRO-1 mission in December 1990.*

flights and was to become the pattern again. There were no major problems at any stage of the mission, which provided a workmanlike reintroduction of the shuttle to space. Before the end of the year Atlantis flew the STS-27 mission (December 2-6) which was dedicated to classified Department of Defense activity. The launch supplied a reminder of the problems presented both by nature and technology. There was a delay of one day because of weather conditions and the orbiter sustained more than normal TPS tile damage, probably due to "ablative insulating material from the right solid rocket booster nose cap hitting [it] about 85 seconds into the flight."

Launch of Magellan

Of the five missions flown in 1989, two launched much-delayed interplanetary spacecraft, one a further TDRS satellite and two were dedicated to DOD activity. STS-29 (Discovery) launched another TDRS during its almost five-day mission which lasted from March 13 to 18. The STS-30 Atlantis mission from May 4 to 8 launched the Magellan radar mapper probe to Venus, which had the distinction of having first been approved in 1982, been cancelled and then reinstated and, upon its successful launch, of becoming the first US planetary mission for 11 years. Columbia returned to space for the first time in three-and-a-half years during the STS-28 mission from August 8 to 13, which was a DOD flight, as was the STS-33 mission of Discovery from November 22 to 27. In between them, an event of major importance was the launch from Atlantis during the STS-34 mission (October 18-23) of the Galileo spacecraft to Jupiter. Because of the loss of the Centaur stage, the combined orbiter (which is to study the solar system's largest planet and its moons for at least two years) and probe (which will take direct measurements of the Jovian atmosphere before it is destroyed by heat and pressure) will take six years to make the journey and require three gravity assists: one from Venus and two from Earth.

While five missions had been accomplished in 1989 with some success, only one – Columbia in August – had been launched to schedule. STS-29 was delayed three weeks from February 18 to March 13 because of shuttle main engine problems, and STS-30 for six days from April 28 to May 4 for another SSME problem and a vapor leak. Yet another SSME fault as well as weather slipped STS-34 six days from October 12 and the year ended with a two day slip resulting from an SRB fault for Discovery's STS-33 mission.

Hubble Space Telescope

The following year – 1990 – proved difficult, too, although there were successes. Columbia's STS-32 mission slipped from a launch on December 18, 1989, to January 9, 1990, mainly because of the need to complete the modification of Pad 39A, which had not been used for four years. Once in space, however, the crew launched a communications satellite and successfully retrieved the Long Duration Exposure Facility (LDEF) which had been in orbit for almost six years. As a rehearsal for considerably longer duration flights to come, the crew of three men and two women, commanded by Daniel Brandenstein, had been in space for almost 11 days when they landed on January 20 and thus beat the previous record set by the STS-9 Spacelab mission.

The illness of the commander (John Creighton), weather and a computer fault slipped the DOD-dedicated STS-36 mission of Atlantis six days to February 28 with a return on March 4. The following month, however, witnessed a much awaited and historic event with the launch from Discovery at a record orbiter altitude of

3

well over 300 miles (483km) of the Hubble Space Telescope. During this STS-31 mission astronauts Bruce McCandless and Kathryn Sullivan were prepared to conduct EVAs should the deployment of the HST run into difficulties but, while the launch was not totally smooth, this activity was not required. The launch of this mission (which finally took place on April 24) was unusual in that NASA on two occasions actually brought the scheduled launch time *forward*, only for a faulty auxiliary power unit (APU) to cause a delay of two weeks in between. Discovery's landing at Edwards AFB on April 29 was the first operational test of new carbon brakes, which had replaced the earlier design using beryllium. It was hoped that the improved mechanical and thermal characteristics of the new brakes would in due course permit some scheduled returns to the run-ways at the Kennedy Space Center, although there was a continuing debate on the need for fitting in addition a drag parachute (which was being installed on Endeavour).

In May of 1990 the STS-35 mission of Columbia was due at long last to carry the ASTRO-1 payload, with UV and X-Ray telescopes mounted on Spacelab pallets, into orbit after a wait of four years. However, at Pad 39A a hydrogen leak was discovered which necessitated the return of the vehicle to the Vehicle Assembly Building. In the meantime a hydrogen leak had also been found on Atlantis, which was to fly the DOD STS-38 mission scheduled for July 1990. A NASA report subsequently stated:

"Shuttle launches were interrupted for five months after hydrogen leaks were detected on Columbia and Atlantis during external tank loading operations. The problem on Atlantis was isolated to the seals associated with the 17-inch disconnect area. After replacing the disconnect, Atlantis passed a tanking test …. Columbia also suffered from a leak in the 17-inch disconnect area which was replaced but a separate leak in the main propulsion system remained elusive. A special leak team … isolated the problem to a seal in a prevalve of the main propulsion system which had been damaged during installation after Columbia's January mission. A tanking test on October 30 verified that all problems with Columbia had been resolved."

Discovery was mercifully free of the problems, and the STS-41 mission from October 6 to 10 was successful in launching another much delayed mission, the ESA-built Ulysses spacecraft, which would enable scientists for the first time ever to study the polar regions of the Sun. A less important but nonetheless intriguing event on this mission was the first use aboard an orbiter of television cameras activated and directed by the astronauts' voices.

With the hydrogen problems overcome, the Atlantis STS-38 mission finally launched from Pad 39A on November 15 with a classified payload that may have been a satellite providing intelligence about the situation in the Kuwait/Iraq region of the Middle East. Return to Edwards was planned for November 19, but high winds caused delays. Approval was then given but a sudden shift in wind direction at the base caused the flight controllers to call off the attempt with a reported 90 seconds only to go before Atlantis would have temporarily lost communication through the TDRS, and would therefore have been committed to a landing at Edwards. The weather at Kennedy on November 20, however, was acceptable and the orbiter successfully made the first landing there since April 1985 after a mission lasting just under five days.

Columbia was finally launched on December 2. Under the command of Vance Brand, who at 59 years became the oldest man to fly in space, were a pilot and three mission specialists who had all flown in space before, together with two astronomer payload specialists. This was the first seven member crew, and also the first with payload specialists, to fly since the Challenger accident. A Soyuz spacecraft with two cosmonauts and a Japanese TV journalist aboard was launched to Mir on December 2, so, together with the two cosmonauts already aboard the Soviet space station, a record-breaking twelve men were in space at the same time.

The STS-35 mission was the first to be devoted totally to one scientific discipline (astrophysics) and, following the practice set by the first Spacelab mission in 1983, the crew divided into two shifts to maximise the returns. Once again there was also very close interaction between the crew and investigators on the ground: indeed, STS-35 was the first flight during which three control centers operated: Mission Control at Houston; the Spacelab Mission Operations Control Center at Marshall SFC (responsible for coordinating science activities); and a special Payload Operations Control Center at Goddard SPC, which operated the Broad Band X-Ray Telescope by remote control.

Unfortunately problems developed both with the European built IPS (Instrument Pointing System) and other electronic units and displays of ASTRO-1 (which had cost an estimated $150m), and this limited the crew's ability to point the telescopes with the required accuracy. Cooperation between ground and astronauts overcame much of the difficulty but it was reported that about one third of the planned observations were lost because of the malfunctions. An unusual event took place when astronomer and mission specialist Jeffrey Hoffman gave a live TV astronomy lesson for students from orbit, dressed in a conventional shirt and tie, almost certainly another first.

Columbia landed at Edwards AFB just before midnight on December 10, 1990 – one day early because of a bad weather forecast.

By the end of 1990, therefore, some lost momentum had been recovered but long term problems remained. The earlier targets of 50, 24 or even 12 launches a year had simply not been achieved. Only five of the nine launches scheduled in 1989 had taken place and only six of the same target number planned for 1990.

Another Observatory

A conservative target of seven missions was set for 1991 and six were accomplished. The STS-37 mission of Atlantis from April 5 to 11 launched the second of

NASA's "great observatories" – the Gamma Ray Observatory (GRO) which subsequently had the prefix Compton added in honor of a physicist of that name. The first US EVAs since 1985 were already planned for this mission but an extra one had to be conducted by astronauts Jerry Ross and Jay Apt to free an antenna boom on the satellite that had stuck. Discovery flew the STS-39 unclassified Department of Defense mission from April 28 to May 6 although it had originally been scheduled for launch in March. The delay was caused by the need to deal with cracks on the orbiter's external tank door drive mechanism. Several technical problems (including a potentially serious one involving a damaged fuel line sensor) delayed the launch of Columbia on the STS-40 mission from May 22 to June 5 with a return after nine days. The main task of the seven astronaut crew was to conduct twenty major investigations in the first Spacelab Life Sciences mission, which – looking toward prolonged stays in space aboard space station

Freedom – was described as the most concentrated study of humans' reaction to the space environment since Skylab in 1973/4.

When Atlantis lifted off on the STS-43 mission on August 2, 1991, with the fifth TDRS spacecraft in the cargo hold the orbiter weighed heavier than on any previous mission by the vehicles in the fleet – 253,000lbs (over 114,000kgs). Once the satellite was successfully launched a battery of technological and biomedical experiments was conducted until return on August 11 after a mission that went like "clockwork". The STS-48 mission of Discovery from September 12 to 18 was significant on a number of counts. Liftoff took place six weeks earlier than originally planned; the placing in orbit of the Upper Atmosphere Research Satellite was the first major step in the Mission to Planet Earth programme; and on the fourth day in orbit the orbiter had to be maneuvered to avoid a derelict Russian booster that would have passed within the 2km safety zone that had been established in mission rules – the first occasion on which such evasion had proved necessary.

By extraordinary coincidence a similar maneuver proved necessary during the final mission of 1991 – that of Atlantis on the military but unclassified STS-44 flight from November 24 to December 1. The vehicle's

3 Post Challenger missions paid attention to the amount of damage sustained by the external fuel tank when the SRBs separated part way into the mission. This photograph, taken during the STS-29 mission in March 1989 after Discovery separated from the tank, shows a burn scar above the forward attach point of one of the SRBs. The study eventually concluded that the scarring was not a hazard.

4 The STS-32 mission in January 1990 beat the endurance record established during the Spacelab 1 flight and retrieved the Long Duration Exposure Facility, shown here with the Sun beyond, which had been in space for six years.

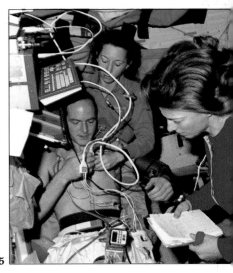

5 Medical experiments and studies are constantly taking place on shuttle flights. STS-32 astronaut Bonnie Dunbar makes notes here, while Marsha Ivins fits the sensors of an echocardiograph on colleague David Low's chest.

1

1 *In May 1992, the STS-49 crew aboard the replacement orbiter Endeavour during its first flight overcame major difficulties to retrieve a stranded Intelsat communications satellite and successfully fit a new motor to it. In their final attempt, astronauts (left to right in this photograph) Richard Hieb, Thomas Akers and Pierre Thuot literally grabbed the satellite before anchoring it in the cargo bay and fitting the new motor. This spacewalk was the longest in space history, and also the first time that three crew members had conducted an EVA at the same time.*

2 *A couple of months later a camera operated by an STS-46 crew member over the Philippines took this splendidly clear view of Mount Pinatubo, which had erupted in June 1991, hurling an enormous amount of debris into the atmosphere. Ash from the eruption surrounds the mountain and all of the rivers draining the area were found to have been widened and clogged with ash. Debris and ash flows down the rivers continued as a result of tropical rainstorms. Significant increases of mudflow deposits in the area could be charted in successive shuttle remote-sensing images taken after the eruption.*

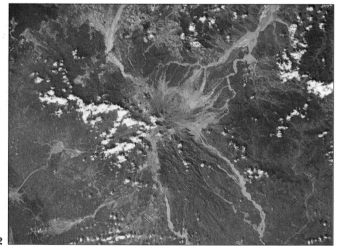

2

orbit was lowered by 2km to avoid a rocket body which eventually passed 32kms away. The crew deployed a Defence Support Satellite (DPS), which was designed to monitor rocket plumes from launches on the Earth's surface, but had to reduce the length of the mission by two days when a navigation unit failed. Atlantis landed on a new runway at Edwards AFB, where it coasted to a halt without braking in a test of the new surface.

Endeavour to the Rescue

Eight missions were planned for 1992 but – by the late summer – NASA was said to be thinking of adding a ninth. Discovery flew the STS-42 mission from January 22 to 30 with seven astronauts operating experiments on behalf of 200 scientists and six international organisations in the Spacelab module during the first flight of the International Microgravity Laboratory. Experi-

ments of a different kind – primarily studying the Sun's effect on the Earth's atmosphere and climate – were flown during the Atmospheric Laboratory for Applications and Science (Atlas-1) mission of Atlantis from March 24 to April 2. One member of the STS-45 crew was Michael Foale, a mission specialist who was the first British born career astronaut to fly in space.

The STS-49 mission and first flight of the new orbiter Endeavour from May 7 to 16 was described by NASA's new Administrator Daniel Goldin as bringing "magic back into the [space] program". The prime task of the crew commanded by veteran shuttle astronaut Daniel Brandenstein was to capture an Intelsat communications satellite that had been stranded in the wrong orbit by a Titan III booster in March 1990 and fit a new motor so that it could be commanded to its correct orbit. The task proved far more difficult than originally envisaged but was ultimately tackled successfully. In so doing demands in skilled rendezvous piloting and EVA work were made of the crew that were described as the most intense in the 30-year history of spaceflight. The mission was the first NASA flight on which four EVAs were conducted and the first flight by any space power on which three crew members conducted an EVA simultaneously. The third spacewalk during which astronauts Richard Hieb, Pierre Thuot and Thomas Akers succeeded in capturing the Intelsat VI satellite and fitting a new motor to it lasted for eight hours and 29 minutes – the longest in space history, a record previously held by Eugene Cernan and Harrison Schmitt of Apollo 17 for almost twenty years. A two day extension was required to complete mission tasks but this presented no problem as Endeavour performed "flawlessly" – a heartening performance as its main engines had to be changed after malfunctions occurred in a firing test conducted one month earlier.

Columbia had been modified for longer orbital flights using an Extended Duration Orbiter kit before it lifted off on the STS-50 mission which lasted from June 25 to July 9. This was another flight dedicated to microgravity experiments but unlike the STS-42 flight these were all domestic US experiments – hence the US Microgravity Laboratory (USML)-1 designation. Bad weather delayed Columbia's return but when it was diverted finally to a landing at the Kennedy Space Center the mission had lasted for 13 days 19 hours 30 minutes and four seconds – almost three days longer than the longest previous shuttle mission.

Atlantis returned to space for the STS-46 mission that lasted from July 31 to August 8. Another "first" was established in that astronaut Claude Nicollier – who operated the remote manipulator system arm to deploy ESA's Eureca materials science platform from the cargo hold – was the first ESA astronaut to have been trained as a mission specialist at the Johnson Space Center, where he had been working since 1980. Although problems were encountered, Eureca in due course became fully operational, but the same success did not attend the other major experiment of the flight – that of greatly expanding knowledge of the behaviors and properties of two spacecraft located at different orbital altitudes and joined by a lengthy tether. The TSS-1 (Tethered Satellite System) spacecraft was the result of more than ten years work by NASA and Italian scientists and engineers working together but unfortunately a malfunction of the tether winding mechanism resulted in the satellite only being unwound for a mere 840 feet (256m) and not the maximum potential of 12 miles (20kms).

Future of the Shuttle

What does the future hold for the shuttle? A contract has been awarded for the development of an advanced solid rocket motor (to replace the redesigned SRBs), which should be operational in the mid 1990s and which, besides being regarded as safer, would increase the orbiters' payload performance. Reports from NASA indicate that attention continues to be given to improving the quality, and therefore the reliability, of such critical items as the shuttle main engines, although there are some criticisms that still more needs to be done in this direction.

Lockheed – the contractor at the Kennedy Space Center with responsibility for the processing of shuttle hardware – has been under constant pressure to improve the efficiency and reliability of its management and work practices. And another potential threat to shuttle operations was removed at the end of 1990 when NASA took delivery of a second Boeing 747 aircraft to ferry orbiters back to Kennedy from Edwards AFB. As for astronaut performance, it should be noted that of the 97 career astronauts who flew from the resumption of launches in September 1988 until the end of 1991 only four belonged to pre-1978 (i.e. pre-shuttle) groups. Moreover, some at least had far shorter waits for a first mission than had been customary – two group 12 (1987) mission specialists aboard the STS-41 (Ulysses) flight, for example, going into space little more than three years after joining NASA.

All in all the shuttle has shown itself to be a highly versatile system, but one which continues to make enormous demands on the groups, and ultimately people, who service and process it. It is complex, and elements of the system have shown themselves to be unreliable, with an inevitably harmful effect on schedules and costs,

even though emphasis on the shuttle's role as a research and development vehicle, rather than a commercial launcher, and safety rather than schedules, has been a major result of the Challenger tragedy. Moreover, NASA itself has looked soberly at the possibility of future tragedies and has stated that each flight has a 1-in-78 chance of "catastrophic failure."

At the end of 1990 a distinguished Advisory Committee, under the chairmanship of Norman Augustine, reported to the Administrator of NASA on The Future of the US Space Program. It was critical of the failings of the shuttle but recognized that it existed, and stated that it "is absolutely essential to America's civil space program for the next decade or more. Necessary steps to assure the viability of Space Shuttle operations in this decade [i.e. an adequate supply of spares] should therefore proceed." At the same time, the Committee pressed for the development of a new, unmanned but potentially man-ratable, launch vehicle with an increased payload capacity which should take over shuttle missions, "except [for those] where human involvement is essential or other critical national needs dictate."

This sense of disappointment with the performance of the shuttle system results to a considerable degree not from what it has achieved (for its achievements have been many) but because of what it was presented as being originally. NASA itself presented the shuttle as the space equivalent of the DC3 aircraft: a workhorse that would fly repeatedly and therefore at much lower cost into orbit. Going into space, ran the argument which gave

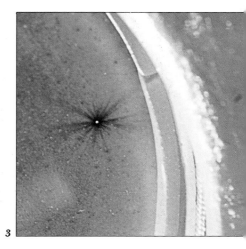

3

3 *There have been a number of cases of shuttle windows being impacted by space debris or micrometeoroids. Endeavour received such an impact on its very first mission; this photograph taken by STS-49 mission commander Daniel Brandenstein shows a section of one of the orbiter's forward, port side windows. Such events are always thoroughly studied afterwards to establish what lessons can be learned.*

4

rise to great expectations, would become routine. And there was the problem, for, until our technology is far more advanced, going into space will never be "routine": the space environment, in human terms, is too harsh, too unforgiving. In seeking to build the shuttle, if NASA did not realize that fact then it was gravely at fault. If it did realize it, but glossed over the difficulties in order to seek presidential, political and public support for a new and major program, then it was even more at fault – although there must be sympathy for any organization forced to deal with the tortuous US governmental funding process, particularly in an area capable of arousing so much emotion and controversy as spaceflight.

Unfortunately, NASA is still having to live with the consequences.

4 *Landing procedures receive constant attention, and the brand new orbiter Endeavour in May 1992 deployed a drag 'chute when it landed in California. This was the first part of a "detailed test objective" (DTO) which would study the drag 'chute system during future missions.*

12

INTERPLANETARY EXPLORATION FROM 1975

USSR: TO VENUS – AND HALLEY'S COMET

As cosmonauts were making flights to the Salyut stations, the Soviets were also rebuilding their unmanned planetary program. The 1974 Mars failures had been a major blow. As in the 1960s, it was decided to concentrate on Venus, Mars being out of reach both in terms of rocket power and technical capability. A new lander/orbiter was designed. The orbiter was based on the Mars probe but the lander used a new design. The launch vehicle was the D-1e.

Venera 9, the first of the second generation Venus probes, was launched on June 8, 1975, and its twin, Venera 10, followed six days later. Venera 9 arrived at the planet on October 22. The lander was cooled down and separated while the main spacecraft went into orbit around Venus. The lander descended on parachutes until it reached an altitude of 31 miles (50km). The parachutes were then released, but the atmosphere was so thick that the drag was enough to slow the lander to touchdown speed. The main body of the lander was a 3.3ft (1m) diameter sphere. At its base was a "doughnut"

1 *Launch of Vega 2 by a Proton booster on December 21, 1984. Vega stood for VEnera-GAllei (Halley). The two Vegas dropped balloons into the Venusian atmosphere and then continued on to fly-by Halley's Comet early in March 1986.*

2 *Success at Venus came for the Soviets in the fall of 1985. Veneras 9 and 10 both put landers down, and the first black-and-white images were transmitted from the surface. This Venera 10 image reveals a landscape described as being a plateau with slabs of rock.*

landing pad to cushion the impact. Atop the sphere was an aerobraking disc that also acted as an antenna reflector. On the disc was a helical antenna which was wrapped around the parachute housing.

The lander touched down successfully and the camera system returned the first images from the surface. These showed a rock-strewn and soil landscape; analysis indicating that the rocks were basalt. Contrary to expectations, the surface lighting was bright enough for one Soviet scientist to compare it to a cloudy June day in Moscow. (The Sun was 54 degrees above the horizon.) Data were relayed to Earth via the orbiter for 53 minutes and the transmission ended, possibly when the orbiter passed out of range rather than because of lander failure.

Venera 10 landed on October 25 some 1,367 miles (2,200km) from the other Venera site. The images showed a plateau with slabs of rock; an area more weathered than that around Venera 9 and confirming that Venus had a varied landscape.

After the accomplishments of the 1975 flights, the next window must have been a disappointment. Venera 11 was launched on September 9, 1978, with Venera 12 following five days later. The two spacecraft would fly past the planet rather than orbit, since this would allow longer contact with the landers. Venera 12's lander was first to arrive on December 21, with Venera 11 landing on Christmas Day. In both cases, the landers suffered malfunctions that prevented imaging of the surface, but they did return atmospheric and other data (95 minutes for Venera 11 and 110 for Venera 12). This included radio bursts that may have been caused by lightning and the reverberation lasting fifteen minutes from a huge thunderbolt.

Venus in Color

The 1981 Venus window saw two more landers: Venera 13 (October 30, 1981) and Venera 14 (November 4, 1981). The landers incorporated a number of improvements, including two color cameras to provide a complete panorama, a penetrometer to measure soil strength, and a soil-analyzing device. Venera 13 landed on March 1,

was also carried. Venera 15 was launched on June 2, 1983, and Venera 16 on June 7. They arrived at Venus on October 10 and 14 respectively, and the two probes scanned the surface from 30 degrees N to the north pole. The radar images showed impact craters, hills and ridges, together with possible volcanic areas and rift systems. The mapping mission was completed in July 1984.

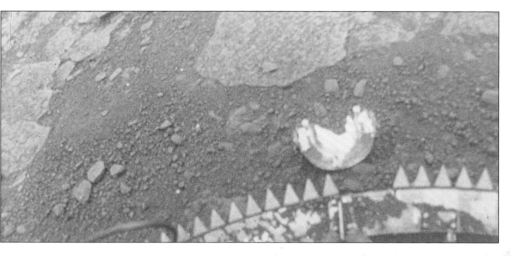

3 *Testing of a Venera lander on Earth: a still from a film. While the success of the 1975 missions to Venus was encouraging, disappointment followed during the 1978 window. Veneras 11 and 12 reached the planet but malfunctions prevented any images being obtained.*

4

5

1982, while Venera 14 arrived four days later. (The landing sites were based on US Pioneer Venus orbiter radar data.) Their images showed orange-brown rocks under an orange sky. Due to a mirage effect, the horizon appeared only about 330ft (100m) away. Venera 13 landed on a plain with flat slabs poking out of granular soil. Venera 14 rested on a 1,640ft (500m) hill, with several small foothills in the distance. The surface resembled a lava flow. Venera 13's soil analyzer found basalt with a high potassium content (rare on Earth). Venera 14 also found basalt but (as on Earth) it had low potassium content. Venera 14's penetrometer unfortunately landed on the jettisoned camera cover. The spacecraft itself lasted only 57 minutes, whereas Venera 13 lasted more than twice as long: 127 minutes.

A different approach was adopted for the 1983 window. The lander was replaced by a large, rectangular radar antenna called Polyus V. Additional propellant

The Venera flights of 1975-83 rebuilt the Soviet planetary program after the earlier Mars failures. Now the Soviets were ready to stage their most ambitious mission: to Venus and to Halley's Comet.

Vega 1 (from VEnera-GAllei [Halley]) was launched on December 15, 1984, and Vega 2 on December 21. The Vega 1 lander touched down on Venus on June 11, 1985. As it descended, a balloon was released to float amid the clouds. Vega 2 followed on June 15, also releasing a balloon. Because the balloons had to be released over the night side, the landers carried no cameras. Vega 1 lander's soil analyzer equipment failed, but that aboard the Vega 2 lander found anorthosite-troctolite, a rock type found in the lunar highlands. A French-Soviet gas chromatograph measured atmospheric composition as the lander descended. The two balloons lasted two days, providing pressure, temperature and wind data.

6

4 *Color comes to Venus: Veneras 13 and 14 in 1981 were versatile spacecraft, with a color imaging capability as well as soil analyzers. This Venera 13 scene, which includes the bottom of the lander, shows a surface not unlike that around Venera 10, with slabs of rock prominent. The life of the lander was 127 minutes and its soil analyzer found basalt with a high potassium content.*

5 *The flight control center during the launch of Vega 1 on December 15, 1984. Release of a picture like this would have been unprecedented a few years before.*

6 *A false-color representation of the nucleus region of Halley's Comet obtained by the Vega 2 probe on March 9, 1986. The most detailed images were obtained by ESA's Giotto spacecraft, but the Soviet spacecraft, which preceded Giotto, played an important part in helping to refine its fly-by of the comet.*

1 *The Voyagers discovered a faint ring round Jupiter. This Voyager 2 image shows the ring as two light orange lines extending out from the limb (edge) of the planet. Two exposures were made through orange and violet filters, and the multiple images of the limb are evidence of Voyager's movement during the long exposures.*

2 *A magnificent Voyager 1 view of Jupiter from 17.5m miles (28.4m km). To the wealth of detail of the Jovian atmosphere can be added the moons Io (against the disc) and Europa (at right).*

The main spacecraft flew past Venus and on to Halley's Comet. Vega 1's closest approach came on March 6, 1986, at a distance of 5,525 miles (8,890km). Vega 2 passed at a distance of 4,991 miles (8,030km) on March 9. Although dust protection had been added, the basic design prevented a closer approach, but the two Vega spacecraft played a critical role in acting as pathfinders for Giotto's 376-mile (605km) approach on March 14.

The decade of Venus exploration rebuilt the unmanned Soviet program. With the lengthy US abandonment of planetary exploration (for example, no US mission was flown to Halley's Comet) the Venus flights gave the Soviets a monopoly. Soon it would be time to try Mars again.

the space shuttle, they are products of the late 1960s and early 1970s. Each Voyager is controlled by six computers (three of them backups), with memories totalling a mere 32 kbits: tiny in comparison to modern-day home computers. But the ability to load new software provided an unprecedented flexibility and allowed the craft to be re-programmed to meet changing situations. That flexibility was supported by an equally essential ingredient for success: navigational accuracy. The flybys had to be controlled to within fine tolerances for each planet's gravity to swing Voyager towards its next target. Even with course corrections en route, a small error at Saturn in 1981 would have made Voyager 2 miss Uranus four-and-a-half years and 3,045m miles (4,900m km) later, some 1,800m miles (2,900m km)

3 *An engineering mock up of Voyager. Prominent features are the instrument platform at the end of the short boom; the large antenna dish; and the 43ft- (13m) long magnetometer boom.*

USA: ODYSSEY OF THE VOYAGERS – AND REBIRTH

The scientific value of the Voyager twins – NASA's two most successful deep space probes – cannot be over-estimated nor, it must be admitted, can their political value to the agency during lean times. While the scientific community criticized what they perceived as an over-emphasis on manned space programs in the 11 years from 1978, when America did not launch a single probe, the Voyagers continued returning data. While the cost of interplanetary missions rose in step with their complexity, Voyager proved to be a bargain-basement investment, costing less than $700m for six planetary encounters and an abundance of scientific discoveries. Only days before the space shuttle exploded in January 1986, Voyager 2 provided mankind's first close-up views of Uranus. The juxtaposition was ironic: all planned planetary missions – Galileo, Magellan, Ulysses and Mars Observer – now depended on the manned vehicle for their first steps into space. Challenger's loss severely delayed them and underlined the scientists' reluctance to be involved with the hugely-expensive, manned program. Current proposals for future missions now call for "expendable" unmanned launchers and there is an associated drift to simpler, cheaper missions.

The success of the Voyagers is all the more remarkable when the technology is taken into consideration. Like

from Earth. In only three decades of the space age, engineers had learned how to guide a probe to arrive only 12.5 miles (20km) off target at the 2,800m mile (4,500m km) distance of Neptune. No Soviet mission has ever come close to that level of achievement, or even attempted to penetrate beyond Mars, which alone has proved sufficiently difficult for them. Not only were Soviet spacecraft too short-lived for such ventures, their navigation was also inadequate. Until NASA provided precise information on the movements of the outer planets, Soviet controllers could not even have attempted the missions.

Versatility of Voyager

The simple Pioneers had already blazed the trail to Jupiter and Saturn, of course, but they provided only the aperitif to the Voyagers' main course. Officially the project's aim was to skim past the two gas giants (reflected in the original Mariner-Jupiter-Saturn name), but the rare planetary alignments of the late 1970s proved too tempting. Voyager was designed with sufficient flexibility, backup systems and propellant for course corrections should NASA and Congress (holding the purse strings) approve continuation beyond Saturn. In fact, the $85m extra funding required for the Uranus phase was not won without a fight. Like Viking before it, Voyager was threatened with an irreversible switch-off to save a modest sum in contrast to the money already spent. Apollo provided a chilling precedent: the science packages left on the Moon were shut down in 1977 to save $2m a year. It would have been a scientific tragedy in the case of Voyager 2, for no other mission was or is planned for the outer planets.

On each of the two Voyagers, the TV imaging systems, with 200mm and 1500mm focal length lenses and several ultraviolet and infrared remote sensing instruments, were mounted on a scan platform essential to avoid picture smearing during the rapid flybys. Other detectors for monitoring radiation, radio emissions and magnetic fields were located elsewhere on the 1,820lb (825kg) spacecraft. The four magnetometers had to be carried on a 43ft (13m) long boom to avoid interference from Voyager's own magnetic field. Also boom-mounted, this time to prevent interference from the magneto-meters, were three plutonium generators providing up to 450W of electricity in the cold, dark, outer reaches of the solar system. That supply is too low to power even a one-bar electric fire, yet, dwindling at 7W each year as the plutonium decays, it ensured some of the most important astronomical discoveries in history.

Voyager spends most of its time with the 12ft (3.7m) diameter radio dish pointing at Earth for transmitting data or receiving instructions. The 23W transmitter working through it meant that terrestrial radio tele-scopes during 1989's Neptune encounter had to lock on and decipher a signal with 20,000 million times less power than a digital wristwatch! The value of carrying backups was demonstrated in 1978 when Voyager 2's main radio receiver failed before even reaching Jupiter, prompting the central computer to switch in the spare unit. This computer also controls two others: one handles the complex movements of the scan platform, while the other collects the science data and prepares it for relay to Earth. All of these electronics are housed in Voyager's central body below the radio dish, where they can be protected against radiation; particularly Jupiter's potentially lethal environment. Inside the bus is a spherical propellant tank to supply the 16 small thrusters, each exerting less force than holding an apple in the hand, but still sufficient for subtle course corrections out to Neptune.

Like the Pioneers, the two Voyagers are now heading out of the solar system, carrying messages for alien cultures should they ever be found. A copper disc holds recordings of greetings in 60 languages, and one-and-a-half hours of music and natural terrestrial sounds, together with a playing needle and instructions. Although interstellar space is almost empty, the record sits under an aluminum sleeve for protection against cosmic dust impacts.

To the Outer Planets

The Voyagers were launched by the Titan 3E-Centaur combination (America's most powerful unmanned vehicle until the introduction by the Air Force of the Titan 4-Centaur in the mid-1990s) but still required a solid-propellant kick stage. Voyager 2 actually departed first, on August 20, 1977, because its partner was to follow a slightly faster route to Jupiter and would overtake it in the Asteroid Belt. Both passed the Moon's orbit in only 10 hours, and Voyager 1 turned its cameras back two weeks later to capture the disks of Earth and Moon in the same image for the first time.

Voyager 1 entered the Jovian system in early February 1979, but the planet's retinue of satellites is so spread out that it took until March 5 to arc in towards the closest point of almost 174,000 miles (280,000km) above the cloud tops. Even at that distance the radiation was so intense that Voyager's electronics were swamped,

and some pictures and other data were lost before it could recover. The dose was sufficient to kill a human a thousand times over. The first major surprise came a day out, when a long camera exposure discovered the faint traces of a debris ring little more than 18.5 miles (30km) thick and stretching down 35,420 miles (57,000km) into the atmosphere. True, it paled besides Saturn's splendid necklace but, unlike Uranus and Neptune, there had been little expectation of finding such a feature.

The best opportunities for studying the four large, so-called Galilean moons came in the days after close encounter. For the first time they emerged as worlds in their own right. The first, Io, proved by far the most astonishing. Voyager's 11,800 mile (19,000km) approach

revealed a surface covered in sparkling orange and brown sulphur, thrown out from what later analysis showed to be active volcanoes – the first ever seen away from Earth. Voyager found eight on what is now regarded as probably the solar system's most geologically active body. The energy is believed to be generated by the distortions induced by Jupiter's strong gravity as Io orbits in a close, eccentric path.

Voyager 1's trajectory kept it over 466,000 miles (750,000km) from moon Europa, but it was still close enough to show a puzzling, pale world encircled by strange streaks. Voyager 2 ventured three times closer in July 1979 and revealed Europa to be a smooth, bright

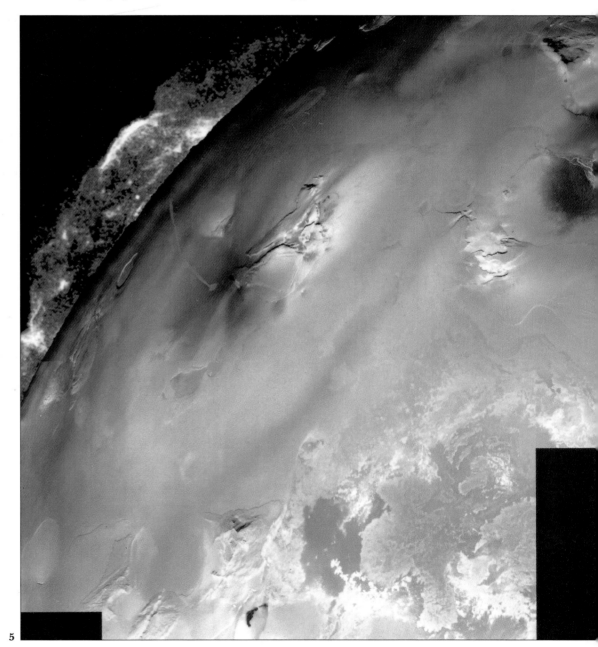

4 Two of the stars of Voyager planetary encounter media conferences at JPL: Dr. Bradford Smith (left), head of the imaging team, and USGS geologist Dr. Larry Soderblom, his deputy.

5 This USGS high-resolution enhancement of the moon Io probably presents a more accurate record of its surface color, and on the limb considerable detail is revealed of the cloud of debris from the eruption of the volcano Pele.

1 *An early color-image reconstruction of the Jovian moon Io, with evidence of a volcanic eruption above the limb. (Compare with that on the previous page.) Io was the first solar system body apart from the Earth where active volcanism had been observed.*

2 *A full disc record of Io from 300,000 miles (862,000km). The surface, which was likened to that of a pizza, revealed the signs of extensive and current volcanism.*

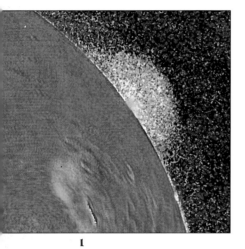

and frozen world. The streaks were simply cracks in the thick ice mantle. Deep below the ice, warm water oceans might be harboring complex living organisms.

The next port of call was Ganymede, the solar system's largest moon; its 1,640-mile (2,640km) radius qualifying it as a planet larger than Mercury. It turned out to look like Earth's Moon, as did the densely-cratered Callisto, except that a huge impact had gouged out a 465 mile (750km) radius basin. Subsequent analysis of the 19,000 images from the 98-day encounter added two moons (Metis and Thebe) to Jupiter's known thirteen. Voyager 1 had also, of course, intensively studied the atmosphere, its composition, intricate banding and circulation, and the Great Red Spot.

Rings and "Shepherds"

Voyager 2 followed only four months later, with a modified observing schedule to take advantage of its predecessor's discoveries, looking for more Io volcanoes (two had apparently fallen dormant) and studying the ring more closely. Each craft's trajectory had been carefully shaped to bend towards Saturn, 497m miles (800m km) and twenty and twenty-five months respectively further on. Spectacular views of Saturn's rings had been expected, but no-one was prepared for what Voyager found. Instead of the handful of known rings, there were thousands of bands of every description.

3 *Saturn imaged from Voyager 2 on July 12, 1981, from a distance of 27m miles (43m km). Exposures through ultraviolet, violet and green filters were used to build up the picture, which is a false-color enhancement.*

4 *The Voyagers each carried a message for aliens who might find them: recordings of the languages and sounds of the world.*

Some were intertwined, and small moons were discovered to be "shepherding" them as they orbited the planet. Voyager 1 found three new moons, but many more are believed to lie in the gaps between the rings. The probes showed the rings' composition to range from 32ft (10m) water icebergs to dust specks – probably debris from the solar system's creation and not, as previously proposed, a shattered moon. "Spokes" seen rotating with the rings continue to present the theorists with problems: although believed to be fine dust perhaps electrostatically levitated above the ring plane, they are difficult to explain.

Voyager 1 flew within camera range of all Saturn's major satellites, finding them to be heavily cratered and probably made largely of water ice. Tiny Mimas exhibited a crater with a radius almost a third of its own; the impact must have come close to shattering the body. Ice-covered Enceladus proved to have the brightest surface seen in the solar system until Voyager 2 reached Neptune's Triton in 1989. But Voyager 1's principal target was Titan, a moon considered to be so important that reaching it prevented the probe from flying on to Uranus. Not much smaller than Mars, it is the only moon with a dense atmosphere, believed by scientists to be similar to Earth's before life appeared. Voyager revealed a mainly nitrogen blanket, together with methane and ethane, and a surface pressure more than one-and-a-half times Earth's, albeit at a chilly -274°F (-170°C). The images were disappointing though, an orange chemical smog proving impenetrable. Voyager only whetted the appetite for more information on Titan: the Cassini mission plans to drop a capsule into the atmosphere in the year 2004.

Voyager 2's path through Saturn's system in August 1981 was constrained by the need to fly on to Uranus, but it still returned better pictures of some moons than its twin did. The Voyagers have continued to provide work for the scientists. Thus in 1990 Dr. Mark Showalter at NASA Ames Research Center identified Saturn's smallest known moon, with a diameter of just under 12 miles (19km). Some 30,000 images were searched in an attempt to locate a body that would explain regular disturbances in the A-ring around Encke's Division, and the moon was found in eleven.

Voyager 2 achieved most of its objectives, but the scan platform produced a major problem for the engineers and encounter sequence planners. When it emerged from behind Saturn after the 62,760 mile (101,000km) close approach of August 26, 1981, the cameras and other instruments were found to be pointing uselessly into space. It took two days before the jammed mechanism was freed, and even then it could only swing its instruments slowly. The Uranus and Neptune encounters had to be meticulously planned around this fault, but it again demonstrated the probe's flexibility.

Many New Satellites

As Voyager 2 left Saturn behind in September 1981 the original mission had been been successfully com-pleted, exceeding all expectations. The first spacecraft was heading up above the planets' orbital plane, never to return, but the second immediately began breaking new ground by heading towards mankind's first inspection of Uranus. Voyager 2's last two targets are so far out that several radio telescopes had to work in unison to detect the spacecraft's faint radio whisper, even though the probe helped by cutting its transmission rate to a quarter of that used at Jupiter. The distance of the Sun, with the subsequent fall-off in light, also meant that exposures had to be lengthened, which threatened image smear problems with the sticking scan platform. Instead, JPL controllers devised a system of swinging the entire vehicle as it passed targets, rather like a camera-toting tourist turning his whole body because of a stiff neck. It all had to be worked out to every last detail, because controllers could provide little help when they were two-and-a-half hours away at the speed of light.

Planning was further complicated by a major peculiarity of Uranus: the system is tilted almost on its side, so that the moons orbited at right angles to Voyager's path. Instead of speeding along the plane of the system, with plenty of picture-taking opportunities, it headed in like a projectile at a dartboard. It could therefore

3

4

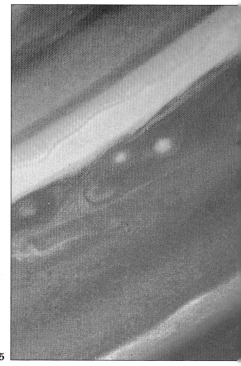

5

look at only one moon in any detail. This encounter with Miranda from over 18,000 miles (29,000km) away nonetheless returned a unique panorama. Planetologists found examples of almost every type of geological feature known in the solar system, all on a moon only about 300 miles (480km) across. By contrast, Umbriel showed a dark and cratered surface, suggesting a great age. Voyager added no less than ten satellites, the largest only 53 miles (85km) in radius, to the known retinue of five. Two of them, acting as "shepherds," help to keep one ring in place. Voyager increased the number of known coal-black rings, apparently composed of icy boulders, from nine to eleven.

The planet itself presented a bland, bluish face. There is little heat so far out from the Sun to churn up the hydrogen-helium atmosphere, although Voyager did detect wind speeds up to 490mph (790kph). The average cloud-top temperature was -351°F (-213°C). Curiously, the magnetic field is tilted at 55 degrees to Uranus' axial spin, so that its tail spirals out from the Sun like a 6.2m mile (10m km) long corkscrew.

Long Time Exposures

Voyager 2's operational difficulties at Uranus in January 1986 were multiplied by the time it reached Neptune in August 1989. The Sun, around 2,800m miles (4,500m km) away, was now so distant that the average image exposure had to last for 15 seconds. Earth's radio telescopes were again linked electronically to pick up Voyager's signals, and new onboard software had been devised to compress the data before transmission, allowing the broadcast of up to 220 pictures daily. Neptune's distance introduced another problem: the planet's position in space was known only to within a

6

few thousand kilometers, yet the probe planned to home in with an accuracy of a few hundred kilometers. "Optical navigation" was therefore employed; a technique in which Neptune was imaged against the star background at intervals, so that small trajectory modifications could be made as the planet neared.

Voyager flew within 3,115 miles (5,015km) of Neptune on August 25, 1989, a mere 12.5 miles (20km) off target after a journey of over 4,400m miles (7,100m km) and twelve years! Surprisingly, it found an active, blue-green atmosphere, apparently driven by internally-

5 Detail in Saturn's atmosphere: this false-color image obtained on August 19, 1981, by Voyager 2 shows complex weather patterns, including three spots. The ribbon-like feature was a jetstream traveling at speeds approaching 330 mph (150m/s).

6 The Voyagers showed that Saturn's ring system was comprised of hundreds upon hundreds of separate rings.

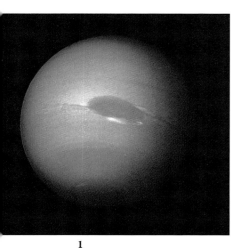

1

generated heat, and a feature christened the Great Dark Spot after Jupiter's famous marking. Methane ice cirrus clouds cast shadows on the clouds up to 47 miles (75km) below, and the solar system's strongest winds of over 1,240mph (2,000kph) were tracked. Several partial ring arcs had been detected from Earth, but Voyager resolved them into five continuous bands, with the inner one stretching down to the atmosphere. Four small moons were added to Triton and Nereid.

The program's final encounter lay with the moon Triton. Its unique face capped everything Voyager had seen throughout its long journey and provided a fitting climax to such a distinguished career. The surface was a vista sculptured by frozen nitrogen, the coldest landscape yet found at -391°F (-235°C). Nitrogen geysers spewing out gas and dust created an atmosphere only 0.0014 percent as dense as Earth's, but nonetheless impressive for such a small body. One geyser alone reached a height of five miles (8km).

Like its twin, Voyager 2 is fleeing the solar system

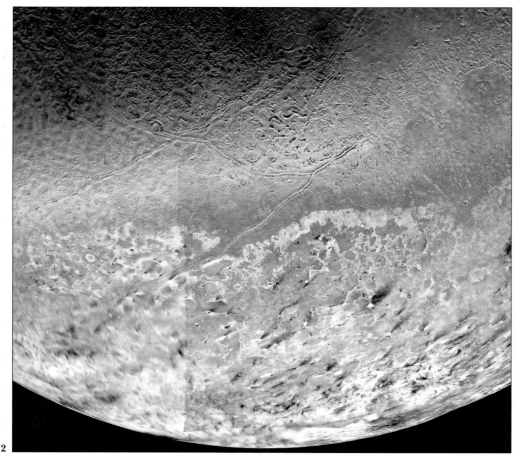

2

1 Neptune's atmosphere revealed much detail. Once again false-color images help to emphasize different features, the most prominent of which in this Voyager 2 image is the aptly named Great Dark Spot.

2 Possibly the strangest world that the Voyagers visited: Neptune's moon Triton, with a landscape of frozen nitrogen shaped into exotic forms, with geysers of nitrogen and dust.

3 This representation of the surface of Venus was secured because NASA's Pioneer Venus Orbiter carried radar instrumentation which could see through the cloud deck. Lower elevations were color coded blue, medium elevations green and the higher yellow. The large, high "continent" at top is named Ishtar.

in what is now officially designated the Voyager Interstellar Mission. Dwindling power and increasing distance will possibly break contact around the year 2020. The pair and their Pioneer predecessors are being tracked very precisely in case any trajectory deviations reveal a tenth planet. During February 1990, and before its imaging system was closed down, Voyager 1's unique vantage point high above the solar system was used to capture six of the planets (Mercury and Mars were lost in the solar glare and Pluto was too dim) in a composite of 64 frames.

The spectacular phase of the Voyagers' lives is now over, but they continue to return radiation and magnetic field data. Scientists hope they will break out from the Sun's magnetosphere before contact is lost and that they will become the first man-made probes to reach true interstellar space. Voyager 1 will fly within 1.6 light years of an unnamed star in the constellation of Camelopardalis in about 40,000 years' time. In 296,000 years, Voyager 2 will pass within 4.3 light-years of the brilliant star Sirius. Who knows, perhaps mankind will be there to retrieve it!

Return to Venus

NASA's only planetary mission launched in the twelve years after Voyager was Pioneer Venus. The agency had tended to concentrate on the more terrestrial-like Mars, leaving the veiled planet to a successful series of Soviet attempts. Only three Mariners had made brief visits to Venus, the third, Mariner 10, en route to more protracted work at Mercury. The Pioneer Venus project was conceived by NASA's Ames Research Center as America's first lengthy global investigation of the planet from orbit, and also the agency's first use of atmospheric penetration probes. As with Pioneers 10/11, they were relatively simple, spin-stabilised spacecraft with fixed instruments, paving the way for more sophisticated investigations at a future date. Built by Hughes, they resembled that company's series of communications satellites; another way of holding down costs.

In addition to Pioneer Venus Orbiter's typical retinue of remote sensing instruments as well as radiation and magnetic field detectors, the spacecraft carried a small radar to penetrate the dense overcast and produce the first surface maps of any detail. Resolution was only about 47 miles (75km), but it would reveal the general topography and help planning for future missions. The landing site for the Soviets' Venera 13, for example, was moved to a more interesting location on the basis of PVO's findings.

Pioneer Venus 2's career, on the other hand, was destined to be considerably shorter. It carried four heatshielded entry probes that would plunge into Venus' atmosphere, returning data on composition, temperature, pressure, density and cloud structure. Even the bus itself acted as a probe, sampling the very high atmosphere with two instruments before its unprotected body succumbed to the searing heat.

Both spacecraft were launched by the venerable Atlas-Centaur, but their very different launch dates illustrate the compromises in planning for planetary missions. The 1,212lb (550kg) Pioneer Venus Orbiter departed from Cape Canaveral on May 20, 1978, to follow a slow route. This meant its arrival speed at Venus was low, requiring a much smaller retro-motor to provide the 2,360mph (3,800kph) braking into orbit. The almost-2,000lb (904kg) Pioneer Venus 2 followed on August 8 by a faster route because, as it was going to plunge into the atmosphere anyway, its arrival speed was not so important. If PVO had pursued the same path, its braking motor would have accounted for half of its weight.

As seen from Earth, PVO slid behind Venus on December 4, 1978, and automatically fired its 4,050lb (1,835kg) thrust solid motor, becoming the third Venus orbiter and eventually settling into a 92.5x40,000 mile (149x65,983km) polar path with a 24 hour period. The orbit was carefully chosen to satisfy a list of requirements. Not least important was that its period of one Earth day allowed its controllers to work regular shifts. The radar mapper operated around the orbit's low point for greater clarity. Much lower and atmospheric drag would have threatened PVO's survival. The distant high point was ideal for the ultraviolet camera to build up a full-disc picture of the clouds. Remember that Mariner 10 had discovered intricate banding: these regular images taken by PVO over a long period would help to unravel the mysteries of the planet's atmosphere. Also, the constant swooping between the high and low points provided a comprehensive slice through Venus' immediate environment for PVO's other instruments.

Pioneer Venus 2 arrived only five days later or, more correctly, the five separate parts of it encountered the atmosphere at 26,100mph (42,000kph). The bus had released the large, 695lb (315kg) Sounder Probe and the three identical 207lb (94kg) North, Night and Day Probes in November so that they would scatter like pellets from a shotgun to enter the atmosphere at different points. The Sounder hit over the equator in daylight, its thick heatshield providing protection during the 40 second deceleration to 435mph (700kph). The science instruments were protected inside a pressure vessel, peering out through eight sapphire windows and one bearing a 13.5 carat diamond. The data were transmitted directly to Earth by a small, battery-powered transmitter as the Sounder descended under a small parachute. The Venusian atmosphere becomes so syrupy near the surface that any lander can cut free of its parachute and fall slowly under air drag. The Sounder plopped on to the surface after a total descent time of 57 minutes. None of the probes were designed to survive and the Sounder

4 *Uranus seen by the Voyagers as the human eye would see the planet close up (right) was a bland, almost totally featureless world. The blue-green color results from the absorption of red light by methane gas in Uranus' cold and deep, yet remarkably clear atmosphere. The false color and high contrast view at left emphasizes very slight differences in the atmosphere in the region of the planet's south pole. These reveal a dark polar "hood" surrounded by a series of progressively lighter concentric bands. The doughnut shapes are shadows cast by dust in the spacecraft's camera optics.*

5 *Uranus' moon Miranda was another exotic world, with two major, distinct types of terrain which included a bright chevron feature.*

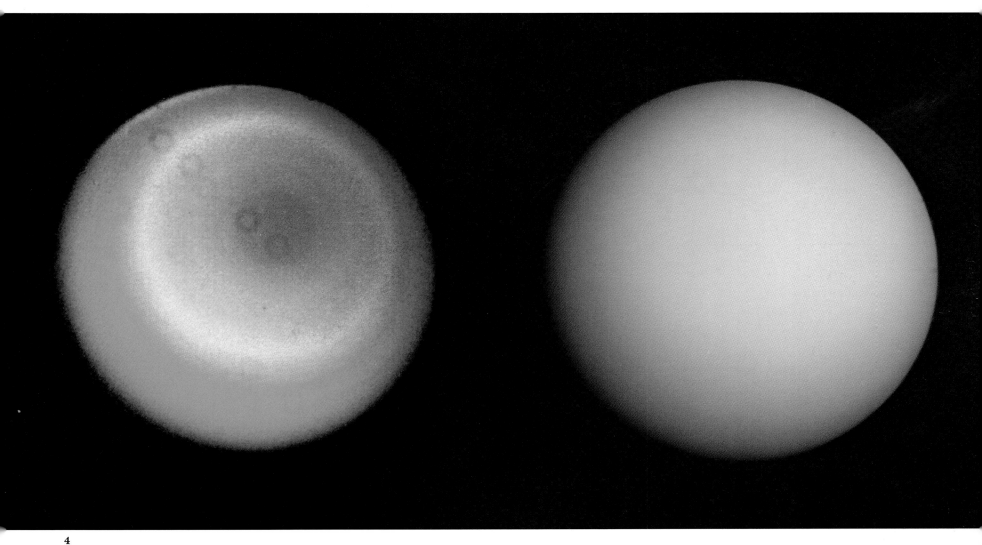

4

succumbed to a combination of the pressure, heat and impact damage.

The three smaller probes enjoyed their brief moments of glory minutes later, and one, the Day Probe, survived for 68 minutes on the surface. The bus was far shorter lived: it sampled the upper atmosphere for 63 seconds before burning up at an altitude of around 75 miles (120km).

Pioneer Venus 2 returned a wealth of data on the atmosphere, revealing its layers, sulphuric acid clouds (which possibly accounted for instrumentation problems) and its stagnant lower level. It confirmed that the 96 percent carbon dioxide content is trapping the Sun's heat in a runaway "greenhouse effect," maintaining the 842°F (450°C) surface temperature. One esoteric

measurement, of the deuterium hydrogen isotope, suggested that Venus had been covered in rolling water oceans at some stage before the greenhouse effect boiled them off.

Meanwhile, Pioneer Venus Orbiter was only just beginning its work. The simple radar mapped more than 90 percent of the surface within two years, revealing a more uniform surface than Earth's. The planet is covered in large, undulating plains punctuated by the large "continents" of Ishtar and Aphrodite, which are about the size of Australia and Africa respectively. Maxwell Montes, almost seven miles (11km) high, is the loftiest feature on the surface. Rift valleys and volcanoes point to an active interior.

PVO settled down to routine monitoring over the

5

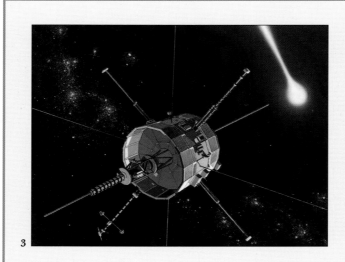

3 *An artist's impression of ICE – International Cometary Explorer. The US missed the opportunity to send a fly-by or orbiter spacecraft to study Halley's Comet in 1985-86. Not to be entirely outshone by other countries' efforts, NASA directed a probe already in space (one in the series of International Sun-Earth Explorers) by a complex series of maneuvers to fly through the tail of Comet Giacobini-Zinner in September 1985. Although the renamed ICE had no cameras, other instruments aboard were able to provide data on the make up of the comet's tail and nucleus.*

3

NASA had to pull out of missions to Halley's Comet because of lack of funding, but it was able to upstage the international fleet to some degree at a bargain basement cost of $3m with a spacecraft launched only four days after Pioneer Venus 2. The International Sun-Earth Explorer (ISEE) was designed for a completely different purpose: monitoring the interplanetary environment from around a point somewhat under one million miles (1.5m km) from Earth, where solar and terrestrial gravity balances.

Its controllers at NASA's Goddard Space Flight Center near Washington, D.C. began a series of maneuvers in 1982 that compared with Voyager's in complexity. No less than five lunar flybys in 1983, the last skimming within 75 miles (120km) of the surface, modified the trajectory towards a meeting with Comet Giacobini-Zinner on September 11, 1985. Lacking cameras, the newly-named International Cometary Explorer (ICE) sped through the comet's tail about 4,850 miles (7,800km) behind the nucleus itself at almost 13 miles/second (21km/sec). ICE discovered water to be the principal ingredient, supporting the long-held theory that comets are akin to "dirty snowballs." Its instruments were not designed for probing a comet, but they warned scientists and engineers what to expect at Halley in March 1986. In fact, ICE itself came within 25.5m miles (41m km) of the famous comet on March 28 1986.

ICE's path was again modified so that, on August 10, 2014, it will be captured within the Earth-Moon system, from where it might be retrieved for analysis of its cometary dust coating. Washington's National Air & Space Museum certainly hopes so, for in that case it will be passed to the museum for display once the scientists have had their fill.

1

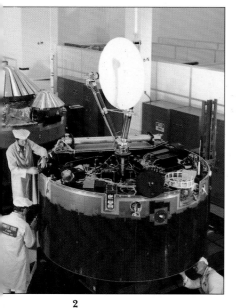

2

1 *August 8, 1978: launch of Pioneer Venus 2 by an Atlas-Centaur booster from the Cape Canaveral Air Force station.*

2 *The two Pioneer Venus spacecraft, the Orbiter in the foreground and the Pioneer V 2 probe partially in view at top left, being prepared in the El Segundo, California, facility of Hughes Aircraft.*

following years, observing changes over the Sun's 11-year cycle. One puzzle remaining to be explained is the atmosphere's sulphur dioxide content. The level has been consistently falling, prompting the exciting conclusion that PVO arrived on the scene soon after a volcanic eruption. Pioneer even made a significant contribution to the intensive Halley's Comet studies over 1985-86, showing how the icy body was shedding water, gases and dust as a result of solar heating.

PVO's longevity proved as remarkable as Voyager's. It was still returning data in 1991 but, unlike Voyager, its days were to end in a fireball. The low point of the spacecraft's orbit was affected by the Sun's activity and PVO began heading inexorably towards the Venusian atmosphere where, in October 1992, it was dragged down by the outer fringes to its fiery fate. Before that happened, however, the spacecraft might well have provided valuable information on the atmosphere to mission controllers planning to lower the Magellan probe's orbit (see below) as a means of securing higher resolution images and gravitational data about the planet – a plan at risk from the fiscal year 1993 funding cuts made in the Congress.

Magellan – Greater Economy

The modest radar of Pioneer Venus provided a tantalising glimpse of what lies below the planet's clouds. The Soviets' Venera 15 & 16 spacecraft followed in 1983-84, mapping north of latitude 30 degrees at 0.6–1.2 mile (1-2km) resolution. Even then, NASA planetologists warned that it would be dangerous to extrapolate global conclusions on the planet's origin, evolution and geology from such a limited survey. Instead, they were waiting for the Magellan orbiter and its powerful radar now systematically to scan the planet at 1,100ft (360m) resolution.

Magellan was conceived in the 1970s as the Venus Orbiting Imaging Radar (VOIR), but as its cost soared to a projected $630m it fell to the financial axe in 1982. However, it was regarded as so fundamentally important to understanding Venus that pressure from the scientific

4

community resulted in the Magellan scaled-down version emerging. Money was saved by using a 12ft (3.7m) radio dish left over from Voyager and some electronics already being built for the Galileo Jupiter project. VOIR was to enter a circular orbit, observe the surface with its radar dish and transmit the data to Earth. Instead, Magellan entered a very elliptical 183x5,260 mile (294x8,463km) polar orbit on August 10, 1990. The mapping in 15.5 mile (25km) wide bands was performed for 37 minutes around the low point in the orbit using the ex-Voyager dish, and then the stored image data were transmitted from the digital recorders through the same antenna for 114 minutes around orbital high point when Magellan was turned to point at Earth. These repeated attitude changes made Magellan's thermal protection the most challenging engineering aspect of the mission, and the elliptical orbit made it more difficult to process the raw

5

data into images, but the approach was still cheaper than VOIR's. The same dish also maps surface temperatures and a separate altimeter measured global heights to within 100ft (30m).

Launch was planned from the shuttle in the Venus window of April 1988, the first time an interplanetary probe had been launched from a manned spacecraft. Challenger's loss in January 1986 created havoc in the shuttle schedule, of course, but the Magellan, Galileo and Ulysses probes suffered more than most. They were to have flown on the powerful Centaur stage but, in the concern over safety in Challenger's aftermath, NASA decided that it was simply too risky to carry a cryogenic propellant upper stage in a manned vehicle. Centaur was duly banned from the shuttle and the waiting missions had to settle for the less powerful IUS solid stage. The next available standard window to

Venus fell in October 1989 which, by unfortunate coincidence, was also the Jupiter window that Galileo and Ulysses were already competing for. NASA decided instead to launch Magellan in spring 1989 on 1.5 circuits of the Sun, effectively storing it in space until Venus was in the desired position.

Successful Cycles

Magellan was lofted into low Earth orbit aboard the orbiter Atlantis on May 4, 1989, and released, attached to its two-stage IUS boost vehicle. The crew watched over Magellan until it opened its two stubby solar panels safely, but once the IUS first stage ignited there was nothing else they could do. The 7,400lb (3,353kg) spacecraft separated on May 5; the first US planetary probe to be sent on its way since Pioneer Venus 2 in 1978. Its mass was reduced by about two-thirds when

4 *This was the first image in ultraviolet obtained by the Cloud Photo-polarimeter aboard PVO on December 5, 1978. It shows clouds illuminated by the Sun during the early morning of the Venusian day. The amount of detail would increase as lighting and spacecraft viewing angles improved.*

5 *The difference that more than a decade makes: a global view of the surface of Venus released in October 1991. Synthetic aperture radar mosaics from the first mapping cycle by the Magellan spacecraft were projected onto a computer-simulated globe to produce this image.*

3 *Another three-dimensional perspective view of Venus prepared from Magellan radar and altimetry data. The major feature shown here, from a viewpoint located almost 350 miles (560km) to the north and at an elevation of one mile above the terrain, is Maat Mons. This is a five-mile-(8km) high volcano, and lava flows extend for hundreds of miles across the fractured plains in the foreground.*

4 *Galileo in the laboratory – the probe is in front.*

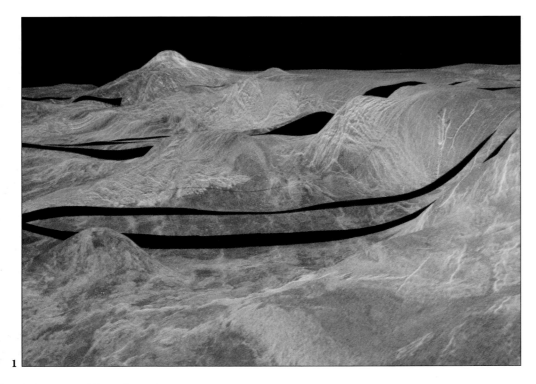

its own solid motor fired (and then separated) in August 1990 to place Magellan in orbit around Venus. The minimum goal was to keep mapping for a full Venusian year (243 Earth days), but by the spring of 1991 the success of the mission – with 84 percent of the planet's surface already mapped and the quality of the radar images transmitted back having generated much excitement amongst scientists because they included features seen nowhere else in the solar system – led to a decision to commence a second 243-day mapping cycle. This second cycle ended on January 15, 1992, and increased the total coverage to 95 percent with particular attention having been paid to the high southern

lati-tudes of Venus. The third cycle ended in September 1992 and concentrated on taking stereo radar images as well as increasing mapping coverage to 99 percent and recording gravity data over target areas. The fourth cycle – concentrating on securing global gravity data – was scheduled to end at mid-May 1993 with additional cycles planned until May 1995. However, Congressional funding cuts in 1992 made it possible that NASA would be forced to end the mission in the spring of 1993.

Galileo – Delay after Delay

The Galileo Jupiter orbiter and atmospheric probe is NASA's major deep space mission of the decade, now heading towards the gas giant for a 22-month tour of the Jovian system. The four Pioneer and Voyager flybys provided brief glimpses of the 88,980 mile (143,200km) diameter gas giant, its faint rings and retinue of at least 16 moons. A long-lived orbiter and a protected atmospheric probe can, by contrast, make detailed studies of many satellites in a series of flybys, map the strongest and most complex planetary magnetic field and probe the primitive hydrogen-helium atmosphere.

But Galileo suffered such a tortuous series of delays, eventually departing more than seven years late, that the rising cost of keeping the program going almost caused its cancellation. The program began in 1978, and by 1983 the estimated cost was more than $600m: that figure has since been revised to around $1,500m. If the actual cost had been evident from the outset then Galileo would never have been approved by Congress. Some would argue that only the sums already invested prevented its cancellation after Challenger's destruction. It could be suggested that the mission has also sucked funds from other projects, preventing the vastly cheaper Lunar Observer from flying.

Many of the problems cannot be attributed to Galileo itself, however. The shuttle's delayed debut and the

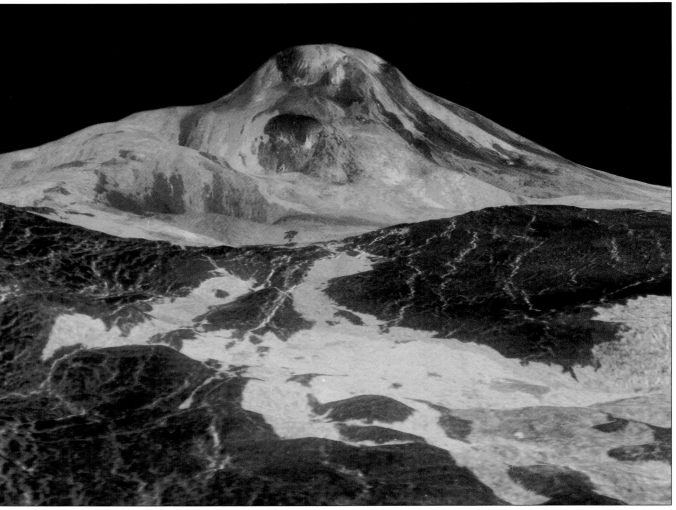

1 *A three-dimensional perspective view from Magellan radar data of the highland region of Ishtar Terra and Lakshmi Planum in the northern hemisphere of Venus. Precipitous slopes and high mountain ranges abound; Danu Montes, one of the ranges, rising (top left center) to over one mile (1.5km) above the plateau, which is itself at an altitude of 1.5 to 2.5 miles (2.5 to 4km).*

2 *Magellan being readied for launch from the shuttle.*

decision on whether it would be boosted by the Centaur-G or less powerful IUS upper stages (neither even built while Galileo was being designed) produced a near-farcical series of changes. At one stage, separate orbiter and atmospheric probes were to be used. The on-off decisions by Congress about the building of Centaur kept forcing drastic changes, with attendant delays and cost increases. Finally, in summer 1982, Congress approved the large liquid stage (Centaur), and Galileo, Magellan and Ulysses engineers could at last proceed with confidence.

May 1986 was to have seen two Shuttle-Centaur launches: Challenger with Ulysses followed five days

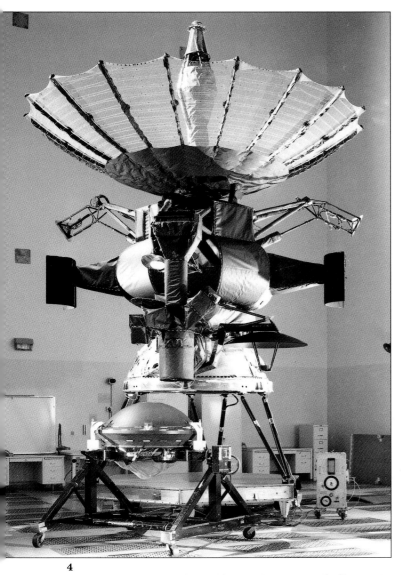

4

later by Atlantis with Galileo. Of course, Challenger's loss four months before halted everything and, as described in the Magellan section above, the potentially explosive Centaur was subsequently banned from use with the shuttle, forcing a return to the IUS! NASA also decided not to attempt two complex launches in the same month, and agreed with ESA that JPL's craft would take precedence for the October 1989 Jupiter window and that the European solar probe, traveling via Jupiter, would wait a year.

Fly-by of Earth

The energy generated by the IUS was insufficient to fling Galileo directly to Jupiter, but JPL's mathematicians were able to work out a path using one Venus and two Earth gravitational swingbys, effectively using the planets as upper stages. The 2.56 metric ton space-craft was released into Earth orbit by Atlantis and her crew on October 18, 1990. The 2.22 ton orbiter section looks very much like Voyager, with a large, 16ft (4.8m) radio

dish, central bus housing the electronics, a boom carrying the nuclear power packs and a scan platform for the scientific instruments. Galileo carries a more sophisticated camera than Voyager, and for the first time includes a "Near-Infrared Mapping Spectrometer." NIMS will be able identify the mineral make-up of Jupiter's moons by recording their tell-tale infrared reflections of sunlight.

Galileo carries a 1.2 metric ton propulsion module (provided by Germany) that will not only brake the spacecraft into orbit about Jupiter but also change its path for repeat visits to the four major moons. Completing the configuration is the 763lb (346kg) probe that Galileo will release into Jupiter's atmosphere.

The spacecraft's one Venus encounter came on February 10, 1990, when it skirted past at 10,000 miles (16,100km) to pick up an almost 5,000mph (8,000kph) increase in speed. The opportunity was not wasted and the clouds were imaged, while other instruments probed the atmosphere at the highest resolution yet and at new wavelengths. Most of the data were stored in Galileo's recorders, however, because the main antenna was still furled. (As the spacecraft was originally designed to head away from the Sun, and not towards it, one of the measures to protect it against high temperatures was to keep the antenna closed, like an umbrella, until May 1991.) The stored information was transmitted through a small, general-purpose antenna in November 1990 as Galileo neared for the first Earth fly-by: it swung within just under 600 miles (960km) on December 8. The cameras made one exposure each minute for a day, from which a 1,500-frame movie sequence was to be assembled, showing Antarctica from a unique perspective. The second Earth fly-by followed exactly two years later, adding new observations of the Moon's north polar region. NIMS was used to look for signs of frozen water in permanently shadowed areas of the Moon, a discovery that would revolutionize the future of space exploration. Apart from providing oxygen for humans to breathe, it could be processed into rocket propellant, which accounts for a major proportion of the cost of getting into space.

Antenna Problem

Following the successful encounter with Earth, Galileo ran into a major problem. In April 1991 the furled high gain antenna was ordered to open but jammed less than half-way open. It was concluded that several of its 18 ribs were still in their launch positions on the central mast – with insufficient lubricant a major cause. It was hoped that exposure to higher and colder temperatures during the months ahead would solve the problem but it was still unresolved by the end of 1992. However, gloom at the prospect of a severely restricted flow of data if all attempts to free the antenna failed eased at that time with the announcement from JPL that the majority of objectives could be met by using the low-gain antenna – but increasing the data bit-rate ten times, by compressing the data and by capturing a greater portion of the antenna's transmissions by maximising the sensitivity of the Deep Space Network antennae on Earth (as was done at the end of Voyager 2's exploration). It was still hoped, however, that the high gain antenna would be freed by procedures planned to be taken around the time of the second fly-by of Earth in December 1992.

Disappointment at the high-gain antenna problem was countered by the success which attended the first ever close encounter with an asteroid occurring on October 29, 1991. Ground based observations of the asteroid Gaspra had located it to within 124 miles (200km) and Galileo's own navigation images narrowed this

5

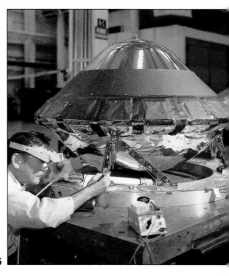

6

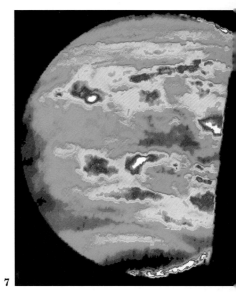

7

5 *Summer 1983 in New Mexico: the Galileo probe is dropped from a height of 30km to test heat shield separation and parachute deployment.*

6 *The probe being worked on by technicians.*

7 *A Galileo view of Venus in the near-infrared obtained on February 10, 1990. The false-color image records radiant heat from the lower atmosphere shining through the sulfuric acid clouds; the thinnest appearing red and white.*

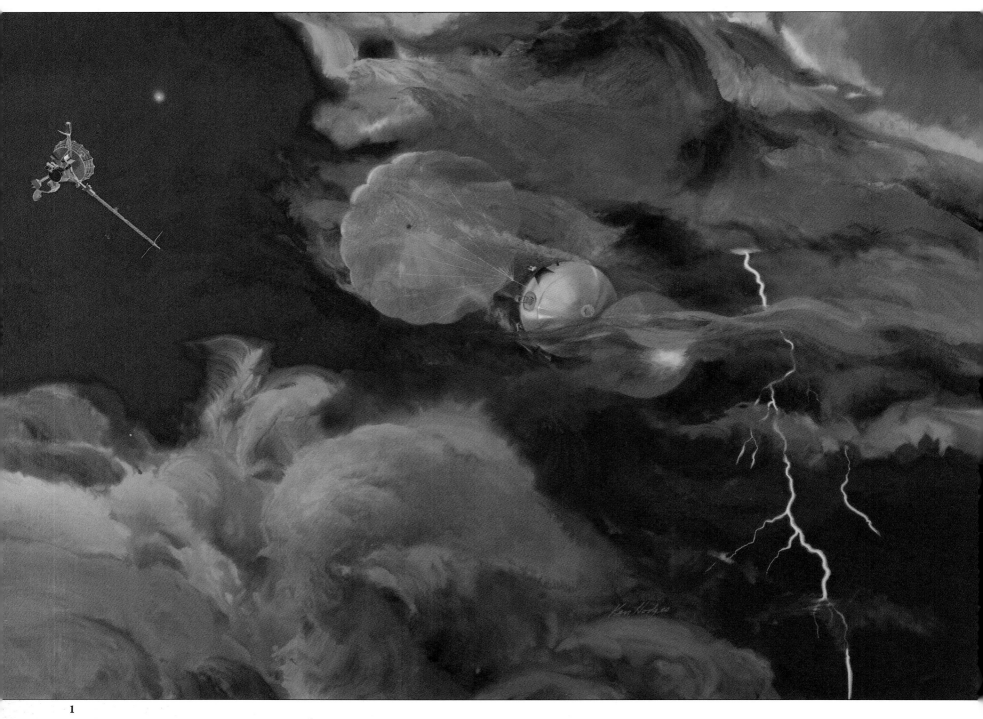

1

1 *An artist's rendition of the Galileo probe's entry into the atmosphere of Jupiter. With the Sun a point of light beyond the orbiter at top left, the heat shield is below the probe with the parachute already deployed. According to current plans, this event should occur in December 1995, when the peak acceleration of the probe will be 115,000mph (185,000kph). The artist here emphasises the violence of the Jovian atmosphere.*

2 *Another probe, another planet, another time: an impression of the Cassini Orbiter, with Saturn in the background, as the ESA-built Huygens probe begins its descent into the atmosphere of Saturn's moon Titan. Unless delays occur, this should take place in June 2005.*

down to about 31 miles (50km) although, because of the possibility of damage from debris around the asteroid, the probe approached no closer than 994 miles (1600km). Encounter took place over 400m km from Earth and a total of 150 images together with other scientific data were obtained and recorded for playback during the 1992 Earth flyby. One low resolution colour image was trans-mitted via the low-gain atenna in November 1991 and revealed a 19x12x11km irregularly shaped body with craters of up to 2km in diameter.

December 1992's precision Earth flyby, at an altitude of only 200 miles (320km), added over 8,000mph (13,000kph) to Galileo's speed, launching it straight towards Jupiter in that month's normal window. En route, it will make a small diversion as it shoots through the Asteroid Belt to come within about 620 miles (1,000km) of a second asteroid, Ida, on August 28, 1993. Heading straight at Jupiter, Galileo will release its probe in July 1995 and then make a small burn to avoid burning up in the atmosphere itself. The probe will hit Jupiter's atmosphere in December 1995 at a staggering 115,000 mph (185,000kph), warding off the 13,532°F (7,500°C) temperature with a thick heat-shield. Peak deceleration will be around 400 times Earth gravity. Releasing its protective covering, the remaining 260lb (118kg) module

will descend below an eight foot (2.5m) diameter canopy, taking readings very much like those of the Pioneer Venus probes. Overhead, the main craft will receive the data for relay to Earth. Unlike Pioneer, however, the Jupiter probe is not a pressure vessel, and it is expected to be crushed after 60-75 minutes when it has penetrated to a pressure of 20-25 Earth atmospheres.

Soon after, the remaining section will fire its main engine to become the first man-made orbiter of the giant planet, beginning 10 looping orbits over 22 months for repeated visits to Callisto, Ganymede and Europa, using each swingby to set up the next. Voyager's navigational feats pale beside those planned for its successor. Galileo will have only the one chance for a close approach to the fascinating inner moon Io: although the spacecraft is protected against radiation, it would find repeated journeys through Jupiter's intense inner radiation belts too damaging.

Cassini to Saturn

There was a proposal at one stage for assembling a second Galileo from the spares and using it as a Saturn orbiter. The suggestion was lost in the program's rising costs and, anyway, would have represented old technology by the time it could have flown. Instead, NASA preferred to design a new spacecraft, although it still bears a resemblance to its predecessors. This Mariner Mark II design is intended for the more demanding missions beyond Mars, but its cost was a major stumbling block despite the use of equipment from Voyager and Galileo wherever possible. The agency asked for funding year after year, only to be turned down by Congress. JPL responded by combining the work on two very different missions, as a result cutting the total $2,100m cost by $500m. The tactic worked and Congress approved the projects in 1989.

One of the missions approved was the Cassini Saturn orbiter (named for a feature of the planet's rings, in turn named for its astronomer discoverer) that is now scheduled for launch in October 1997. Some months after reaching Saturn more than seven years later it would release an ESA-built Huygens probe (named for another astronomer) into Titan's tantalising atmosphere. Cassini is to be launched by the energetic Centaur upper

2

stage which means departure on the USAF's new Titan 4 launcher and not the manned shuttle – but even then it will need gravity assists from Venus and Earth.

As currently planned, Cassini will flyby Venus twice – in April 1998 and June 1999 – and Earth in August 1999. If fuel allows, it may examine an asteroid before swinging by Jupiter in January 2001. Although Cassini will come no closer than about 2.3m miles (3.7m km) to the planet, it will be able to join Galileo – if the aging orbiter is still working – in joint studies, perhaps returning stereo views. The Jupiter swingby is necessary to bend the flight path for Saturn arrival in November 2004. ESA's 423lb (192kg) Huygens entry probe will not be fired into the gas giant's atmosphere but instead into the more interesting atmosphere of the moon Titan, which Voyager 1's brief encounter showed to warrant a spacecraft of its own. Rather like Galileo's probe, Huygens, when it begins its descent in June 2005, will brake in the thick nitrogen atmosphere, analysing its chemical composition, cloud structure, wind speeds, temperature and pressure. It is possible that the probe will continue transmitting after splashing into a methane-ethane sea. The Cassini element of the mission is not scheduled to end until about the end of 2008.

Constant budgetary pressures have resulted in modifications to the original Cassini Mariner Mk II concept. There has been a substantial weight reduction – in case the improved solid rocket motors for the Titan 4 are not ready – and the established practice of having science instruments mounted on a steerable platform has been abandoned in favor of fixed instruments that will be pointed by spacecraft attitude changes. In addition, the scheduled flybys of Titan by the orbiter have been reduced by around one-third in number and the total cost of developing the spacecraft and associated systems has been cut by $250m.

But at least Cassini survived after the initial two-mission Congressional approval in 1989. The second did not. CRAF (Comet Rendezvous/Asteroid Flyby) was an imaginative and exciting project in which a Mariner Mk II spacecraft would be launched in 1996 to rendezvous in 2003 with Comet Tempel 2. Unlike the armada that was sent to Comet Halley during the last visitation, CRAF would accompany the comet from near the time of aphelion (its furthest point from the Sun) until after it had rounded the Sun. By the conclusion of the mission in 2005 our knowledge of comets – which it is generally believed date from the beginning of the solar system and thus provide a unique insight into its evolution – would have been increased enormously.

Alas, NASA was forced to abandon the mission in January 1992 after the White House deleted it from the FY 1993 funding request. The reason was to safeguard Cassini at a time when the politicians were voicing growing concern at the cost of ambitious planetary ventures.

No more planetary missions have been approved beyond Cassini, although there is no shortage of suggestions. As described earlier, NASA is planning a lunar orbiter to launch, possibly in 1997, to map the Moon's resources from polar orbit. 1992's Mars Observer (also see earlier) could be followed by MESUR, intended to pepper the surface with a number of small landers. Rover and sample return missions are very attractive, but are likely to suffer from limited budgets. The Near-Earth Asteroid Rendezvous (NEAR) mission for close study of one of these Earth-approaching bodies is under serious consideration.

Early on, NASA's next planned Mariner Mk II mission was scheduled to be Rosetta – a joint mission with ESA which both sides had been investigating since 1984. This was to have gone one better than the CRAF plan by landing in 2007 on a relatively fresh comet, collecting a range of samples, storing them at below -166°F (-110°C) and delivering them to Earth in perfect condition three years later – a daunting task! In 1991, however, NASA decided to support an alternative major mission – PF/NO or Pluto Flyby/Neptune Orbiter. ESA was left with the prospect of endeavoring to go ahead on its own with a more limited mission to an asteroid or comet; hoping to re-initiate collaboration with NASA on a comet lander mission; or to shift emphasis to another "Cornerstone" mission that had been a rival of Rosetta: this was FIRST – the Far Infrared Space Telescope.

Other proposals abound. Solar Probe would penetrate to within a scorching 2.5m miles (4m km) of the Sun and TAU would travel ten times further than Voyager on a 50-year trip into interstellar space. It is all a question of money – and it is money that lies at the root of project cancellations and program changes. In the case of NASA in particular, this can have a major effect on the health of its relations with other national space agencies such as ESA.

3 *The Cassini-Huygens joint NASA/ESA mission to Saturn and its moon Titan was the only formally-approved future interplanetary NASA mission as the Bush Presidency ended. Mission concepts abound but cost is a major problem, although imagination is not. The TAU mission (Thousand Astronomical Units) would be launched from the space station and travel almost 100 billion miles (160 billion km) in an astrometric survey of the stars.*

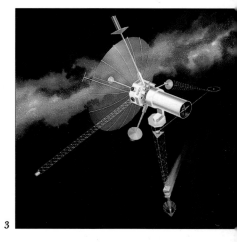

3

4

4 *An artist's impression of the Mars Observer spacecraft in orbit. Launched in September 1992, the spacecraft was scheduled to arrive at the Red Planet one year later; America's first visit there since 1975. The Observer carries instruments and cameras to investigate the Martian surface, interior, climatology, atmosphere and magnetic field over one Martian year – 687 Earth days.*

13

MIR AND MARS: THE ERA OF GLASNOST

On February 20 1986 – a few days after the Challenger tragedy – the Soviets launched the Mir space station. Mir (the Russian variously for "peace," "community" and "world") was a modified Salyut, the forward docking port having been replaced by a multiple unit (up, down, right, left and forward). It was designed to be the core of a modular space station.

The launch came at a time of change for the Soviet Union. In the late 1970s and early 1980s, Brezhnev and the other Politburo members were becoming increasingly old and feeble. When Brezhnev died in November 1982, he was replaced first by Yuri Andropov and then Konstantin Chernenko. On Chernenko's death in March 1985, Mikhail Gorbachov was named

1 *Yuri Romanenko (foreground) and Alexander Leveykin work on the Mir-Kvant 1 docking mechanism. When the module was launched on March 31, 1987, a hard dock could not be effected and an EVA was required to solve the problem.*

Communist Party General Secretary. At 54, he was the youngest to hold that position since Stalin. He moved quickly to end the years of decline and corruption. One of his reforms was called *glasnost* (Russian for "openness"): the Soviet press could now discuss (within limits) such formerly classified subjects as fires, accidents – and the Soviet space program.

Glasnost was reflected in Soviet press coverage: for example, Mir's launch was not only televised, but detailed information was provided on its design. This

continued with the launch of Soyuz T-15. The date, time, and the crew of Leonid Kizim (commander) and Vladimir Solovyov (flight engineer) were announced the day before. When launch came on March 13, 1986, it was televised live. Before this, details of launches and crews had almost always not been announced until after spacecraft had reached orbit. (The only other Soviet launches that had been televised live were Soyuz 19 [ASTP], Soyuz T-6 [France] and Soyuz T-11 [India].)

Kizim and Solovyov had a complex mission involving both Mir and Salyut 7. After docking with the new station, they began activating its systems and conducting experiments. Progress 25 and 26 brought up supplies and fuel. Mir had been launched into an orbit which paralleled Salyut 7's. On May 3, it was announced that Soyuz T-15 would leave Mir and maneuver over to Salyut 7, and the transfer took place two days later. Kizim and Solovyov were to complete some of the experiments planned for the shortened Soyuz T-14 flight. On May 28 they conducted a spacewalk to assemble a long frame, which was unfolded, then folded back. A second spacewalk was made on May 31: this time instruments to monitor the vibration of the frame were mounted on it.

While the cosmonauts were aboard Salyut Mir was also busy, and Soyuz TM-1 docked with the station on May 21 at the beginning of a nine-day unmanned test flight. The Soyuz TM contained much new equipment: an improved rendezvous system, computers, inertial guidance units, fuel tanks, electrical and hydraulic systems, main engine and abort rocket, together with a lighter parachute and new spacesuits for the crew.

On June 25, Soyuz T-15 undocked from Salyut 7 and maneuvered back to Mir. Once aboard, the crew reactivated its systems and remained aboard until July 16. Both the final undocking and landing were covered live. The flight had lasted 125 days. Because the add-on modules were not yet ready, the Soviets announced that no more manned flights to Mir would be made for the rest of 1986. Salyut 7/Cosmos 1686 was boosted into a high storage orbit, thereby removing any possibility that it might interfere with Mir, and allowing possible re-manning and/or recovery at a future date. (In the event, the combined vehicles made an uncontrolled re-entry into the atmosphere on February 7, 1991, with some debris being scattered over parts of Argentina and Chile.)

There was no choice other than to have Romanenko and Leveykin make an unplanned spacewalk to attempt a repair, which took place on April 11. The crew left one of the ports and moved to the rear of the station, where ground controllers extended the probe to allow an inspection of the port. The crew saw that a hard dock was being prevented by a white plastic bag that had become caught in the port. The bag had been loaded on Progress 28 and, when the supply craft had separated, the bag had been caught in Mir's hatch. Once it was pulled free, Kvant 1 could complete the docking. The new module had a rear port to allow a Soyuz or Progress to dock and where fuel could also be transferred. On April 12 the tug separated and was boosted into a higher orbit. (It had been planned to de-orbit the tug but the second docking used too much fuel.)

Kvant 1's arrival took place at an opportune time. In February a supernova had been discovered in the Large Magellanic Cloud (a satellite galaxy of the Milky Way) and Kvant's instruments observed the supernova on June 9, with X-rays from the supernova first being detected on August 10. The cosmonauts made space walks on June 12 and 16 to add a third solar panel to

2 A sure sign of glasnost: a picture ("declassified" according to the caption) showing the assembly of Mir space station sections at the Krunichev machine building plant. Such a release would have been unheard of a few years before. At left is the incomplete docking adapter.

3 The Mir complex photographed in the summer of 1988, when Soyuz TM-5 was docked to the Kvant 1 astrophysics module. The multiple docking adapter and large solar arrays can be seen clearly at the other end, together with an unidentified unit equipped with its own, much smaller solar arrays. Because of the angle of the Sun, these arrays cast two deep shadows which run the whole length of the station.

3

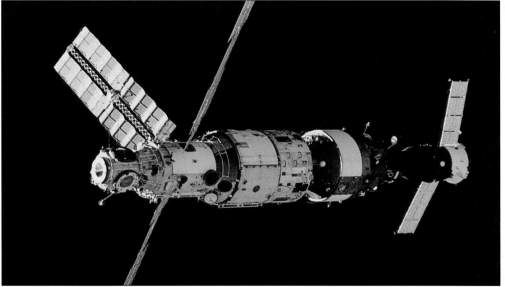

4

5

Manned flights to Mir resumed with Soyuz TM-2 on February 5, 1987, and the crew of Yuri Romanenko (commander) and Alexander Leveykin (flight engineer) would spend 11 months in orbit. Soviet information policy was a mixture of NASA-style openness and traditional Russian secrecy. The crew was named on January 28 and the launch date was announced the day before. The crews' preparations, final press conference and the lift-off were all carried live on TV. On the other hand, it was a year before the Soviets acknowledged that Romanenko and Leveykin were the backup crew: Vladimir Titov and Alexander Serebrov had been the prime crew, but Serebrov failed his pre-flight physical.

The first step in the expansion of Mir began on March 31, 1987, when a D-1 booster placed Kvant 1 (Russian for "Quantum") into orbit. It carried X-ray telescopes and other astrophysical experiments from Great Britain, West Germany, the Netherlands and the European Space Agency. Kvant was attached to a tug which maneuvered it towards Mir. On April 5 it was nearing Mir, but at a distance of 656ft (200m) the thrusters failed to slow it down and Kvant 1 sailed past. Soviet engineers analyzed the situation and decided to make a second docking attempt which took place on April 9, when Kvant made a soft docking at the rear port. When the docking probe was retracted, however, a hard docking could not be completed.

Mir. It was on a collapsible frame and did much to ease an electrical power shortage, since Kvant had no panels to supply its experiments and had to draw power from Mir.

The crew played host to two crews. The first was Soyuz TM-3 on July 22, 1987, which carried the first Syrian cosmonaut, Mohammad Al Faris. The two other crewmen were Alexander Viktorenko (commander) and Alexander Alexandrov (flight engineer). Originally it was planned that the Soyuz TM-3 crew would stay together. During the early summer, however, minor heart problems had been noted in Leveykin and ground control decided to bring him home, replacing him with Alexandrov. The crew returned in the Soyuz TM-2 spacecraft on July 30. Leveykin, looking pale from his 174 days in orbit, was the last one removed from the Descent Module. Ironically, post-flight examination indicated the fears had been groundless and Leveykin was returned to flight status.

The second flight (and Romanenko and Alexandrov's replacement) was launched on December 21, 1987. Soyuz TM-4's crew was Vladimir Titov (commander), Musakhi Manarov (flight engineer) and Anatoli Levchenko (researcher). Titov and Manarov were to take over Mir and spend a year in space, while Levchenko was a test pilot with the Soviet shuttle program, his eight-day flight (like that of Volk on Soyuz T-12) being intended to provide

4 Mir is here seen from a different angle and against the blackness of space. Attached at one end is Soyuz TM-3, which docked with the station in July 1987. One of the long-stay cosmonauts, Alexander Leveykin, already aboard Mir at that time, appeared to have developed heart trouble (fears later proved groundless) and returned on July 30 in Soyuz TM-2, from which this picture was taken.

5 A close up of a section of Mir's solar arrays.

ternal politics of the Soviet space program. Development began in mid-1976, only two years after cancellation of the N-1. In terms of payload and thrust, Energiya and the N-1 were similar: both could orbit about 100 metric tons. Like the N-1, Energiya was developed by what had been called the Korolev Design Bureau, which had been renamed NPO Energiya in 1974 and was now run by Valentin Glushko. What is doubly ironic is that, after years of rejecting it as a propellant, Energiya's designers chose liquid hydrogen for the core stage. The N-1, which Glushko had opposed and interfered with, was Korolev and Mishin's rocket. Energiya was Glushko's rocket.

During the final preparations for Energiya's launch, Gorbachev toured Tyuratam and gave a space policy speech in which he said:

experience and orientation. Launch and docking took place without mishap (unlike Titov's first two flights). The spacecraft were switched and the returning astronauts came down on December 29. All three crewmen spent different times in space: Romanenko 326 days, Alexandrov 160 days and Levchenko just under 8 days. Romanenko's total time in space was now 430 days. Sadly, Levchenko died on August 6, 1988, from a brain tumor.

Energiya Booster

Not all Soviet space activities during 1987 took place aboard Mir. On May 15 came the long-awaited first launch of the Energiya booster. The superbooster consisted of a central core stage with four strap-on boosters. The payload rode on the back of the core stage in the same way that the US shuttle orbiter was mounted on the external tank. This allowed Energiya to serve both as a heavy-lift booster and as launch vehicle for the Soviet shuttle.

In addition to being a counter to the US shuttle program, Energiya also gives an insight into the in-

"We have grown lax in recent years and we have put up with disorder, a slackening of discipline and backwardness. This has resulted in great losses. This must not happen again …. There are shortcomings and oversights and it is therefore necessary to make good the lost ground …. We must switch more boldly from experiments and research work to the planned and wide-ranging application of existing opportunities in the interests of the country's socio-economic development. We expect considered and viable proposals on expanding the application of the achievements of space technology in the national economy."

It was a clear indication that prestige was no longer the sole reason for space activities: *applications* were now to be stressed.

Energiya's launch came at 7:30 pm Moscow time on May 15, 1987: 30 years and a half hour after the first SS-6 launch. Liftoff, strap-on separation and core stage shut-down were normal. Still short of orbital velocity, the payload separated and prepared to fire its own engines. The payload, a long, tube-shaped object weighing 100 metric tons, was designed by the Chelomei

1 *Energiya, a heavy-lift launcher capable of orbiting 100 metric tons. In this picture, the Soviet shuttle ("Buran" – snowstorm) is mated to Energiya prior to launch in November 1988.*

2 *Energiya on the pad prior to its first flight in May 1987. For the first time liquid hydrogen was used as a propellant.*

3 *Buran being prepared.*

Bureau. Rather than being merely ballast, it was a large space station core called Polyus (Russian for "pole") which had taken two years to develop. Although the core lacked life support, it did have a complete attitude control system. Unfortunately, a faulty circuit in an on-board instrument prevented Polyus from reaching orbit. It made a spectacular re-entry and burned up over the Pacific. Despite the disappointment, the Soviets showed films of the launch the following day. (It had taken two decades for the Soviets to show photos of the D-1 booster!)

Commercial Program

During this time, the Soviets were moving into the commercial market. Following the loss of Challenger, as well as the Delta, Titan, and Ariane failures in 1986, the Soviets offered to provide launch services. This included not only launching satellites but also flying experiments and even Western cosmonauts to Mir. A special ministry, called Glavkosmos, was set up to handle the marketing efforts. This attempt to break into the marketplace required the Soviets to release the number of launch failures of the D-1 booster. Between 1970 and August 1986, 97 attempts had been made with seven failures. Later, the Soviets released data for other boosters:

Name	Successes	Failures
A-1 (Vostok)	88	1
A-2 (Soyuz)	554	12
A-2e (Molniya)	179	10
C-1 (Cosmos)	317	14
F-2 (Tsyklon)	61	2
J-1 (Zenit)	13	0

(A-2 January 1972 through December 1987; J-1 1986 to August 1988; all others January 1970 through December 1987.)

The Soviets' commercial launch program ran into an initial roadblock: the US Government was unwilling to give export licenses to satellite manufacturers. As long as the launches were made from Soviet soil, the US feared that satellite technology would be stolen, and it would be several years before Soviet efforts would find a way around the problem.

Kvant – and Afghan Visitor

Meanwhile, Titov and Manarov settled in aboard Mir in late December 1987. Observations of the supernova continued with the Kvant 1 (later in their mission the cosmonauts would have to make an EVA to work on one of the X-ray telescopes) and other medical, scientific and Earth resources investigations were also carried out. The need to switch the Soyuz TM (and political factors) meant three guest flights in 1988. In two cases, it was a repeat for the country involved. First was Soyuz TM-5 on June 7. It carried the second Bulgarian cosmonaut, Alexander Aleksandrov (no relation to the Soviet cosmonaut of the same name), who had been on the backup crew for the aborted Soyuz 33 flight in 1979 when Bulgaria had been the only guest country to have its mission fail. The flight lasted a total of 10 days, a record for an Intercosmos flight, and activities comprised a mixture of experiments and speeches before the return to Earth on June 17.

In February 1988, the Soviets and Afghanistan signed an agreement to fly to Mir. With the normal 18-month training cycle, this meant a flight in July 1989. However, the Soviets soon began withdrawing their 100,000 troops. All would be gone before the Afghan flight would be made, and it was not clear whether the Soviet-installed

regime would survive that long. The Soviets therefore decided to accelerate the schedule, the Afghan cosmonaut flying after only six months of training. The launch of Soyuz TM-6 came on August 29. Its crew was made up of two Russians: Vladimir Lyakhov (commander) and Dr. Valeri Polyakov (research cosmonaut), and the Afghan crewman was Abol Mohmand, an Air Force pilot with combat experience. Polyakov was to stay aboard Mir when Lyakhov and Mohmand returned to Earth.

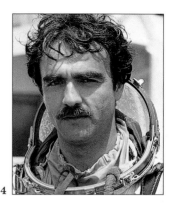

Once aboard, Mohmand read a solemn passage from the Koran and spent much of his time photographing Afghanistan (also, it was speculated, looking for mountain retreats of the anti-communist guerrillas). Because of the speed at which the flight had been planned, there were no Afghan-developed experiments **4**

5

on board so Mohmand had to use Soviet or Bulgarian equipment.

On September 6, the eight-day flight neared its end. Lyakhov and Mohmand boarded Soyuz TM-5 and separated from Mir. Two hours before retrofire, the Orbital Module was separated. This procedure was first used on the Soyuz T flights and saved 10 percent of the fuel needed for the re-entry burn. The retrofire was scheduled to take place at 5:19 am Moscow time over the South Atlantic. Orbital sunset took place about three minutes before this, and the setting Sun confused the primary and backup infrared horizon scanners aboard the spacecraft which were used to orient it correctly for retrofire. Because of the confusion, the computer prevented retrofire. Seven minutes later, the confused readings caused by sunset cleared up and the computer fired the rocket. By this time Soyuz TM-5 was 2,175 miles (3,500km) down the orbital ground track and would have landed in Manchuria, so the crew manually shut down the engine after three seconds. Ground control ordered them to try again after two more orbits. **6**

The second attempt was made at 8:24 am Moscow time. The horizon scanners were disengaged and orientation was set by the gyros. At ignition, the autopilot switched to the backup computer, which performed a pre-programmed maneuver. The engine burned for six seconds and then shut down. Lyakhov quickly restarted it and retrofire continued for about 60 seconds before the navigation computer detected an attitude error and cut the engine. Had the shutdown come a minute or

4 In August 1988 the guest cosmonaut aboard Soyuz TM-6 was Abol Mohmand, a pilot in the Afghan air force.
5 Vladimir Titov (commander) and Musakhi Manarov spent a year aboard Mir during 1988. Here, Titov (right) and his crewmate suit up prior to underwater training.
6 A TM-6 cosmonaut practicing water egress. Unlike the Americans before the advent of the shuttle, Soviet missions always came down on land.

1 November 15, 1988: the Soviet shuttle Buran touches down for a fully automatic landing after two revolutions of the Earth. Despite years of development, the orbiter was far from ready for manned flight.
2 Energiya launches Buran.
3 The first Energiya launch – more than one year before Buran.

1

so later, the crew could have been in grave danger because the re-entry path would have been too shallow; prolonging the heating phase and possibly destroying the capsule.

The crew would have to wait another day before the next landing opportunity, which would occur when the ground track again passed over the Soviet Union. The cosmonauts were in no immediate danger as they had life support for 48 hours. However, they were very uncomfortable. The internal temperature was a cold 45°F (7°C), they only had an emergency food supply and, worse still, the launch and landing spacesuits could not be removed and lacked waste management facilities.

The software problem which caused the original abort was easily bypassed and the third try was made at 4:01 am on September 7. The 230-second burn went normally and the landing was successful. (It was not covered live, however.)

There was criticism of Lyakhov for manually restarting the retrorocket. Clearly, he should have done nothing and let ground control sort things out. But the root problem was the rushed schedule and the lack of thorough planning. The flight plan required the re-entry burn just after sunset, whereas all other Soyuz retrofires had taken place in daylight. Had there been more time, the problems likely to result from this would have been realized. Ironically, the political requirements for the speed-up proved to be incorrect. After the Russian pullout, the Afghan war became a bloody stalemate: the government holding the cities and the guerrillas the countryside.

Buran Flight

On September 29, 1988, a few hours after the launch of the first post-Challenger flight in the US, the Soviets released a photo of the Russian space shuttle on the Energiya booster. Although similar in shape to the US shuttle, it was not a direct copy. Its main engines, for example, were on the booster: this changed the shape of the rear fuselage. The Soviets took the proven aerodynamic design of the US shuttle and adapted it to the Energiya booster.

The Soviets announced that the first test launch of Buran (Russian for "Snowstorm") would take place on October 29. Despite 12 years of development, orbital

3

and sub-orbital tests of the heat shield and flight tests of a jet-powered aerodynamic test aircraft, it was not ready for a manned spaceflight. It had no life support system and the computer displays had not even been designed yet. The guidance control programming allowed only two orbits and a single landing site. Clearly, the all-important electronic systems had minimal maturity. The launch was a political decision: a flight demonstration being ordered even though engineers and shuttle pilots opposed the flight as premature. But, after 12 years of expense, the program had to show results: the risk of a program cancellation must have been as high as that of a launch failure.

The October 29 attempt ended in disappointment. At T minus 51 seconds, the launch computer detected that the crew access arm had not retracted (at launch the arm would be hit by the booster, destroying the rocket) so it stopped the count. The Soviets postponed the launch for four hours, then cancelled it. The second attempt was made on November 15. Buran lifted off, made a quick turn (like the US shuttle) and headed downrange. Two and a half minutes later, the four strap-ons separated in pairs. The core stage burned for eight minutes before shutting down. The shuttle made two burns of the Orbital Maneuvering Engines to place itself into orbit, then made two revolutions before re-entering and making a fully automatic landing on a runway.

2

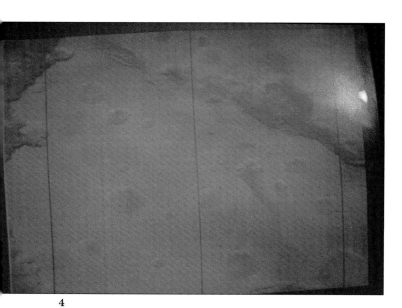

4

5

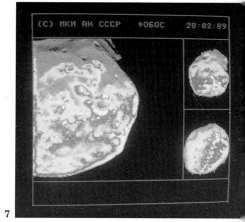

7

they were really attacking was their own (then secret) shuttle program.

Mars and Phobos

Ironically, the one disappointment of the Soviet space program at this time was in the unmanned sphere. The Soviets had resumed Mars flights with the launch of Phobos 1 on July 7, 1988, and of Phobos 2 five days later. In early September, contact was lost with Phobos 1 due to ground control error. An erroneous signal was transmitted which shut down the attitude control system: the solar array drifted off the Sun, the probe lost power and died. Phobos 2 continued and entered orbit around Mars on January 29, 1989. It maneuvered several times until it was in an orbit 218 miles (350km) above that of the Martian moon Phobos, where it imaged both the moon and the surface of Mars.

Phobos 2 then began closing in on its namesake

4 Phobos 2 had a TV camera system aboard and this picture shows an area of the Martian surface. Raw data can usually be improved, so it may be assumed that the detail eventually to be found in images from the Phobos spacecraft will be considerably better than that shown here.

5 A Phobos spacecraft being prepared. The technician affords a good size comparison.

6 Mars' satellite Phobos seen against the planet. This image, although appearing in "conventional colors," was obtained in the near infrared with a stated resolution of under 1,100ft (330m).

7 Three images of the moon Phobos obtained on February 28, 1989. The large, false-color image has a stated resolution of 230ft (70m) and was obtained in the visible band. The two smaller pictures were obtained in the infrared with a resolution of about 1,200ft (370m). All of the pictures emphasize the characteristically rugged surface of Phobos, with evidence of many impacts from meteorites and larger bodies.

6 [caption on image: (C) IKI PHOBOS - FREGAT 28:02:89]

The success brought out some of the internal conflicts and politics of the Soviet space program. Roald Sagdeyev, the retiring Director of the Soviet Space Research Institute, condemned both the US and the Soviet shuttle programs as having "absolutely no scientific value." He added that the two shuttles were "an outstanding technological achievement" but a costly mistake. "We have put too much emphasis on manned flight at the expense of unmanned efforts that produce more scientific information at lower cost," he wrote.

Sagdeyev's comment was a reflection of the beliefs of academic space scientists in the West. (It was a philosophy similar to that of President Eisenhower, too: that space is a matter of scientific interest only, with no political, social or military importance.) It was also a question of control: the space scientists saying, in effect, that only they should decide the goals and direction of spaceflight. Manned programs, on the other hand, were engineering efforts: science had to take second place: it was the Apollo debate all over again. But Buran caused controversy even among Soviet manned spaceflight advocates. This was because it was set apart from the "real" Soviet program. The shuttle pilots were selected separately from the cosmonauts, for example. At the same time, Buran competed for available space funding. For several years, Soviet spokesmen and cosmonauts had been rejecting the concept of a reusable spacecraft like the US shuttle but, in retrospect, it is clear that what

moon. It was intended that the spacecraft would hover 164ft (50m) above the moon, where it would secure images and measurements. It would then drop two small landers onto the surface of Phobos, one of which was mobile and able to flop around the surface. However, on March 27, before this could be accomplished, Phobos 2 fell silent.

Unlike past practice, the Soviets engaged in a public debate over whether it had been a meteorite strike or a failure of onboard systems. It was noted that the designers had to "comply with a set of restrictions relating to funds and the weight and size of spacecraft." The Soviets added: "we are determined to rectify the defects we have identified, notably, those relating to power supply, operational errors and the possibility of probes functioning in an automatic mode."

Spacewalk by Chretien

During this same time (between fall 1988 and winter 1989), Mir was active even if shadows were growing over the Soviet manned space effort. The third of the 1988 guest launches was made on November 26. On hand at Tyuratam were French President Francois Mitterand, Soviet Foreign Minister Eduard Shevardnadze, Western journalists, and a member of the rock group Pink Floyd. The Soyuz TM-7 crew was Alexander Volkov (commander), Sergei Krikalev (flight engineer) and Jean-Loup Chretien (research cosmonaut). Chretien was the

8

8 Roald Sagdeyev, for some years head of the USSR Academy of Sciences Space Research Institute (IKI). A strong advocate of international cooperation in space, he was highly critical of the resources devoted to manned spaceflight.

first guest cosmonaut to make a second flight to a Soviet space station, and launch of the joint Soviet/French crew was covered live. (An added feature was that bane of Western television, commercials, advertising watches and banks.)

The Soyuz TM-7 flight was different from the standard guest flight. It lasted much longer – 24 days – and a spacewalk by Chretien on December 9 was the first by a national of a country other than the US or the Soviet Union. The main experiment was deployment of a collapsible grid structure called ERA. The equipment was successfully attached to Mir, but when the switch was thrown to extend it nothing happened. Volkov and Chretien shook it but it still did not extend, despite the use of much profanity. At this point, Mir passed out of tracking range. When contact resumed via a tracking ship, the cosmonauts reported that ERA had deployed successfully. In fact, Volkov had kicked it several times, a time-honored repair method that was, however, counter to ground control orders. Nonetheless, the kicking had saved a FF50m experiment. Once the work was completed, the ERA was jettisoned and the nearly six-hour-long spacewalk ended. For the rest of the flight the crew conducted various tests and photographed the area of Armenia devastated by a massive earthquake.

Return to Earth came on December 21. Titov, Manarov and Chretien boarded Soyuz TM-6 and separated from Mir, with Volkov, Polyakov and Krikalev remaining aboard. Retrofire was delayed three hours due to software problems. This resulted from a conflict between the new programming (meant to prevent the problems that occurred on the Afghan flight) and the old software. The crew changed the computer program and made a normal retrofire. Only then was the Orbital Module separated (it had been retained to prevent the hardships of the Afghan flight) and re-entry and landing went as planned. Titov and Manarov had spent 365 days, 22 hours and 39 minutes in space, while Chretien had orbited for just over 24 days.

Titov and Manarov suffered no major problems from their year in space. By the fifth day back, they were walking two to two-and-a-half miles (3-4km) and swimming 1,312 to 1,640ft (400-500m) in the pool. The flight also marked the end of the Soviets' steady increase in flight time: Soviet space medicine specialists requested that future flights be kept to six months in length during 1989-91. This would allow them time to conduct ground-based (bed-rest) tests before committing to spaceflight.

Political Opposition

With 1989 came problems that had been building for several years. When Mir was launched in February 1986, it was planned that three modules would be added in the following year: the Kvant 1 short module and two large modules docked to the upper and lower ports. Later, two more short modules would be added on the side docking ports. The Kvant launch and docking were (after a few difficulties) successful. However, a series of problems kept delaying the long modules: 1986, 1987 and now 1988 had passed without their launch.

There were also political problems. Gorbachov did the unthinkable, allowing free, open and contested elections for the new Congress of Soviets, something that had not occurred since the early months of the Russian Revolution. By the time the voting was over, in March 1989, many Communist party candidates had lost, and one of the things opposed by the opposition in the new Congress was the Soviet space program.

Boris Yeltsin, Gorbachov's most important rival,

advocated a five to seven-year-long halt. One Soviet journalist wrote of an emerging "space-phobia."

Both these problems had an impact on Mir. Originally, the first of the long modules was to be launched in late April 1989: the second would follow soon after. With only one module attached, Mir would be difficult to maneuver and would use more fuel. Volkov, Krikalev and Polyakov's training and flight plan were based on this. In due course, they were to be relieved by Alexander Viktorenko and Alexander Serebrov, who would make a spacewalk with a Manned Maneuvering Unit. However, in February 1989 the Soviets announced that the first launch would be delayed until the second half of the year (to increase the likelihood of both modules being launched close together), which meant that the replacement crew would have had little to do.

News of the setback came to the West in an odd way. A team from ABC television was in Moscow to arrange a live television link-up between Mir and the space shuttle Atlantis. The Soviets were uncooperative and finally told ABC that the planned link-up would not be possible as Mir would be unmanned at the time of Atlantis' flight. When it was learned that the Soviets were ending two-and-a-half years of continuous space operations, some Western newspapers and television media reacted as though the entire Soviet manned space effort had been abandoned. There was, however, a little light relief. Krikalev was an Army reservist and for several months he had been sent orders to report for duty. Finally, despite being on duty aboard Mir, he was threatened with jail if he did not report. Tass headlined the story: "Space is no escape from dim-wit bureaucrats, cosmonaut learns."

Mir Unmanned

The Soviets prepared Mir for a long period of unmanned operations. During April, the Progress 41 spacecraft made several engine burns to raise Mir's orbit. These left it in an orbit higher than that of any earlier Soviet space station so as to counter the effects of the solar sunspot maximum, which had greatly increased atmospheric drag. So much fuel was used by the Progress that it lacked the fuel needed to complete its own de-orbit burn. (Normally, Progress spacecraft, as well as Salyuts and space station modules, were de-orbited in a controlled manner over the Pacific, any debris that survived re-entry falling into the Ocean.) The Progress burns had raised Mir's orbit by only 6.2 miles (10km). Clearly, every kilometre possible was needed.

Soyuz TM-7 itself landed on April 27. Volkov and Krikalev had spent 151 days in orbit, and Polyakov 240 days. There were probably several reasons for the recall. Without the two add-on modules, as already stated there would be little for a replacement crew to do. There was also a shortage of electrical power, with the solar panels having lost six percent of their output per year. With new, power-hungry systems added (such as Kvant 1), more was being taken up with housekeeping, leaving even less for experiments. (When they arrived, the two large modules would have their own solar arrays to make up for the loss.) In these circumstances, a manned mission was probably seen as not worth the cost. Moreover, with the fierce, current debate over space spending, a temporary halt in operations might have been regarded as a way to ease the criticism.

Hidden History Revealed

The summer of 1989 was also the 20th anniversary of the Apollo 11 Moon landing. After all this time, the Soviets finally admitted that they, too, had been reaching

1

2

1 French cosmonaut Jean-Loup Chretien during his EVA from Mir in December 1988 – the first spacewalk conducted by a national from a country other than the USSR or USA. Together with mission commander Volkov he spent almost six hours outside the space station, during which time the deployment of a recalcitrant experiment was facilitated by the time-honored remedy of kicking the equipment.

for the Moon. Newspaper articles and a biography of cosmonaut Valeri Bykovsky released details. This included the planned Soyuz 1/2 flight, the N-1 booster failures, Zond plans and the Moon flight profile. A group of US scientists was shown the actual spacecraft which would have made the flight. It was a dramatic climax to a year of unprecedented disclosures. Even so, this information only confirmed much of what had long been suspected by Western observers of the Soviet space program. Using Soviet statements, analyses of their hardware and orbital maneuvers of test flights, James Oberg, Phillip Clark, Nicholas L. Johnson, David Woods, Charles P. Vick, John Parfitt, Alan Bond and others had reconstructed much of the "hidden history" of the Soviet Moon effort. This research had been published in *Spaceflight* magazine and the *Journal of the British Interplanetary Society*.

Indeed, the Soviet disclosures were a direct result of the analysts' efforts. Once the analysts had reconstructed events, the Soviets were willing to give their own account. One early example of this cause-and-effect relationship occurred in April 1986. A year before, in

postponement. Viktorenko and Serebrov continued to conduct Earth resources studies and also installed a new computer on Mir.

The long-awaited Kvant 2 (also called the Re-equipment Module or Module D) was launched on November 26 by a D-1 booster. After reaching orbit, telemetry indicated that the right-hand solar panel had not fully extended, but engineers were able to correct this by turning the panel with its electric motor while rolling the module. Kvant 2 carried a large airlock, the Manned Maneuvering Unit and a shower. A set of gyros provided stabilization and there was a device which split water into oxygen and hydrogen. (Kvant 1 had similar systems.)

Once the solar panel had been extended, maneuvers became easier and Kvant 2 headed for a December 2 docking with Mir, Progress M-1 having separated on December 1. When Kvant had closed to within 12 miles (19km) of Mir, the docking was aborted because Mir's computer had become overloaded. Viktorenko and Serebrov tried to make a manual docking but also failed. The problem was cleared up eventually and a second

2 Many satellites may be seen and photographed from the surface of the Earth with ease. This picture was taken during the late evening of July 19, 1989, as Mir was passing over southern England. During a time exposure the station left a bright trail on the film – to one side of the bright star Vega in the constellation Lyra. Mir was unoccupied at this time. A much fainter trace crosses the track of Mir in the photograph at the same height as Vega: this was left by Salyut 7 as it passed through the field of view six minutes after Mir.

3 Cosmonauts Anatoli Solovyov and Alexander Balandin suit up prior to their TM-9 mission.

4 French cosmonaut Chretien entertains mission commander Alexander Volkov (right) and Sergei Krikalev aboard Mir. Both appear slightly less than enthusiastic!

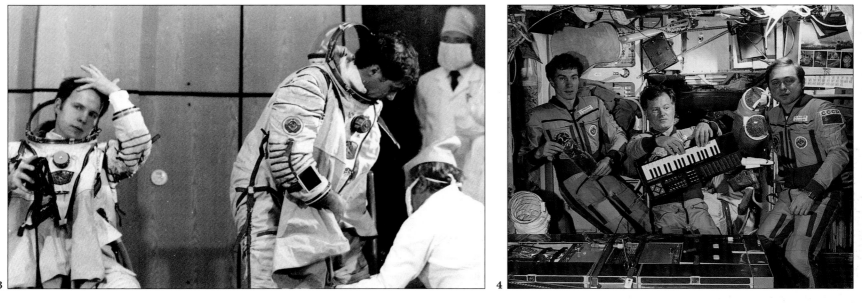

3

4

5

April 1985, Curtis Peebles received a declassified document from the Lyndon B. Johnson Presidential Library. It listed Soviet launch failures from 1957 to 1964. The document was sent to other analysts and published in late 1985. The following April, the Soviets published several articles marking the 25th anniversary of Gagarin's flight. They included details of two unmanned Vostok launch failures which had never before been officially disclosed. Both had been listed on the LBJ document.

Launch of Kvant 2
The halt in space operations lasted through the summer. The first sign of its end was the launch of Progress M-1 on August 23, 1989. This was an improved Progress, with solar panels, an enlarged cargo bay and the ability to transfer excess fuel to Mir. Progress docked at the forward port: something that the earlier version could not do. Soyuz TM-8 followed on September 6 with Alexander Viktorenko (commander) and Alexander Serebrov (flight engineer) aboard. It docked at the rear port after Viktorenko had been forced to take over manual control at a distance of only 13ft (4m) from the station.

Originally, it had been planned that the first of the modules would be launched on October 16. However, a few days before, it was announced that a problem with the module's navigation system had caused a 40-day

try succeeded on December 6: two days later, Kvant 2 was rotated to the upper docking port and the space station assumed an "L" shape.

Progress M-2, launched on December 20, carried the first US commercial payload to be flown on Mir: a protein crystal growth experiment built by Payload Systems, Inc. This was an entirely American experiment with the Soviets having no access to the results. The crew simply turned the package on for 56 days then shut it off. Payload Systems had signed an agreement with the Soviets to fly a further six three-month-long flights for similar experiments. The package was flown on Mir because the station could provide a much longer period of weightlessness than NASA's shuttle.

Five spacewalks were made during January and February 1990. The first two retrieved samples that had been attached to Mir during the French spacewalk. The third (January 26) rigged a mounting bracket for the Manned Maneuvering Unit on Kvant 2. On February 1 and 5 the MMU was flight-tested: Serebrov going first with Viktorenko following. In both cases, the MMU was tethered to Mir. (MMU flights by NASA astronauts have all been untethered because the shuttle could "scoop up" an astronaut in the event of a problem, something Mir could not do.) These tests were the last important activities of the mission.

On February 11, 1990, the replacement cosmonauts Anatoli Solovyov (commander) and Alexander Balandin

5 The scene in the Soviet Flight Center on February 1, 1990, when Soyuz TM-8 cosmonauts Viktorenko and Serebrov were conducting the first spacewalks using a manned maneuvering unit. Unlike use of the American MMU, the crewman remained tethered to Mir. Powered by jets of compressed air, the unit was stored on a mounting bracket fitted to the Kvant 2 module, which had brought up the MMU on December 1, 1989.

(flight engineer) were launched on Soyuz TM-9. As the launch shroud separated, three of the eight insulation blankets on the Descent Module were torn loose. When Soyuz TM-9 docked with Mir, the three blankets were visible on television as they stuck out from the spacecraft. Viktorenko and Serebrov took photographs of the blankets when they undocked in Soyuz TM-8 on February 19: one was sticking out at a 90 degree angle, while the other two were at angles of 60 degrees.

Once the photos were analyzed following the landing, it was decided a spacewalk would be needed to inspect and repair the damage. There were several problems. The blankets protected the Soyuz heat shield from exposure to space so, to avoid any possible damage, the attitude of Mir was controlled to prevent Soyuz TM-9 being exposed to extremes of heat or cold. Another risk was that, when the time came for the return to Earth, the blankets would block the Soyuz horizon and vertical sensors that established the correct orientation during retrofire. The Soviets announced the problem with the blankets to the public on February 23: the equipment needed for the spacewalk would be brought up on the second module.

Despite the problems, the Soyuz TM-9 flight represented another policy change for the Soviet space program. The flight was expected to return a profit: the cost was reported to be 80m rubles, with a planned return of 105m rubles. It was claimed that the Soyuz TM-8 had broken even on its 90m ruble cost, most of the earnings coming from 297 grams of gallium arsenide semi-conductor material (the material used to make computer chips) produced during the mission.

Much of Solovyov and Balandin's time was spent installing new, higher capacity computers on Mir. This took longer than expected, once more delaying the second module launch. During March, the cosmonauts conducted an experiment that was important to the development of future, closed-life support systems (i.e. life support systems where everything is reused). Called Inkubator, it involved the hatching of Japanese quail eggs in weightlessness. The first hatched on March 23 and, by the end of the month, seven had hatched. Sadly, the chicks were frail and unable to feed themselves. It was decided they would have to be put to sleep, but despite the unhappy outcome of the experiment (and an earlier experiment on NASA's shuttle which had also

failed) Inkubator had shown it was possible for an embryo to develop fully in weightlessness. Mir also received two Progress flights: Progress M-2 and Progress 42. The latter was the last of the original version.

Launches from Australia?

That same spring, the Soviets finally seemed to have broken into the commercial launch market. To overcome US opposition required an unusual compromise. The booster was to be the J-1 Zenit ("Zenith"), and the launches would not be from Russia but from Cape York in northeastern Australia. The program would be run by the US company United Technologies, with the satellite and booster checkout being carried out by Australian personnel. The Soviets were to act only as booster suppliers. This arrangement avoided US export restrictions, as the satellite would never leave western control. Approval for the scheme was given by President George Bush in July 1990. The J-1 booster could put 12,919lb (5,860kg) into a geosynchronous transfer orbit: the near-equatorial location of Cape York resulting in it being able to carry a much heavier payload than in a launch from Russia. Construction could begin in 1992, with a demonstration launch of a Soviet satellite in 1994. Commercial launches could begin in 1995, but this would depend on the project proceeding, one problem being the opposition of local aborigines.

Future Soviet plans for commercial launches went

1 *A Progress unmanned supply spacecraft photographed from Mir.*

2 *August 1989: cosmonauts practice EVA procedures outside Mir under water. Although unidentified in this picture, the crewmen were most likely Alexander Viktorenko and Alexander Serebrov, whose Soyuz TM-8 launched on September 6, 1989.*

3 *Another "declassified" picture, this time of the Yuzhmash (Southern Machine-Building Works) in the Ukraine, showing Zenit launchers under construction. The TASS caption recorded that " ... in the not so distant past, workers in one section of the plant were not aware of what was done in the next section – and to mention space rockets outside the gates was regarded as a criminal activity."*

4 *A TV still of Alexander Serebrov making the first EVA using the Soviet manned maneuvering unit, February 1, 1990.*

far beyond the Cape York proposal. During the summer of 1990, the Soviets informally approached the US for permission to launch D-1 boosters from Cape Canaveral in Florida, which would both increase the payload capacity (due to the Cape's proximity to the equator) and avoid US export restrictions on satellites. The Soviets at that time were also considering setting up a D-1 launch site at Cape York or in Brazil but decided that a formal request should not be made until the American government had established guidelines for foreign launchings of US satellites. All of these considerations were very much overtaken by subsequent, dramatic events within the Soviet Union itself.

Docking of Kristall

Back to Mir: the second module, called Kristall ("Crystal"), the name stemming from its materials processing role, was launched on May 31. Originally called Module T (for Technology), it carried four processing units. It was estimated that these could provide 220lb (100kg) of biotechnology materials per year, and the total profit (presumably over several years) was estimated to be 1,000m rubles. At its end were two docking ports: one for the Buran shuttle and another for a small X-ray telescope that would be brought up by the Buran. A third port mounted cameras for Earth observations.

Kristall spent several days maneuvering towards Mir in a fuel-saving approach. On June 6, shortly before docking, one of the small rocket engines began firing for too long. Ground control shut down the system and switched to the backup maneuvering engines, docking taking place successfully on June 10. The following day, Kristall was rotated to the lower docking port. Mir now had a "T" shape but the delay exacted a price: the estimated profit from the flight being reduced by 20m rubles.

As the flight of Solovyov and Balandin neared its close, it was time to make the spacewalk to repair the Soyuz TM-9 blankets. To reach the docked spacecraft, Kristall carried a 20ft (6.1m) long "ladder." The spacewalk was made on July 17, when Solovyov and Balandin spent seven hours in space. They inspected the heat shield and several explosive bolts to ensure they had not been damaged by exposure to space. They then tied down the loose blankets with metal clips. The cosmonauts had to move carefully to avoid damaging the parachute compartment. Once the work was finished, Solovyov and Balandin made their way back to the Kvant 2 airlock only to find that the outer hatch could not be closed, which meant in turn that the airlock could not be repressurized. Finally, they depressurized the entire Kvant 2 module and this allowed them to open the inner airlock hatch. They went through and closed the inner hatch. Once this was sealed, they could repressurize the Kvant 2 and remove their spacesuits. There had been an element of danger in all this because their spacesuits had a limited, self-contained oxygen supply that had to be replenished before they re-entered Mir.

The hatch problem required a three-and-a-half hour spacewalk on July 26. In addition to the hatch repairs, Solovyev and Balandin removed the ladder: their final major task before the conclusion of the Soyuz TM-9 mission.

The replacement crew of Gennadi Manakov (commander) and Gennadi Strekalov (flight engineer) lifted off in Soyuz TM-10 on August 1. Manakov was a veteran test pilot who had made over 300 parachute jumps and was described in Western news reports as a candidate to fly the Soviet shuttle. (He was not, however, a member of the "Wolf Pack" of test pilots connected with the shuttle program.) The Soyuz TM-9 crew

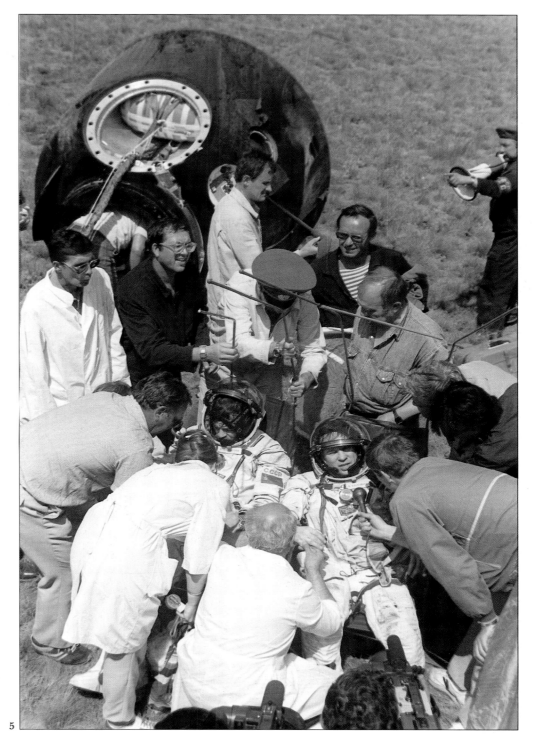

5

returned to Earth on August 9 after 179 days in orbit. Despite concerns in the press over the heat shield, the re-entry went normally.

As 1990 wore on glasnost had the effect of causing flight plans to be announced for several years in advance – in great contrast to the early days of the program. Featured quite strongly in the plans were "guest" cosmonauts who were mostly foreign nationals whose places aboard Soyuz and Mir were secured by hefty payments which went to help balance the heavy cuts in Soviet manned spaceflight budgets which in 1990 were reported to be down 25 percent on 1989. The guests would go up with replacement crews and return with the returning crew from Mir after about eight days. All future Soviet flights were scheduled to be four to six months long (an 18-month flight had been discussed but rejected) and future guests were to come from Japan, Britain, Austria and Germany – with France having no less than five missions on option.

1990 ended with the arrival on Mir of a Japanese TV reporter Toyohiro Akiyama, who accompanied Musa Manarov and Viktor Afanasyev on TM-11 when they launched on December 2 to relieve Manakov and

6

5 August 9, 1990: Soyuz TM-9 cosmonauts Solovyov and Balandin are surrounded at the end of their 179 days in orbit, which included a number of EVAs, one devoted to repair of thermal blankets on the Soyuz which had been loosened during launch and which could endanger a safe return to Earth.
6 Different weather conditions awaited the Soyuz TM-8 crew on their return in February 1990.

Strekalov – the TM-10 crew leaving the space station with their Japanese guest on December 10. It was reported that the eight day stay on Mir had cost Akiyama's employers (the TBS TV station) and other Japanese companies $12m.

Helen Sharman, the first Briton to go into space, followed on May 18, 1991, when she joined Anatoli Artsebarski and Sergei Krikalev aboard Soyuz TM-12. The promoters of the Anglo-Soviet Juno project had run into serious financial difficulties but the Soviets decided that the British woman should go into space as planned. She returned on May 26 with Manarov and Afanasyev in the TM-11 Soyuz. The two TM-12 cosmonauts began a round of EVAs to install outside experiments, retrieve others, repair the Kvant docking antenna and erect a 14m platform on which would be located later an attitude control thruster which would result in fuel economies. On August 20 the Progress M-9 supply spacecraft launched as scheduled – but then events in space were totally eclipsed by what happened back on Earth.

Soviet Union Disappears

A coup against Mikhail Gorbachov was attempted but failed – with Boris Yeltsin playing a prominent role in defeating the coup leaders. Gorbachov's position was so weakened that he eventually lost power and, indeed, by the end of 1991 the Soviet Union has passed into history with the formation of the Commonwealth of Independent States (CIS) – a loose grouping of the former constituent republics of the USSR.

costs. More importantly, it was agreed that the individual states would not impede others from using space facilities in their territories – a clear reference to (largely) Russian space missions being launched from Baikonur in Kazakhstan.

In April 1992 a number of the CIS states formed their own national space agencies but the cosmonaut training centre at Star City remained under the control of the Russian air force – and, while the economic problems were severe, the fact that the Mir, Soyuz, Progress and Buran spacecraft together with the Energiya launcher had been designed by what came to be called NPO Energiya (the Energiya Scientific and Industrial Enterprise) operating under the control of the Russian Ministry of Industry resulted in greater coherence in space activities than might have been feared. Nonetheless there were worrying events – such as a riot among soldiers and three reported deaths at Baikonur resulting from poor food and working conditions allied with rocketing inflation.

Krikalev's Lengthy Stay

On Mir, Artebarski and Krikalev were above the furore both literally and metaphorically – although the latter had known as early as July 1991 that he was to be aboard the space station for an extra tour of duty because of two planned flights being merged into one. Soyuz TM-13 arrived on October 4 with commander Aleksandr Volkov – who was to replace Artsebarski – accompanied by Austrian "guest" Franz Viehbok and Kazakh national

Inevitably the space program was seriously affected by the developments. Amidst numerous and far-reaching organisational changes – and with hyperinflation resulting from attempts at economic reform – "massive confusion" existed in the former Soviet space program, according to a leading specialist in the US Congressional Research Service. But while the problems were many, the fact that most of the major space design bureaux were located in the Russian republic was an advantage, and at the turn of 1991-92 the forces tending to fragment the previously coherent Soviet space program were at least limited to a degree by the setting up of a CIS co-ordinating committee for space activities – the Interstate Space Council – which looked toward some privatisation in the industry and to the states working together on

Toktar Aubakirov, both of whose passages had been purchased. The requirement that a Kazakh pay for his flight was indeed a sign of the (new) times. Artsebarski commanded the TM-12 Soyuz when he returned with the passengers to Earth on October 10.

The Progress M-10 supply vehicle docked with Mir in the third week of October after one approach was aborted because of a computer software problem. Volkov settled in to the routine and together with Krikalev began a lengthy program of observations and experiments in the fields of astrophysics, geophysics, medical studies and materials processing. Both took advantage of a New Year message session to appeal for a continuation of the space program. There was another Progress docking on January 27, 1992, but the early part of the year was chiefly noted for widespread reporting in the West (and particularly in Britain for some reason) that Krikalev – there was little mention of Volkov – was "marooned" in space because the Russians could not afford to fly a mission to bring him back! In fact the two cosmonauts returned to Earth on March 25 aboard TM-13 (as they could have done at any time). Krikalev had spent almost 311 days in orbit and could claim the strange honor of having left the Soviet Union in May 1991 and returned to the independent state of Kazakhstan ten months later when the Soviet Union no longer existed.

The replacement TM-14 crew of Alexander Viktorenko and Alexander Kaleri was launched on March 17 – together with a German fare-paying passenger Klaus-Dietrich Flade who returned eight days later with Volkov and Krikalev. Once again a Progress spacecraft carrying supplies suffered initial difficulties before being able to dock with Mir, and this occurred some days before the TM-15 replacement crew of Anatoli Soloviev and Sergei Avdeiev arrived on July 27 along with French cosmonaut Michel Tognini, who spent two weeks on the space station at a cost to the French space authorities of around $13m. By the time he returned on August 10, 1992, with Viktorenko and Kaleri it was estimated that the total of 23 space launches from the CIS in the year was almost 50 percent down on the annual level of space activity of two years before.

While the future of the CIS in space could not be predicted with any certainty, by the late summer of 1992 things could be considered a little clearer in Russia. The Energiya managers foresaw Mir being operated at least until 1995 – a minimum life span of around ten years. It had been upgraded by the addition of the Sofora thruster package and more modules were to be launched from 1993. Spektr ("Spectra") would have environmental research equipment aboard and Priroda ("Nature") would contain remote sensing equipment. (Expansion of Mir by the additional modules commencing with Kvant in 1987 had led increasingly to a demand for a new means of returning payloads to Earth and this led to the introduction of a small recovery capsule, with a 330lb (150kg) payload, that could be placed in the hatch of the Progress M spacecraft. A Progress, after delivering supplies, retrofired and then ejected the capsule which parachuted to a landing while the main vehicle burnt up in the atmosphere.)

The plans for a Mir-2 station of radically new design were abandoned by Energiya on grounds of cost in favor of simply replacing the central core of Mir-1 with a new and improved version around 1997-98. As an example of the increased costs faced in the space program – particularly where supplies from other republics were concerned – managers at Energiya in August 1992 quoted a computer replacement for Mir-1 which the factory in Kiev (in the Ukraine) was offering for eight times the cost of the original computer supplied for the station.

Before the events of 1991 there had been plans for an unmanned second test flight of Buran by early 1992. However, reports in August of 1992 indicated that while the project had not been cancelled it was continuing at a "low level," which presented problems as it was difficult and costly to keep the shuttles already built in a flyable condition. The Energiya heavy lift launcher was in a somewhat similar position of awaiting missions which could justify its use. The Energiya organisation was in discussion with other CIS states on the possibility of sharing the cost of developing a class "M" Energiya launcher that would be capable of placing around 34 metric tons of payload into a low Earth orbit compared with the existing launcher's 100 metric tons.

Back to Mars

The former Soviet Union's plans for interplanetary exploration were similarly affected by the political events and the resulting economic difficulties. The plans for a return to Mars in 1994 and 1996 had excited considerable international interest. Two improved, Phobos-type probes were to be launched in November 1994 for arrival in Mars orbit the following October. The orbiters would each drop penetrator probes over different areas of the planet. These would relay seismic data and possibly weather data to both the parent orbiters and NASA's Mars Observer, which should be in orbit at the same time. Each (then) Soviet orbiter would also drop a balloon probe. The French-made balloon would be 70ft (21m) high and carry high-resolution television cameras to image the landscape drifting below. At

4 Soviet plans for unmanned missions to Mars in 1994 and 1996 aroused great international interest, despite subsequent events. France undertook to build a balloon that would explore the Martian surface both with cameras and more directly, as depicted in this artist's impression.

5 In the summer of 1992 Russia was marketing its space "know how" world-wide with great energy. Shown here is Professor Yuri Solomonov, of the Moscow Research and Technology Center, at the head of the START project which offered a complete launcher/satellite service into low Earth orbit.

4

night, the lifting gas would cool and the balloon would descend to the surface – the descent stopping when a tail-like "snake" beneath the balloon touched the surface. The snake was to be built in the USA and it was intended that it should conduct soil analyses. It was hoped that the balloons would survive for about ten days and cover a wide area. If the 1994 probes were successful, two more were scheduled to follow during the 1996 "window": one would drop a lander on the Martian moon Phobos and possibly a rover onto Mars. The other probe would continue on to several asteroids.

However, as early as October 1991 it had been decided by IKI (the Russian Institute of Space Research) that only one spacecraft would be sent in 1994 – and that it would drop two small penetrators and two landers. The balloon would be postponed to 1996, when another single spacecraft would take it and a rover-type vehicle to Mars. (A prototype of the rover underwent tests in 1991 and 1992 – the latter tests taking place in Death Valley in the USA).

By early 1992 there were reports from Russia that the 1994 mission might be cancelled because of the economic chaos and resulting lack of progress on the hardware. In the spring, Germany and France were reported to be assisting with funding. IKI stated that the project had "highest priority" but that work was proceeding at an erratic pace because of sharp increases in component prices and because finance was only being provided from the state on a monthly basis. At the time of writing in the late summer of 1992 the fate of the undoubtedly exciting missions was still not clear.

5

14

SPACE FOR ALL:
THE END OF THE SUPERPOWER MONOPOLY

The launch of Sputnik 1 in October 1957 had particular significance for two European countries: the United Kingdom and France. The UK had been conducting scientific research into the upper atmosphere for some time, and during the International Geophysical Year (1957-58) there were no less than seven launches of its highly-successful Skylark sounding rocket from the Woomera range in Australia. Britain was recognized as having the most advanced space technology in Europe at that time, and within a few months was examining the likely success and costs of adapting the Blue Streak IRBM and the Black Knight experimental rocket to form a launcher for placing satellites into orbit. British scientists immediately looked to the US as a means of bridging any delays before a national launcher was available, and their first scientific payload was lofted into orbit aboard Ariel 1 by a Thor Delta vehicle in April 1962; the world's first *international* launch. (The first British-designed satellite as well as payload was Ariel 3 in May 1967 and the sixth and last satellite in the extended Ariel series – all launched by US vehicles – did not fly until 1979.)

The British scientists would indeed have had a long wait for their national launcher. Blue Streak was cancelled as a missile project in 1960, but in 1964 it was decided to develop a three-stage launcher from the Black Knight research rocket. Construction of the vehicle, called Black Arrow, began in 1967, and in October 1971 a demonstration satellite, *Prospero*, was successfully launched from Woomera. Unfortunately, in a manner that seemed all too typical of a vacillating British policy, the program had already been cancelled a few months before. As to gaining independent access to space, the French had secured that no less than six years before.

First French Steps
Charles de Gaulle assumed power in France within a few months of Sputnik's launch and, in a concentrated drive to re-establish a leading position for his country in Europe, it was inevitable that space (linked essentially at first to the development of missile technology as an independent means of being able to deliver France's nuclear deterrent) would be accorded some priority. As a result, when a Diamant launcher (other vehicles bore

1 *About the time of the IGY, the UK was regarded as Europe's most advanced country in space technology. This picture shows the launch by a Black Arrow booster of the Prospero satellite on October 28, 1971. However, the independent British launcher program had already been cancelled and the future for boosters in Europe was effectively handed to the French.*

2, 4, 5 *The UK depended on US vehicles to launch its early scientific satellites – named Ariel. These pictures show the launch of Ariel 5 by a Scout booster on October 15, 1974; of Ariel 2 on March 27, 1964; and of Ariel 1, the first international launch, by a Thor-Delta from Cape Canaveral on April 26, 1962 respectively.*

2

equally delightful names such as "Emeraude," "Saphir" and "Topaze") lifted off from an army base in Algeria on November 26, 1965, and placed the A-1 (Asterix) satellite in orbit, France was the third nation – behind only the two superpowers – to achieve the feat.

The political, economic and technological drive associated with President de Gaulle's policies generally did not mean, however, that building an independent power base for France resulted in an unpreparedness to cooperate – and space was no exception. Within a few weeks of the Asterix launch a US Scout vehicle was launching another French satellite, and by the end of the 1970s about one-third of the French satellites orbited had been launched by US or Soviet rockets. French space policy from the earliest years has been totally consistent: it decided on those areas of space technology where it wished to concentrate and has since devoted the maximum possible national resources to developing them, at the same time as it has succeeded for the most part in convincing other European nations that they, too, should devote resources to exactly the same objectives.

Launcher Developments

Whatever might be possible nationally, it was evident at an early stage that the European countries would need to cooperate if space was not to be left to the superpowers. Beginning in 1960, Britain and France

4

5

3

cooperated in proposing that a European launcher be developed. Talks led to the signing in March 1962 of a convention establishing a European Launcher Development Organization, with Australia (contributing the Woomera range), Belgium, France, Germany, Italy, Holland and the UK as members. Britain was to develop the Blue Streak as the first stage with France and Germany contributing the second and third stages respectively. The other three European members would work on the development of test satellites, guidance systems and telemetry links. The first launch was scheduled for 1966, with the initial program estimated to cost £70m: the equivalent of at least $1,000m in 1992 values. The UK agreed to contribute up to 38 percent of the total cost.

The project did not go well. Individual countries took a long time to ratify the agreement, and integrating the planning and development of the various systems between the member countries, without a prime contractor to exert overall control, was a nightmare. Lengthy delays and increased costs caused the UK to threaten

6

7

3 *A three-stage Skylark research rocket ready for launch from Andoya in Norway. The rocket is still operational and there were more than 400 launches down to the end of 1992.*

6 *France was the third world power after the USSR and USA to achieve an independent space launch capability. This picture shows the Asterix satellite being prepared for launch by a Diamant booster on November 26, 1965.*

7 *The Diamant with Asterix aboard shortly before launch.*

1 *A model of ESA's EXOSAT satellite seen against a projection of the Whirlpool Galaxy in the Canes Venatici constellation. The X-ray observatory operated for three years after its launch in May 1983.*
2 *The European Space Research Organization relied on NASA to supply launch vehicles in the late 1960s and early 1970s. Here a Thor-Delta launches HEOS – Highly Eccentric Orbit Satellite – from Cape Kennedy, 1968.*

1

2

3 *A British Aerospace technician works on the GEOS-2 satellite. It was launched into geostationary orbit – the first European satellite to be placed there – in July 1978 to conduct a study of the Earth's magnetosphere. GEOS was developed for ESA by the STAR consortium led by British Aerospace.*

4 *Washington, D.C., September 24, 1973. NASA Administrator James Fletcher (right) and ESRO Director-General Alexander Hocker sign a communique announcing European participation in the space shuttle program.*

to withdraw from ELDO in 1966. Agreement on a new budget ceiling and a reduction in the UK's contribution to 27 percent staved of the threat for a while, but a review of space policy in Britain during 1968-69 resulted in the government deciding to purchase launchers from abroad (i.e. from the USA). By the end of 1969 its withdrawal from ELDO was complete, despite protests from France and other members that Europe must develop an independent launch capability. The work directed to that end (and ELDO itself) continued until 1973, when the organization was absorbed in a re-structuring of spaceflight research and development activities in Europe. There had been eleven ELDO launches in all – ten of the Europa I vehicle from Woomera and one of the more advanced Europa II from a new French launch site at Kourou in French Guiana – but all had failed to place a satellite in orbit. It was perhaps a sign of what might have been for the future that in every case the British-built Blue Streak first stage performed successfully.

Research in Orbit
Although totally separate, a European Space Research Organization had come into being at around the same time as ELDO. Formed to promote European collaboration in space science and technology, it had ten members – Belgium, Denmark, France, West Germany, Italy, the Netherlands, Spain, Sweden, Switzerland and the UK – with Britain on its formation in 1964 making the single biggest contribution (about 25 percent) to the ESRO budget. In Britain there was a strong body of scientific opinion which would have preferred to channel financial resources into national or bilateral space projects, but the value of ESRO was championed by (among others) Sir Harrie Massey of University College, London, who was one of the leading space scientists and policy makers of the time. ESRO certainly progressed further and faster than ELDO, and during the 1960s a number of projects were developed and various installations were built in member countries such as Germany, Italy, Holland and France. Once again the UK demonstrated its inability to derive maximum advantage from cooperative projects when, as the country contributing the biggest share to the ESRO budget, it contrived to arrange matters so that not one ESRO installation was built on British soil, although a reasonable number of Britons were included in the top management.

Between 1968 and 1972 seven ESRO satellites were successfully placed in orbit by US launchers to study the upper atmosphere, cosmic rays, the Earth's radiation belts, the solar wind and its interaction with the Earth's magnetosphere, in addition to other phenomena. But all was not well with the organization. Scientists complained of vast sums being swallowed up by the creation of technical and engineering facilities – as distinct from going into scientific projects – while costs generally rose. At the same time, there was a strong body of opinion in favor of a unified approach to space in Europe, with a resulting dismay being felt at the lack of success of ELDO in creating a European launcher. Some countries in ELDO reacted to the increased costs resulting from the UK's withdrawal from the organization by reducing their contribution to ESRO. As the 1970s approached policy drifted, and a NASA appeal to Europe to share in some suitable way in post-Apollo developments (including the space shuttle) tended once again to emphasize the differences between the UK and France over the need for an independent launcher. On space policy the continent was indeed in some disarray.

But out of the experiences of the 1960s eventually came a growing awareness of ways in which to proceed

to the future – ways which were charted in some detail at a ministerial meeting held in Brussels in December 1972. In essence it unified launcher development, space applications and space science under one roof, and gave the major countries the chance to develop projects of greatest concern to themselves. A convention setting up the European Space Agency was signed in May 1975 by Belgium, Denmark, West Germany, France, Ireland, Italy, the Netherlands, Spain, Sweden, Switzerland and the UK (who were joined as full members in 1987 by Austria and Norway), with the new body taking over the staff, structure and establishments of ESRO as its initial basis. The first director general of ESA was Roy Gibson of the UK, who had previously been head of ESRO.

All members agreed to contribute to a *mandatory* activities budget, which covered the general budget and the science program, calculated on the basis of their average national income over the previous three years. In addition, they could contribute to optional programs in which they were particularly interested. On the formation of ESA the science budget represented about 15 percent of the total, but it was the optional budget system which eased many of the previous problems. With France making the biggest single contribution, ESA was to support the development of a new, three-stage L3S launcher (which was subsequently to be called Ariane) as an optional program. Germany emphasized a growing interest in manned spaceflight and micro-gravity applications by taking the lead in developing the Spacelab pressurized module and instrument bearing pallets exposed to the vacuum of space that would fly in the cargo bay of NASA's space shuttle. For its part, the UK capitalized on its instrumentation and spacecraft experience by spearheading the development of a maritime communications satellite.

Other countries contributed as they wished to the various optional programs – the investment for all being based on the concept of so-called "just return," with industrial companies in the member countries being expected to benefit according to the national contribution to ESA finances. (It was not intended that ESA itself should develop and build spacecraft, although its engineers and scientists would be responsible for the definition stages of a program.) As a result of ESA as well as national space contracts, over the years a number of companies became recognized as leaders in aerospace technology – for example, Aerospatiale and Matra in France, Dornier and MBB in Federal Germany and British Aerospace and Marconi in the UK.

From COS-B to Giotto
The lengthy processes of ratification meant that ESA did not formally come into being until 1980, but the work done by ESRO caused the new agency's science program to move forward powerfully from the very beginning. COS-B – a gamma ray sensing satellite and ESA's first – was launched in August 1975, and two GEOS spacecraft to study the Earth's upper atmosphere and magnetosphere followed in 1977 and 1978. Those same two years saw two more ESA satellites in space: ISEE-2 (ESA's contribution to the joint International Sun Earth Explorer program with NASA) and the International Ultraviolet Explorer (IUE), a cooperative ESA, NASA and UK project which was still operating in 1991 and which by then was claimed to have generated more scientific papers than any other single satellite.

There was then a gap of some years before the launch of the European X-ray Observatory Satellite (Exosat) in 1983, which built on the work of NASA's HEAO series of spacecraft, followed by the lift off in

1985 of the Giotto mission to Halley's Comet. The latter spacecraft brought ESA perhaps its highest level of publicity ever. It performed admirably when it flew past the comet on March 13/14, 1986, at a distance of just under 600kms (yielding the first clear images as well as much data of the comet's nucleus) and participated in an excellent demonstration of international co-operation in the furtherance of major scientific goals. Nor was that all. The Giotto mission was extended and in July 1992 the space probe flew to within 200km of the nucleus of Comet Grigg-Skjellerup at a distance of over 130m miles from Earth. By so doing, Giotto established a number of records – including being the first deep space spacecraft to be placed in hibernation and reactivated; the first to return from deep space and perform an Earth flyby before being re-targeted to a cometary encounter; and being the only probe to encounter two comets – the second flyby being the closest to be achieved by any spacecraft to date.

work of a trail-blazing spacecraft launched ten years before and which will also, like that earlier mission, be based on extensive international cooperation.

Success of Ariane

This was an impressive performance in space science by ESA – and there was an equally powerful showing in both the critical area of launcher development as well as in the building of applications satellites. In the former, France showed its determination by financing around two-thirds of the costs, and it was rewarded with success at the end of 1979 with the first launch of an Ariane 1, a three-stage vehicle capable of putting 4,078lb (1,850kg) into geostationary transfer orbit. Eleven were to fly in the period to 1986, being then replaced by more powerful variants. The 1990 standard vehicle was the Ariane 4, which had an optimum GTO capability of 9,480lb (4,300kg). Ariane was declared operational in January 1982 and production and launch operations

5 A solar array deployment test of the ECS-1 communications satellite being conducted at the Ariane launch site at Kourou, in French Guiana. The spacecraft was launched in June 1983 and subsequently became operational under the control of the European Telecommunications Satellite Organization (Eutelsat). British Aerospace led a consortium in the construction of the ECS satellites and the company played a leading role in developing communications satellite technology generally for ESA. The picture gives an excellent demonstration of the growth in size of satellites since the early days, although by the early 1990s many were considerably bigger than ECS.

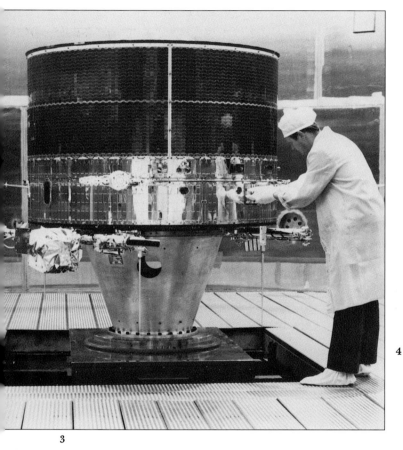

3

4

Also of importance was Hipparcos (High Precision Parallax Collecting Satellite), which was launched in August 1989 to conduct the first space-based stellar survey. It failed to reach geostationary orbit because of the failure of a boost motor, but skilful work by ESA engineers and controllers enabled a modified mission to proceed with a high level of success. ESA supplied the Faint Object Camera as well as the solar arrays on the Hubble Space Telescope which was finally launched from the shuttle in the spring of 1990 – the intital contacts over the project having begun when ESRO was still in being. Later that same year another greatly delayed joint ESA-NASA mission, that of the Ulysses spacecraft, which was to investigate the polar regions of the Sun for the first time, was also launched from the shuttle. The spacecraft successfully flew by Jupiter in February 1992 and gathered scientific data of great importance as well as having its orbital inclination modified by the gas giant so that it would fly over the solar poles as planned. ESA's next major scientific satel-lite to be launched (hopefuly in 1993) will be its Infra-red Space Observatory (ISO), which will build on the

passed from ESA to a specially formed company, Arianespace, which was charged with developing Ariane as a commercial space launcher. The vehicle enjoyed considerable success at a time when US launchers were in disarray following the Challenger tragedy, and in the twelve years to end 1991 there were 43 successful launches and five failures. A measure of its confidence in the future was Arianespace's decision in 1988-89 to place a single order for the supply of no less than fifty Ariane 4 launchers over a period of eight years at a fixed cost of about $2,750m.

Communications Satellites

What was to become known as British Aerospace led a consortium of European companies in developing and building communications satellites for ESA in a program which saw the handing over of the satellites, once operational, to organizations such as the European Telecommunications Satellite Organization (Eutelsat) and the maritime system Inmarsat. The program did not get off to an auspicious start as the first OTS (Orbital Test Satellite) was lost in September 1977 when its US Delta launcher blew up shortly after lifting off from Cape Canaveral. A second spacecraft, however, was successfully launched in May 1978 into a geostationary orbit, from which it was not "retired" until January 1991.

The experience gained led to the development of

5

the ECS (European Communications Satellite) system, with spacecraft about twice the size of OTS and basically offering 12,000 telephone circuits with two TV channels. Five ECS satellites were built for eventual transfer to Eutelsat control, and these were launched between 1983 and 1988, with one being lost as the result of an Ariane failure: the cost of the two last versions of the series was $66m each. Three modified ECS satellites were built for leasing to Inmarsat and two were launched in 1981 and 1985, with one being lost because of a launch failure. Both of the remaining two satellites continued to be operated by Inmarsat in 1991 two years before ESA had launched Olympus 1, a new generation of satellite developed again by a consortium of companies led by British Aerospace. This was the world's largest civil communications platform and was scheduled to conduct a five year demonstration test of various functions, including direct TV services, specialized communications channels, and of operations at 20/30GHz. ESA's plans for the 1990s include the development of data relay as well as navigation and land mobile communications satellites.

In applications, ESA has also devoted much attention to the development of meteorological and other remote-sensing satellites. The first Meteosat geostationary meteorological satellite, with Aerospatiale as the prime contractor, was launched in November 1977 as ESA's contribution to the Global Atmospheric Research Program (GARP), and three further Meteosats have been launched since, with two more scheduled to follow in the period to 1995. The system was declared operational in 1983 and, while ESA continues to operate the satellites, overall responsibility for the system passed in 1986 to Eumetsat (European Organization for the Exploitation of Meteorological Satellites). ESA commenced a major activity in the remote sensing of Earth resources in the summer of 1991 with the launch of its ERS-1 (European Remote Sensing) satellite, a 2.3 metric ton, highly-advanced vehicle with a battery of microwave sensors, which was intended to yield data on land, sea and ice formations, with particular attention being directed at improving understanding of interactions between the atmosphere and oceans. Dornier was the prime contractor for the spacecraft, with significant contributions being made by both Matra and Marconi. To ensure continuity in the flow of data ESA plans the launch of a second ERS satellite probably in 1994 – with discussion of a more advanced environmental satellite (POEM – Polar Orbiting Earth Mission) scheduled to launch in the late 1990s still continuing.

Financial Pressures

As the end of the 1980s neared, ESA members discussed major new initiatives to take the Agency into the next century. All, with the exception of the UK, strongly backed a French lead aimed at achieving European autonomy in space, including manned spaceflight. Three main optional programs were approved. A higher performance Ariane 5 would be developed at a cost of approximately $4,200m. This would be needed to launch a manned orbiter called Hermes, which would be developed for missions beginning in the later 1990s at a cost of about $5,300m. And, thirdly, ESA would develop under its Columbus program a pressurized laboratory module to be attached to the space station Freedom as well as an unmanned polar platform and a free-flier laboratory – the last to be serviced by Hermes. This program was costed at around $4,400m. (To gain experience of operating such unmanned platforms ESA developed Eureca, the unmanned European Retrievable Carrier, which was placed in orbit by the

shuttle in 1992 with two other flights possible in 1994 and 1996.)

France proposed to take up over 44 percent of the Ariane 5 budget and just a little less of that for Hermes, with Germany taking 22 percent and 27 percent respectively. Germany was to contribute 38 percent to the cost of the Columbus program, with Italy's share at 25 percent and France's at just under 14 percent. Italy also decided to make sizable 15 percent and 12 percent contributions to the Ariane and Hermes projects respectively. The costs contemplated were high, and it was agreed that the Hermes and Columbus programs would be scrutinized after three years (early 1991), but the Ariane 5 project was to proceed without any such limitation.

The British Government took a firm stand against what it considered to be an unbalanced policy and huge increases in costs. A debate had been taking place domestically in Britain in the period leading up to the ESA decisions. Roy Gibson, the former Director General of ESA, had been appointed to head a newly formed British National Space Center in 1985 and set to work to develop an overall plan which was presented to the Government. The Government never published the plan but it was rumored that Mr Gibson had proposed at least doubling government expenditure on space from the then current modest level (about $215m in 1987). Mr Gibson's proposals appeared to find little support in the Government and he subsequently resigned.

A few months later (at the end of 1987) the House of Lords Select Committee on Science and Technology published its keenly awaited report on UK space policy. It, too, favored increasing state spending on space, but

1 *August 4, 1984: first launch of the Ariane 3 version of the booster. It successfully orbited a French communications satellite and the ECS-2 satellite for ESA.*

2 *Weather forecasting in Europe benefited greatly from the launch of the first Meteosat geostationary satellite in 1977, with three more following and others planned. Aerospatiale is prime contractor for Meteosat.*

only to around $360m after five years. However, it believed that the UK directed too much money to ESA programs (the proportion was about 60 percent), and that over a course of years the national share should be increased to half. On manned spaceflight, the report was adamant: "For the foreseeable future space offers enough opportunity to telecontrolled craft for the involvement of man in space to be an expensive and hazardous diversion. It is not necessary to put a European in space independently of the Americans ...."

On this at least the UK government agreed, and at the ESA ministerial meeting in November 1987, which approved the new drive toward manned spaceflight in Europe, it was represented by Kenneth Clarke, a minister scarcely noted for either tact or diplomacy. He criticized the proposed policy and announced that the UK could not support any of the three optional programs (although, in fact, it subsequently took a 5.5 percent share in the Columbus program, since British Aerospace had worked hard to take the lead in developing the polar platform). The cost of participation to the UK would amount to about $450m over ten years, and the government also stressed the emphasis which was to be placed on extending the domestic infrastructure in Britain devoted to the remote sensing of Earth resources.

For the while, however, the other ESA members pursued the policy laid down in 1987 and the ESA budget for 1991 reflected this. No less than 48 percent of the total budget of around $2,900m was to be devoted to the development of Ariane 5 and Hermes, with 15 percent going to Columbus and allied micro-gravity projects. The scientific share of the budget was just over ten percent, with telecommunications accounting for over nine percent and Earth observations for about 7.5 percent. In descending order France, Federal Germany, Italy and the UK were the biggest single contributors to ESA, but in terms of total state expenditure on space (national spending as well as contributions to ESA), Britain was a very distant fourth. France led with an expenditure of about $1,750m, followed by Germany (over $900m), Italy ($750m) and the UK with $275m. The significant feature here was the large stride being taken by Italy, which was expanding not only in the context of Europe but in developing joint projects with NASA.

Cooperation with Russia

No matter the ambitious program agreed in 1987, economic realities began to intrude as the decade of the 1990s progressed. In particular, Germany – facing the enormous burden of reunification – voiced concern at escalating costs which in the case of Hermes totalled 40 percent by the end of 1991, although to be fair over half of the increase resulted from a deliberate "stretching out" of the program. An ESA ministerial meeting in November 1991 postponed giving the expected, final approval for Hermes and Columbus to go ahead until a further ministerial meeting one year later.

Against a continuing background of what an issue of the British Interplanetary Society's magazine *Spaceflight* described as "space in recession," ESA's director-general Jean-Marie Luton was instructed by the ministers to review the 1987 plan and by the summer of 1992 it appeared obvious that a significant scaling down would be recommended to the ministers the following November, more especially as even France by this time was calling for ESA budget restrictions. The changes were almost certain to include reducing Hermes to the status of an unmanned "demonstrator" spacecraft, with manned flight (if approved at all) postponed to at least 2005. The development of the Columbus pressurised

module to be attached to the Freedom space station was supported but a decision on the man-tended free-flier was postponed. Participation in manned spaceflight was reaffirmed (a European astronaut centre was being set up in Cologne and the selection of six new astronaut candidates was announced in May 1992) but the opportunities in the near term were seen as being aboard the US space shuttle, with the hope of flights to Mir as well. An independent, ESA manned spaceflight capability was thus to be postponed indefinitely. In addition, Soviet experience was to be tapped further by placing contracts with Russian space research institutions and companies to study in detail the possibility of collaboration between the two sides on the Hermes program.

4

While the mandatory scientific share of ESA's budget is not high, a five percent increase per year over the next five years was agreed at the end of 1990 and this was regarded as signalling a go-ahead for the ambitious Horizon 2000 program, which had been conceived some years before. This program consisted of four major "cornerstone" missions. The first, scheduled for launch in 1995, will comprise a single Soho (Solar and Heliospheric Observatory) and four Cluster spacecraft. Soho will conduct an extensive survey of the Sun, while Cluster will essentially be concerned with space plasma physics. The second cornerstone mission will be an advanced X-ray Multi-Mirror (XMM) observatory to follow on EXOSAT, with a planned launch date of 1998. The third mission in the series, which is only in the early definition stages, is Rosetta, a possible joint mission with NASA, which would attempt to return samples to Earth from the nucleus of a comet. The current tentative launch date is 2001. The fourth cornerstone planned is a Far Infrared Space Telescope (FIRST), which would conduct very high resolution spectroscopy at submillimetre wavelengths. Other missions besides these elements in Horizon 2000 are planned, and the most important is another joint mission with NASA: Cassini/Huygens. NASA will build an orbiter (Cassini) to conduct a four year tour of the Saturn system in the late 1990s, with ESA supplying an instrumented probe (Huygens), which will investigate the atmosphere of the Saturnian satellite Titan.

5

The plans are undoubtedly impressive, but concern about costs (particularly in the light of the enormous sums being spent outside the science program) continue to surface from time to time. Thus at the end of 1990 there was speculation as to whether the planned mission

3 An artist's impression of ISO – Infrared Space Observatory – which ESA is due to launch in 1993. The telescope will be placed in a cryostat of liquid helium to keep it at - 270°C. Aerospatiale is designing and building ISO.

4 An artist's impression of the Hermes spaceplane.

5 ESA's Olympus 1 – the world's largest civil communications platform, weighing 5,760lbs (2,612kg) and with a solar array measuring 85ft (26m) across, which was launched in July 1989 for an intended five-year period of tests.

1 *The Infrared Astronomical Satellite – IRAS – prior to launch on January 25, 1983. Weighing 2,365 lbs (1,076kg), the satellite was a joint US-UK-Netherlands project which, over a period of ten months, conducted a survey of the entire sky at infrared wavelengths.*

2 *This image of the Andromeda Galaxy (M31) was produced by combining IRAS data in three IR wavelengths. Blue represents the warmest material, green intermediate and red the coldest. The bright yellow ring of clouds reveals regions where stars are being formed. The nucleus is also bright in IR, but this emission was believed to be due to dust and other material ejected from old stars. The faint object above the nucleus is an elliptical galaxy companion of M31, and the object below and to the left is an optically faint but infrared-bright background galaxy.*

3 *The German Roentgen Satellite (ROSAT) was launched on June 1, 1990 to study cosmic X-ray emissions in a joint German-US-UK program. The spacecraft is shown here being worked on by Dornier technicians shortly before launch.*

4 *Preparation of Japan's first satellite – "Ohsumi" – at ISAS' Kagoshima space center.*

5 *The successful launch by a L-4S-5 booster on February 11, 1970.*

6 *NASDA's main launcher between 1975 and 1982 was the N-I (shown here). Japanese-built but based on McDonnell-Douglas Delta technology, the vehicle could place a payload of about 130kg into a geostationary orbit.*

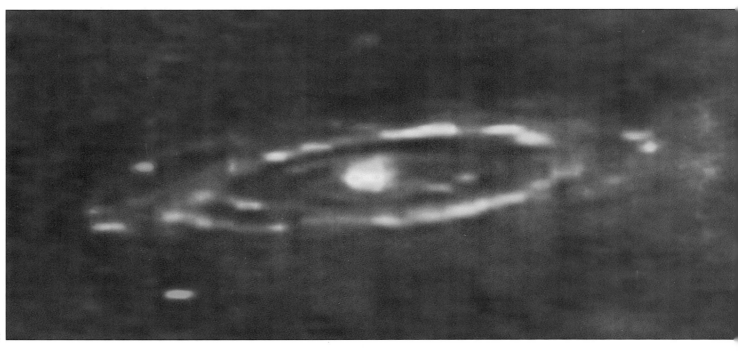

to send Giotto on to another comet – "perhaps the cheapest mission ever offered in the history of space science" – could proceed. It did – with great success.

Relations with NASA

ESA is a classic example of the increasingly international nature of spaceflight and space science. Indeed, outside of the military sphere, it is becoming rare to find a space project which is entirely domestic. Projects shared between individual countries and typical of this approach over recent years included the highly successful Infrared Astronomy Satellite (IRAS) – a joint US/UK/Netherlands mission – launched in 1983 and, more recently, Rosat, which was launched in June 1990. This was an X-ray satellite developed primarily by Federal Germany with contributions from the UK and US. In August 1992 France and the US launched the Topex/Poseidon spacecraft, one of the major objectives of which was to construct the first global, three dimensional model of the world's ocean currents. There were reports at that time that France planned to extend its cooperation in space with the USA – partly because of uncertainty about future developments in the CIS with which, as the former Soviet Union, France had had very close ties in space.

International projects do not always flow smoothly and NASA, beset by the problems posed by the annual funding process and changes in political attitudes to spaceflight, has figured in a number of such situations. The most unfortunate was the International Solar Polar Mission. This was a long standing commitment between NASA and ESA to fly two spacecraft simultaneously to the north and south poles of the Sun. NASA unilaterally abrogated the agreement in 1981 because of budgetary problems, and subsequently some of the ESA member states protested to the State Department in Washington through their ambassadors. The mission was in due course reshaped using the ESA spacecraft (renamed Ulysses), with NASA in 1990 providing the launch, the onboard power unit and tracking facilities. In the context of a future, major project, the constant reshaping in the US of the proposed space station Freedom over the course of some years has presented NASA's international partners such as Canada, ESA and Japan with considerable, and potentially very costly, difficulties. Reports of deteriorating relations were such that early in 1991 ESA issued a statement denying that it was ready to quit the project. No doubt the problems will be resolved, but the situation could well be regarded as a vindication of the French-inspired drive for European independence in manned access to space.

Japanese Progress

Japan's achievements in space over the past two decades have strongly mirrored the country's general performance as a technological leader in the world economy. In February 1970 it became the fourth country to achieve an independent launcher capability (after the USSR, USA and France) and, in February 1977, the third into geostationary orbit. By 1991 annual public spending on space (quite apart from extensive investment by the private industry sector) totalled some $1,300m, thus making Japan one of the leading countries behind the two major space powers. The Japanese space effort, however, continues to suffer from a unique restriction in that the two principal spaceflight organizations may conduct no more than two flights annually from each of their two launch sites, and even these are limited to the periods January-February and August-September – a reflection of the strength of the country's powerful fishing interests. This restricted domestic launch activity in Japan in the period 1970- end 1991 to only 43 flights.

Government space activity in Japan is spread over a number of ministries and institutions, so that effective coordination has assumed great importance. Since 1960 the Prime Minister has been advised on space policy by what is now called the Space Activities Commission – a small group of five "wise men." The Commission has established *Fundamental Guidelines of Space Policy,* which are reviewed at intervals and according to which annual development programs are conceived and executed. Broadly speaking, the principles governing Japan's space activities to date have: (a) emphasized their peaceful nature; (b) aimed at developing advanced technological capabilities so that in effect the country is independent of any other in conducting spaceflight missions; and yet (c) expressed a wish for "harmony" in international relations. The active coordinating role in striving to achieve these goals – as well as responsibility for annual space development plans over the whole of the two decades - has lain with the Science and Technology Agency, which reports to the Prime Minister. It was the STA which in 1964 set up the National Space Development center: a body that five years later became the National Space Development Agency (NASDA) which now accounts for around two-thirds of all state expenditure on space.

But it was not NASDA that first gave Japan an independent launch capability. In the 1950s the Institute of Industrial Science at the University of Tokyo had begun work on sounding rockets, and in 1963 the research and development effort was extended to a solid-fueled rocket capable of placing satellites in orbit. Later that same year the University opened the Kagoshima Space Center. In 1964 the Institute of Space and Aeronautical Science at the University of Tokyo was established, and it was this group which in February 1970 put the first Japanese satellite into orbit. Subsequently (in 1981) ISAS became an independent, inter-university institute – with "astronautical" replacing "aeronautical" in the title – under the Ministry of Education. It was then charged with responsibility for both research in space science and for the development of space satellites, as well as the launch vehicles which were to place them in orbit.

In its approximately 20 successful launches to date, ISAS has partly concentrated on a series of satellites (EXOS) that have studied the Earth's upper atmo-

sphere and another series (ASTRO) conducting X-ray investigations of astronomical objects. The last of the latter series to be launched in 1987, ASTRO-C, also called Ginga (Japanese for "Galaxy"), included instrumentation from both the US and UK and monitored emissions from the supernova which became a prominent object during that year. ISAS launched two spacecraft in 1985 (Sakigake/"Pioneer" and Suisei/"Comet"), which in March of 1986 joined probes sent by ESA and the Soviet Union in a fly-by of Comet Halley. Sakigake was Japan's first ever deep space probe, and Suisei successfully flew past the sunward side of the comet at a distance of 93,827 miles (151,000km). The trajectories of both spacecraft have been modified since to achieve fly-bys of other comets in the later 1990s. Despite some problems, ISAS achieved another significant success early in 1990

4

5

when it sent a test engineering satellite (Muses-A/Hiten) to the Moon. A 40cm diameter sub-satellite weighing 12kgs was released from the main vehicle and this went into lunar orbit – the first man-made object to do so since the USSR's Luna 24 in 1976.

The Institute has ambitious plans for the 1990s, some of which involve participation by international partners. A $33m solar imaging spacecraft – Solar-A (Yokoh/"Sunshine") – was launched in August 1991, and in the following year a study of the Earth's geomagnetic tail commenced following the July launch of the Geotail satellite, a contribution to the International Solar Terrestrial Program (ISTP). The $40m ASTRO-D X-ray mission in 1993 is regarded as a precursor to the launch of NASA's major AXAF observatory, and Muses-B, scheduled for launch in 1995, is a radio observatory which will combine with Earth-based receivers to form a VLBI (Very Long Baseline Interferometry) system. By this time ISAS plans to be operating a new M-5 solid fuel booster developed at a cost of over $130m. Three times more powerful than its predecessor, the M-3SII launcher, the M-5 is intended to be able to place payloads of 4,400lb (2,000kg) into low Earth orbit, send 1,320lb (600kg) to the Moon and 660lb (300kg) to Mars or Venus. This new launcher will be used by ISAS for a

6

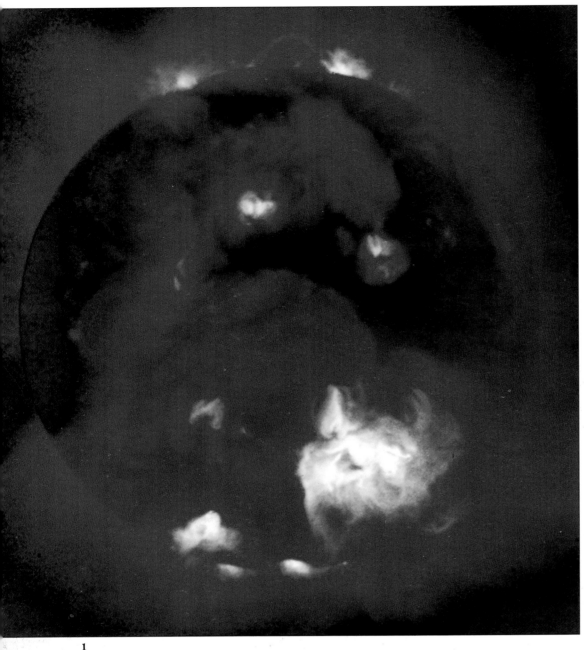

1

2

1 An excellent image of the solar corona taken by the ISAS Yohkoh (Sunshine) spacecraft in the low energy X-ray waveband. Large, dark coronal "holes" and bright points are prominent, but a fainter loop prominence can be seen clearly at center top.

lunar orbiter/penetrator mission due to be launched in 1996-97. Other long term plans envisage a Venus orbiter and a sample return flight to a cometary coma.

NASDA has concentrated on developing liquid fueled rockets and application satellites, as well as on directing Japan's future manned spacecraft activities. The Agency has its principal space center at Tsukuba, to the northeast of Tokyo, and its launch complex at Tanegashima. The first launch from the latter took place in 1975, and since then NASDA has concentrated on the development of three broad categories of satellites.

Engineering Test Satellites (ETS), as the name implies, are used to establish basic technology. Five had been launched up to the end of 1991 and in orbit NASDA gives them the designation Kiku, which means "chrysanthemum" in Japanese. Five communications satellites (CS), designated Sakura/"cherry blossom," and five broadcasting satellites (BS), Yuri/"lily," were launched over the same period. Earth observation satellites have been devoted to different tasks. So far four Geostationary Meteorological Satellites (GMS), Himawari/"sunflower," have been launched and in 1987 and 1990 NASDA launched two Marine Observation Satellites (MOS), Momo/"peach tree," data from which are being received abroad (for example, by ESA, Canadian and Australian stations) as well as in Japan. In February 1992 NASDA added to its remote sensing capabilities by launching an Earth Resources Satellite (JERS) with a synthetic aperture radar payload. An Advanced Earth Observing

Satellite (ADEOS) which will have an ocean color and temperature scanner as well as a visible/near infrared radiometer is scheduled to follow later.

Not included in the general categories of satellites orbited up to 1991-92 were several amateur radio and microgravity experiment spacecraft, launched "piggyback" with major payloads, and also an Experimental Geodetic Payload (EGP) satellite (Agisai/"hydrangea"), sent into orbit in 1986 to provide information on movement of the Earth's crust and other geo-physical phenomena (much like NASA's LAGEOS-1 satellite launched in 1976).

A New Launcher

A second major thrust of NASDA development and engineering activity has been directed to liquid fueled launchers. Although a number of its early satellites were launched in the US, the Agency concluded a licensing agreement with McDonnell Douglas in 1969 for the development of a Delta-based N-I orbital launch vehicle. There were seven successful launches between 1975 (NASDA's first) and 1982, and these were followed in the period to 1987 by eight successful launches using the N-II. This vehicle increased the payload capacity significantly (from 286lb/130kg to 770lb/350kg to geostationary orbit) by improvements in all three stages, and further refinements resulted from the use of a Japanese developed inertial guidance system.

Gaining experience by the use of US technology was a sensible policy for NASDA but the licensing agreement resulted in some limitations on Japanese freedom of action, including, for example, the need for prior US agreement before it could launch a satellite for a third country. A totally Japanese developed launcher was the ultimate objective, and an interim stage was reached in August 1986 with the first launch of the H-I vehicle. The first stage still derived from Delta technology, but the second, high performance stage using cryogenic propellants and the solid fueled upper stage had been developed entirely by the Japanese. The second stage LE-5 engine had a re-start capability and a thrust of 10.5 metric tons, thereby helping to improve the H-I's geostationary payload capacity to 1,200lb (550kg).

The first entirely Japanese launcher will be the H-II, which was approved for development by the Space Activities Commission in 1984. The first stage comprises a new LE-7 engine burning liquid hydrogen and oxygen with a thrust of about 100 metric tons, and with two solid rocket boosters supplying a further combined thrust of 320 tons. The second stage of the two-stage launcher will be powered by an improved LE-5 engine, the LE-5A. The target payload capacity to geostationary orbit is 4,400-4,850lb (2,000kg-2,200kg), a performance broadly similar to the Ariane 4 and Titan 3, though with less launch mass. Initially NASDA hoped that the first trial launches of the H-II would take place in 1992, but development problems with the LE-7 engine in particular led to the first launch being put back initially to 1993 and then subsequently to 1994. The development cost to first launch was estimated early on at around $800m but it seems certain that the delays will result in an increase in this figure. (At mid-1992, NASDA was reported to be working also on a new solid fuel launcher, designated the J-1, to be used for placing small satellites in a low Earth orbit).

Japan has embraced the concept of manned spaceflight, associated with extensive micro-gravity and life sciences experimentation, enthusiastically. A Japanese payload specialist, Mamoru Mohri, participated in a Spacelab mission aboard the shuttle orbiter Endeavour in September 1992 to operate 34 materials processing and life sciences experiments (out of a total of 43) at a total cost of $80m. Japanese experiments also flew aboard the Spacelab International Microgravity Laboratory (IML-01) flight in January 1992.

Manned Spaceflight

A major participation has been planned by Japan in the Freedom space station, and a preliminary agreement was signed with NASA as early as 1985. A three-part Japanese Experimental Module (JEM) attached to the main body of the station is envisaged: a pressurized module will provide a shirt-sleeve environment for the conduct of experiments; another experiment facility will be exposed to the vacuum of space and be reached from within the module by a manipulator arm; and a supply unit containing experiment equipment, samples, supplies and gases will be flown once or twice a year to the station and attached to the pressurized module. Inevitably, delays in the USA over the design and development of the space station must be having an impact on Japanese work on JEM, with a resulting

in the development of a manned, reusable spaceplane is NASDA's proposed H-II Orbiting Plane (HOPE). This is a winged and unmanned space vehicle to be launched by the H-II which will glide back to a runway landing under automatic control. It is envisaged as both a support craft for supplying the Japanese modules attached to the space station and as a means of establishing basic technology that will be required by the future manned spaceplane. Because of the H-II development problems it seems unlikely that test flights will take place by the mid 1990s as planned, and another project delayed by those problems is the Space Flier Unit. This is an unmanned, reusable free-flying experiment platform (similar to ESA's Eureca), which in part was intended to verify the design of the exposed facility planned for attachment to Freedom as part of JEM.

Unlike Europe and China, Japan has hitherto shown little interest in entering the commercial launcher market. Its licensing agreement with McDonnell Douglas and also the severe limitations on launch activity in Japan have largely accounted for this, but once the H-

2 An artist's impression of Yohkoh: it was launched on August 30, 1991.

3 The first Japanese scientist has already flown aboard the space shuttle and more experience is to be gained by a select band of Japanese astronauts and technicians in the next few years, leading up to the launch of the space station Freedom. The HOPE spaceplane is seen as a future supply vehicle servicing the Japanese module located in the station.

increase in costs, but total investment in the station was envisaged in 1988 at about $2,200m, with an annual operating cost approaching $300m. In the meantime, NASA in the United States responded to these initiatives by inviting Japanese candidates to join the next mission specialist class starting training at the Johnson Space Center in summer 1992 – and also NASDA engineers to work in Mission Control to gain experience of flight operations, which will be essential in planning missions aboard the space station.

Japan aims ultimately at the development of a manned spaceplane, but while some proposals have referred to flights in the first decade of the next century (at a development cost of around $18,000m) NASDA, ISAS and other groups involved in developing the new technology have adopted a careful, step-by-step approach. ISAS has been researching an unmanned suborbital delta wing vehicle, and of major importance

II launcher is in operation Japan might wish to enter the market, perhaps from a launch site to be developed elsewhere in the Pacific area. It could be tempted in this direction by its experience in losing two communications satellites in February 1990 and April 1991, as a result of Ariane and Atlas Centaur booster failures respectively – and the formation of a new Rocket Systems Corporation (a consortium of 70 companies) might be a sign of things to come. It is more likely, however, that the country will choose to exploit its prowess in other high technology areas, such as communications satellites. It has concentrated on developing satellites operating at frequencies of 20/30GHz (most satellites have hitherto operated at lower frequencies where the wavebands have become crowded), and in a world which appears to have an insatiable appetite for communications Japan may well choose to exploit its technological prowess in this new direction commercially.

4 Like other space nations, Japan is exploiting the value of remote sensing from space. It has launched four geostationary meteorological satellites, one Earth resources spacecraft with a synthetic aperture radar system and two Marine Observation Satellites – the first of which is shown here in the laboratory. MOS-1 operated a multispectral radiometer, a visible and thermal band infrared radiometer and a microwave scanning radiometer.

5 ISAS' M-3SII solid fuel launcher – shown here – will be replaced in the mid-1990s by the more powerful M-5 booster, which will be capable of sending 660lb (300kg) to Mars or Venus.

1 *China has mounted a powerful campaign to win business from foreign countries for its launcher industry. Probably because of that, it has been prepared to release far more and far better quality pictures of its space activities than the former Soviet Union did for many years. This photograph shows China's second weather satellite (Fengyun 1) being launched from the Taiyuan center, south of Beijing, on September 7, 1990. This satellite replaced one launched in 1988 but which quickly failed.*

China launched 13 satellites, and five of the launches were believed to have been failures or partial failures. (The Chinese also subsequently reported a major explosion at one of their launch pads on January 28, 1978, resulting in some serious injuries and burns.) The progress in the second decade was steady rather than spectacular in numerical terms – 15 launches – but with only one failure reliability was obviously improving. China's busiest space year to date was 1990 – with five launches including that of a powerful new launcher derivative as well as a commercial communications satellite for a foreign customer – although there was only one successful launch in 1991. Over the period the launch vehicles were steadily developed, with most bearing the prefix initials "CZ," for Chang Zheng, or Long March. The workhorse has been the CZ-2, an ICBM derivative using hypergolic propellants with a first stage thrust of over 625,000lb (284,000kg).

Since the first launch of a Chinese satellite there has been a strenuous national attempt to develop the country's economy. The plants engaged in the space program have in some cases at least been required to play their part in this work, and in 1985 a US delegation visiting a factory near Shanghai making CZ-2 and 3 boosters (as well as ICBMs) noted that it was also producing refrigerators. Nonetheless, while the Chinese performance in space has not been intensive it has demonstrated a methodical sense of purpose.

Cooperation – and Problems

There has been a continuing interest in orbiting recoverable spacecraft which have probably been devoted to reconnaissance as well as microgravity experiments and remote sensing of Earth resources. (China was the first country after the Soviet Union and the USA to master the technique of recovery.) In 1981 a triple payload, which included a solar observatory, was launched, and three years later came the launch of a test communications satellite into geostationary orbit, which was followed in subsequent years by a number of operational comsats. The first Chinese sun-synchronous meteorological satellite (FY-1) was launched in 1988 but quickly failed, and a replacement (FY-1B) was orbited in September 1990. An experimental geostationary metsat system (FY-2) is currently being developed with launch expected in the mid 1990s.

The different launch requirements for the various types of satellites necessitated the construction of new facilities. Thus to the earliest site at Jiuquan in the Gobi Desert at 41.2°N (used to place satellites into orbital inclinations of between 50 and 70°) was added in 1984 Xichang in the south-east at 28°N (for geostationary missions launched by the CZ-3) and in 1988 Taiyuan, south of Beijing, at 38°N (for sun-synchronous – polar orbiting – metsat and remote sensing missions launched by the CZ-4 vehicle).

As the 1980s progressed visits were exchanged between US and Chinese delegations on a number of occasions. The Chinese stated that the openly available literature on US launchers had been of value to them, and at least as early as 1984 began to publicize the availability of the Long March as a commercial launch service. The French aerospace company Matra paid for a launch of microgravity units aboard a recoverable spacecraft in August 1987, and the Chinese met the potential criticism of possible "technology transfer" to a communist country by accepting and returning the units to Matra unopened. The drive for commercial, foreign payloads continued, and in 1987 the Chinese organization responsible – the China Great Wall Industry Corporation – set up several offices in foreign countries.

2 *Foreign visitors have commented on how China's space industry operates in a rural setting. That impression is suggested by the background to this picture, which is described simply as the preparation of a CZ-3 booster.*

3 *Taken at the Jiuquan launch site on May 28, 1991, the photograph shows "scientific and research staff … conducting a final check of a return technical exploration satellite …."*

4 *A night view of a CZ-3 launcher, taken at the Xichang center in 1987. The type of launcher and the location suggests that this could have been the launch of a geostationary satellite.*

Chinese Launchers

China became the fifth nation, after the Soviet Union, the USA, France and Japan, to develop a launch capability when it orbited a satellite on April 24, 1970. Information about the country's space program is not extensive and much of it has been shrouded in military secrecy. Rocket propulsion was included in a scientific and technology development plan introduced by the ruling Communist Party in the late 1950s and artificial satellites were added as a development task in 1965. It is intriguing to note how often in various countries the early stages of building a spaceflight capability appear to have centered on individuals, and China is no exception. A major role in developing the country's long-range military rockets and space launchers is usually attributed to Dr Chien Hsue-shen, who was expelled from the US in 1955 during the McCarthy anti-communist campaigns when he was working as a research engineer at CalTech.

In its first decade as an independent space power

There were numerous reports at this time of US companies (effectively without a domestic launcher since the Challenger tragedy) entering into provisional agreements for launches aboard Long March boosters.

There were problems, however. Visiting US specialists warned of the extensive bureaucracy and military control with which foreign customers would have to deal, and also of the strangeness of a very rural and "old-world" setting that required the evacuation of nearby villages when launches were to take place. More demanding were the requirements of the US Government before it would issue licences for the export of spacecraft built by US concerns or involving some US technology.

4

In 1988 three such licences were sought for Hughes Aircraft satellites: AsiaSat 1 (the refurbished Westar 6 spacecraft retrieved by a shuttle crew in 1984 and subsequently purchased by an international concern based in Hong Kong) and two comsats for Australia: Aussats B 1 and 2.

Before issuing the licences President Reagan insisted that agreement be reached over protecting the technology of each satellite while it was in China, and also on "fair trading" in the provision of launch services – a reference to the fear of US launcher companies that the Chinese would undercut the going price for launches. He was satisfied with the assurances he received (including a Chinese agreement not to carry more than nine foreign satellites in the period to end-1994) and the licences were granted, although President Bush suspended them for some time in 1989 as a result of the crushing of the Tiananmen Square uprising in Beijing. AsiaSat was launched by a CZ-3 vehicle from Xichang in April 1990, but there was a setback in March 1992 when a CZ-2E booster, which was due to launch the Australian Optus B-1 communications satellite, shut down after ignition start at the launch pad. This followed the failure of a domestic satellite launch in December 1991. However,

the Australian satellite was not damaged and was launched successfully in August 1992. Moreover the problems did not stop the Intelsat organisation signing an agreement with the Chinese in April 1992 for the launch of one of its comsats – scheduled tentatively in 1995-96.

The world launcher market is fiercely competitive and it seems almost certain that criticism of pricing policies by non-US concerns will continue to be made. Thus in May 1990 it was being alleged by US companies that China had broken its pledge of 1988 to ensure that the market was "both open and fair" by agreeing to launch an Aerospatiale-built Arabsat communications satellite for $25m – about half the going rate. (In fact,

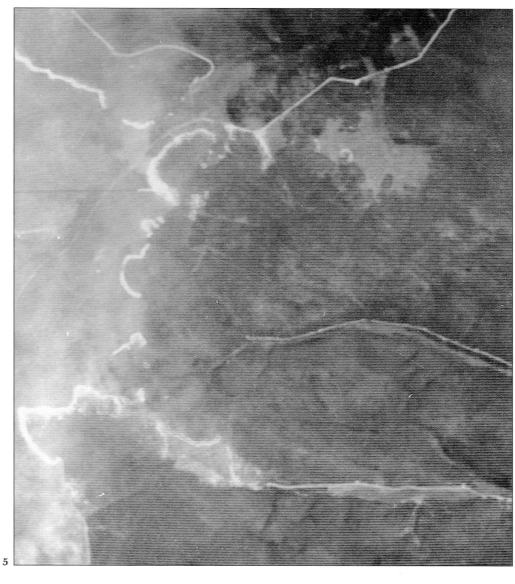

5

Arianespace won the contract.) The Chinese position will be strengthened and the competition intensify further if the CZ-2E booster is further uprated by the mid-1990s as planned, to rival the most powerful version of the Ariane 4. Significantly, Chinese and ESA representatives met in Paris in December 1990 to discuss the "fair exploitation" of launcher technology worldwide: it was agreed that the discussions "should be continued in a framework to be defined."

But the traffic has not been all one way. In 1987 China purchased a station for the reception of Landsat data at a cost of $11m, with an annual reception fee of $600,000 being paid. It has entered into a number of agreements for cooperation in space technology, and in the mid 1980s there were reports about a Chinese satellite being launched from the US space shuttle with a Chinese payload specialist aboard, but this came to nothing. There have been suggestions also of a 1993 visit to Mir.

5 An image taken from a TV screen of forest fires burning in northeast China during the spring of 1987. The thin, white lines away from the pall of smoke are fire breaks. It is not clear whether the scene was taken from a domestic Chinese remote sensing satellite or from a Landsat, but higher resolution would be expected from the latter.

6 Launch of the Australian Optus B-1 comsat on August 14, 1992.

6

Compared with ESA and Japan, China's position on manned spaceflight has been obscure. As early as 1979 Chinese astronauts were stated to be in training, but the economic situation caused a postponement of the announcement of the plans for at least the remainder of the decade. However, in 1987 Chinese specialists were talking of manned spaceflight in the 1990s. Work on a space suit and a "Gemini class" spacecraft was reported to be proceeding, but during a visit to the Kennedy Space Center late in 1990 a senior Chinese figure was reported to have talked of plans for a four-man spacecraft to be launched into Earth orbit by a Long March CZ-2E. The Chinese space program may not have the depth and versatility of the European and Japanese programs but it would not be surprising if its representatives were put into orbit over the next few years, perhaps in a less advanced spacecraft than those contemplated by ESA and NASDA.

Many years of cooperation with NASA included Canada developing and producing the shuttle remote manipulator arm used in deploying and retrieving satellites, and an early agreement to develop the Mobile Servicing System to be used on the proposed space station. A Canadian astronaut flew aboard the shuttle in 1984 and another aboard Discovery during the micro-gravity mission flown in January 1992 – with a third following on Columbia in October 1992. The various activities of the Canadian government in space were drawn together under the Canadian Space Agency in 1989 and investment in 1990 was running at a rate of about $200m, with over 40 percent being devoted to the space station project.

Indian Way to High Technology

India has taken the lead among the so-called developing countries in developing spaceflight applications. A number of such countries have taken the step of opening stations to receive valuable natural resources data from the Landsat and Spot satellites, including Argentina, Brazil, Ecuador, Indonesia, Pakistan and Thailand, but India has gone much further and has endeavored to seize the opportunity (in the words of the chairman of the Indian Space Research Organization) of bypassing an intermediate stage by leap-frogging directly from the status of a developing country into high technology. Its satellites have been launched by the US, the Soviet Union and Europe since 1975, but in July 1980 India became the seventh nation to achieve an independent launcher capability, although subsequent progress in this direction has proved difficult.

A number of satellites combining communications and remote sensing packages gave India experience of the new technologies and in 1988 the country's first operational remote sensing satellite (IRS-1a) was launched from the Soviet Union, from where, a few years earlier, an Indian Air Force Major had left on a seven-day Soyuz/Salyut mission. A second remote sensing satellite was launched by a Soviet Vostok booster in August 1991 followed by the successful domestic launch

2

1 *An Indian SLV-3 lift off. It became the seventh nation to achieve an independent launcher capability in July 1980.*

2 *An artist's impression of Canada's Anik-B domestic comsat.*

3 *An Indian technician at work on a SROSS satellite – one of a series designed to carry remote sensing and scientific payloads.*

4 *Dr. Fred Whipple, father of the "dirty snowball" theory of comets, speaking (left) at a press conference held at ESA's European Space Operations Center in Darmstadt, Germany, after Giotto had flown past Halley's Comet in March 1986. At right is Prof. R. Lust, then Director General of ESA.*

5 *Tests being completed on the Giotto spacecraft at British Aerospace's Bristol facility in 1984.*

Early Canadian Moves

Two countries on opposite sides of the world have shared in the advances in spaceflight. With its extensive land mass, Canada moved very early to benefit from the space applications technologies offered by communications and remote sensing satellites. In the early 1970s it began operation of the world's first domestic commercial space communications system with its Anik ("brother" in the Eskimo language) satellites and was negotiating with NASA to receive data from the first ERTS (Earth Resources Technology Satellite, later re-named Landsat) years before it was launched in 1972. If all goes well, Canada's own Radarsat spacecraft with a high resolution synthetic aperture radar system aboard will be launched in early 1995, and valuable advance experience in this area was gained as a result of co-operation between Canadian corporations and ESA during the preparatory stages of the latter's ERS-1 satellite. (Canada has a formal agreement with ESA and contributes to its general budget.)

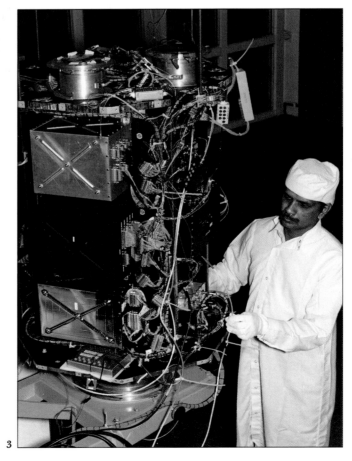

3

of a scientific satellite in May 1992. Subsequently the Indian Satellite Research Organisation (ISRO) declared the objective of applying its growing expertise in satellite construction on the world market. State expenditure on space was over $200m in 1990-91, with launcher development taking 45 percent and satellites 32 percent.

Brazilian Plans

With an enormous land area in South America, Brazil, like India, has been working to apply communications and remote sensing satellite technology to the development needs of the country and to establish an independent launcher capability. INPE (the Brazilian Institute for Space Research) replaced an earlier organization in 1971 and is developing four satellites under a program begun in 1983. Two of the spacecraft (SCD 1 and 2) will collect environmental data from ground based platforms, while the other two (SSR 1 and 2) will be 4

5

6

imaging remote sensing satellites. The first SCD was scheduled for launch in 1992 (by an Orbital Sciences Pegasus air launched booster) but the control center was commissioned as early as 1989.

Brazil's Ministry of Aeronautics has responsibility for developing the VLS launcher system and the equatorial launch site at Alcantara. The original cost of the entire satellite/launcher program was estimated at $1,000m, but should the launcher system not be ready in 1992 INPE has already stated that it will seek a foreign launcher. In 1988 the country signed an agreement with China (CBERS) to develop two further remote sensing satellites, with Brazil meeting $45m and China $105m of the total cost. Long March vehicles were to be used and the first launch was planned for 1994. In the meantime, Brazil has reception stations for taking data from both Landsat and Spot satellites.

A satellite based communications system has progressed faster than remote sensing in the country. Over 120 ground stations access two US designed but Canadian built Brasilsat satellites launched by Ariane in 1985 and 1986. Two new generation comsats are scheduled for Ariane launches in 1993 and 1994.

Goodwill from Science

In the first three decades of spaceflight there has been both rivalry and cooperation. National programs, whether of the "super powers" on the one extreme or those striving to hasten economic development by the application of high technology on the other, have often had very materialistic goals. Sometimes, however, space technology has saved lives, as in the case of the SARSAT rescue system operated by Canada, France, the USA and the USSR. At other times, basically scientific missions have provided evidence of goodwill and a preparedness to cooperate which has transcended any problems. A fine example of the latter was the manner in which nations, banded together in the Inter-Agency Consultative Group (IACG), combined to maximise the scientific returns from the peaceful armada of spacecraft which encountered Halley's Comet in the late winter of 1986. There was perhaps something uplifting in the spirit of friendship and cooperation with which humans greeted this most celebrated of celestial visitors more than 87m miles (140m km) out in space, where literally as well as metaphorically national boundaries have no meaning.

7

6 *The most detailed image ever obtained of a cometary nucleus (a composite of six images in fact): Halley's Comet as seen by the Multicolor Camera aboard ESA's Giotto probe.*

7 *Academician Roald Sagdeev, leading Soviet space scientist, was present at the Giotto encounter gathering held at ESOC in March 1986.*

15

SCIENCE AND APPLICATIONS IN ORBIT

1 *Launch of an OGO satellite aboard an Atlas Agena-B booster from the Kennedy Space Center. The Orbiting Geophysical Observatory satellites (of which there were six in the period from 1964 to 1969) were directed primarily at studying the dynamics of the Earth's magnetosphere.*

1

2 *Explorer 10 was a small satellite – shown here being examined by two NASA specialists – which was one of a number of spacecraft in the Explorer series (as well as other NASA programs together with Soviet satellites) which studied the interaction between the solar wind and the Earth's magnetosphere. The mechanism became well understood, and over the ensuing years could be compared with that at other planets in the solar system as they were visited by interplanetary spacecraft. Explorer 10 was launched on March 25, 1961.*

Both unmanned and manned space missions have conducted wide-ranging operations in the area of science and applications. The early satellites were intended to provide information on the Earth's upper atmosphere and magnetic fields; research referred to as "fields and particles" or "sky science." The first example of this was Sputnik 1. By analyzing tracking data as its orbit and that of its booster core stage decayed, it was possible to calculate the density of the upper fringes of the atmosphere. Six years of tracking data from Vanguard 2 revealed changes in this density caused by solar activity. The increased energy from the Sun heated the atmosphere, causing it to expand outward, and this in turn caused satellites' orbits to decay faster.

The most important discovery of the early space age was the Van Allen radiation belts. Explorer 1 carried a Geiger counter, and this showed that the radiation level increased with altitude, as had been expected. But when Explorer 1 reached 300 miles (483km), the Geiger counter failed and the same thing happened with Explorer 3. Dr. James Van Allen and his staff realized that the Geiger counters had been jammed by radiation too intense for them to measure. This radiation had been trapped within the Earth's magnetic field, forming the radiation belts that were subsequently to be named for Van Allen. A more sophisticated system was flown on Explorer 4 and showed the shape of the belts, which resembled two parentheses bracketing the Earth and open over the poles; a shape like that of a bar magnet's field.

The size of the belts was first shown by the Pioneer lunar probes. As Pioneer 3 climbed away from the Earth it recorded two peaks of radiation: the first at an altitude of 2,000 miles (3,218km) and a second at 10,000 miles (16,090km). Beyond this, the radiation dropped until it faded to a constant level at 40,000 miles (64,360km). Thus, there seemed to be two belts: an inner and an outer one.

Both Pioneer 4 and the Soviet Luna missions showed the source of the radiation. They detected a flow of protons and electrons from the Sun – the solar wind – the speed and strength of which depend on solar activity. The speed of the solar wind ranges from 190 miles per second (306kps) to 555 miles per second (893kps) at times of high solar activity, and because the particles are electrically charged they become trapped within the Earth's magnetic field. It was soon realized that the solar

2

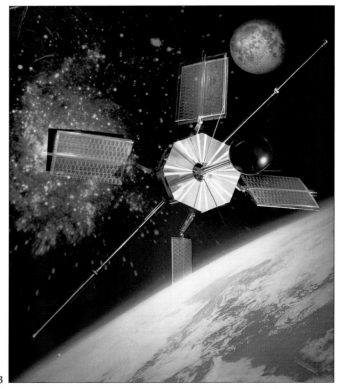

3

wind and the radiation belts were the causes of auroras. The particles, which are funnelled into the atmosphere above the poles by the magnetic field, interact with the thin gases of the upper atmosphere and cause them to glow. Solar flares also set up currents within the magnetic field, which disrupt the ionosphere, often causing blackouts in long-range radio communications.

A large number of satellites were launched between 1958 and 1966 to study the interaction between the solar wind and the Earth's "magnetosphere." Among the missions were Explorer 7, 10, 12, 33 and 44, Pioneer

which extends as far as the Moon. The shape of the magnetotail is similar to that of the airflow around an airplane at supersonic speed, and it also seems to wiggle.

Many of these features had been theorized beforehand (the solar wind, for example, was first suggested by Olaf Birkeland in 1896), but it was not until the advent of satellites that they were confirmed. Moreover, satellites brought the realization that what were once seen as separate phenomena were actually interrelated. The Earth was not isolated in space but was directly affected by the Sun.

3 *The IMP spacecraft – Interplanetary Monitoring Platforms – were described as "physics laboratories" exploring the near-Earth environment. This is an artist's impression of the compact, 136 lb (62kg) spacecraft IMP-2, which was also designated Explorer 21 and was launched in October 1964. Because space science and the spacecraft were still relatively new, NASA's caption described IMP-2 at considerable length. "[It] has an octagon-shaped base, eight inches (20cms) high and about 28 inches (71cms) in diameter. All but two of its experiments are mounted on this base. (IMP's) ... most distinguishing physical feature is the manner in which its magnetometers, which measure magnetic fields in space, have been mounted to avoid interference from the weak magnetic field created by the satellite itself. Solar cell arrays are mounted on four panels ... and the satellite's tiny, four-watt transmitter ... can transmit data to NASA's world wide network of tracking and receiving stations at distances of over 126,500 miles (204,000km)."*

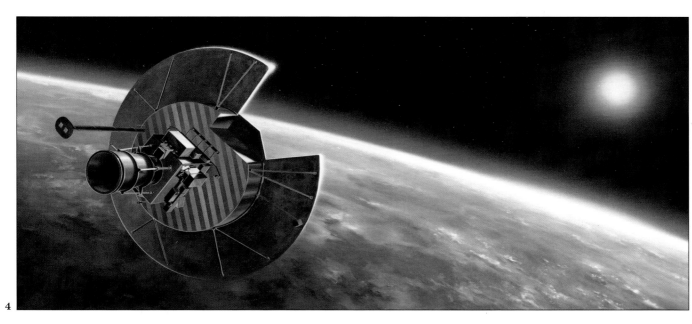

4

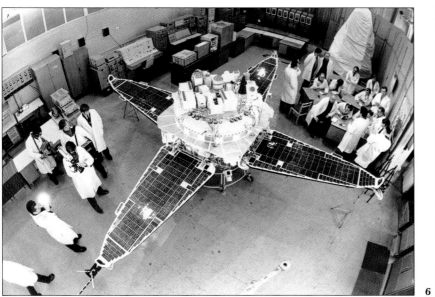

5

6

5 and 6 and the Interplanetary Monitoring Platform (IMP) 1, 2, 3 and 4 satellites. Soviet missions included Sputnik 3, Cosmos 1, 3, 5, 7, 26 and 49, and the Elektron 1/2 and 3/4 flights. They showed the complex structure of this interaction. When the solar wind strikes the Earth's magnetic field a bow wave is formed. Behind this is a region of turbulence where the solar wind is slowed, and beyond is a cavity within the solar wind: the Earth's magnetosphere. The location of the magnetosphere's boundary (called the magnetopause) is dependent on the Sun's level of activity. When there is a solar flare, the magnetopause is pushed back towards the Earth. There is a sudden increase in auroras, a disruption in the ionosphere and a drop in the number of cosmic rays near the Earth. When solar activity is low, the magnetopause is farther from the Earth and the reverse occurs. As the solar wind flows around the magnetosphere, it forms a long tail (the magnetotail)

Effects of Solar Activity

With the basic structure of the magnetosphere understood, efforts shifted to a study of its mechanisms. Satellites launched for this purpose during the 1960s and 1970s included the Orbiting Geophysical Observatory (OGO) series, Pioneer 7 through 9, a large number of Explorer and IMP flights, the West German Azur, Aeros 1 and 2, Helios 1 and 2, the French Tournesol and Aureole 1 and 2, the Japanese Denpa, Taiyo, Ume, Kyokko (Exos A) and Jikiken (Exos B) satellites, the Indian Aryabhate satellite, and the Soviet Prognoz, Intercosmos and many of the Cosmos satellites.

The years 1976 through 1979 were designated the International Magnetospheric Study. This was a study of the effects of solar activity on Earth's weather, magnetosphere and atmosphere. NASA and ESA launched three International Sun-Earth Explorers (ISEE) during 1977 and 1978. ISEE-1 and 2 were placed into highly-

4 *NASA's Solar Mesosphere Explorer (SME) – shown in this artist's impression – was launched in October 1981 with the primary objective of measuring the effect of the Sun's ultraviolet radiation on the Earth's ozone layer.*

5, 6 *Many Soviet spacecraft have been devoted to the study of the near Earth environment. Both of these pictures show the Prognoz 10 (Intercosmos 23) satellite, which was launched in April 1985 to continue studies of the interaction between the solar wind and the magnetosphere. As a relatively recent spacecraft, its large size causes no surprise.*

1, 2 *On August 20, 1979, a comet collided with the Sun – an event not detected on Earth but recorded by a satellite studying the Sun's corona. This was the first instance of a comet being discovered by satellite instrumentation – even if the data were not recovered until the fall of 1981! The US Naval Research Laboratory had its Solwind coronagraph aboard the Department of Defense satellite. This instrument studies the outer atmosphere of the Sun by occulting the bright solar disc to create an artificial eclipse.*

1

2

Images are taken routinely and the two reproduced here show the comet (belonging to a type normally called "sungrazers") approaching the Sun and then disintegrating into a dust storm. The bright spot at top right in 2 is the planet Venus.

3 *Activity on the Sun recorded by a soft X-ray imager aboard the Solar Maximum Mission spacecraft on September 2, 1988.*

4 *Final preparations for the launch of SMM by a Delta booster from the Kennedy Space Center on February 14, 1980.*

5 *The Solar Maximum Mission spacecraft undergoing vibration tests at the Goddard SFC in August 1979. SMM is mounted on top of a Multi-Mission Spacecraft: a module designed to provide power, control and communications to a variety of packages mounted on it.*

elliptical orbits, while ISEE-3 was placed at the L-1 Lagrangian point some 900,000 miles (1.5m km) from Earth. This allowed variations in solar activity over time and distance to be recorded. The ESA GEOS 1 and 2, the Japanese Exos A and B and the Soviet spacecraft Prognoz 4 and 5, Intercosmos 14 and 18 and Venera 9 and 10 were also part of this study. In addition, on August 3, 1981, NASA's Dynamic Explorers 1 and 2 were launched together on a single Delta booster into differing orbits to study the interactions between the ionosphere, the magnetosphere and the Earth's atmosphere. Soon after, on September 21, the Soviets launched the French Oreol 3 satellite to study the magnetosphere and ionosphere.

Pioneer 10 and 11 and Voyager 1 and 2 also provided data on the magnetic fields of Jupiter and the other outer planets. These showed similar structures and effects, though on a vastly larger scale. As these spacecraft leave the solar system they will search for the heliopause: the region where the Sun's magnetic field fades out and interstellar space begins.

Interest in magnetospheric studies remains high for the 1990s, with ESA planning to launch the Soho-Cluster project. The Soho satellite will observe the Sun, its corona and the solar wind, while the four "cluster" satellites will study the reaction of the magnetosphere to changing solar activity. The Japanese Geotail spacecraft will, as the name suggests, look at the effects of the solar wind on the magnetosphere's tail.

Astronomical Satellites

Space is the ideal site for astronomy. There is no weather or turbulent atmosphere to distort images and, more importantly, instruments in orbit can be used to study the ultraviolet, infrared, X-ray, gamma ray and radio wavelengths of the electro-magnetic spectrum, which on the surface are largely absorbed by the atmosphere. Astronomical studies in space fall into several categories. In many ways, solar studies overlap with those of the solar and Earth's magnetic field. Indeed, several of the satellites mentioned earlier also carried instruments to study the Sun. The Explorer 37 and 44 satellites measured solar radiation, OGO 2 studied solar ultraviolet and X-ray emissions, the Soviet Intercosmos 1 carried instruments to find the polarization of solar flares, while Prognoz 3 and 5 studied solar flares and solar X-ray and gamma radiation. The first British satellite, Ariel 1, was designed in part to observe solar radiation in ultraviolet and X-ray wavelengths.

Specialized solar satellites in the Orbiting Solar Observatory (OSO) series were built in the U.S. The first, OSO-1, was launched on March 7, 1962, with a payload to study solar flares. Later missions studied solar ultraviolet, X-ray and gamma rays (OSO-2), the extreme ultraviolet (OSO-4) and the solar corona (OSO-6 and 7). The series ended with OSO-8, launched in June 1975.

By this time, solar studies had also been undertaken by manned spacecraft. The Skylab space station in 1973-74 carried the solar telescopes of the Apollo Telescope Mount, and the Orbital Solar Telescope (OST-1) aboard Salyut 4 in the mid-1970s performed similar studies. On the Soyuz 18 flight in 1975, for example, the Crimean Astrophysical Observatory requested the crew to photograph a solar disturbance that had first been noticed from the ground.

The replacement for the OSO series was the Solar Maximum Mission, better known as Solar Max. Launched on February 14, 1980, the spacecraft, as the name suggests, was meant to study the Sun during the peak of the 11-year solar cycle. Unfortunately, on November

23, 1980, three fuses blew in the attitude control system, leaving Solar Max without a fine-pointing control capability for four of its six telescopes, thereby limiting its usefulness. The fault was repaired in orbit by the crew of the orbiter Challenger in April 1984, highlighting the ability of a human crew to save a valuable unmanned satellite, and Solar Max continued to operate for the next five years until it re-entered the atmosphere. This was not the Shuttle's only involvement with solar studies. The STS-51F (Spacelab 2) mission, launched in July 1985, carried a package of three solar telescopes which included a British X-ray instrument as well as a U.S. ultraviolet telescope.

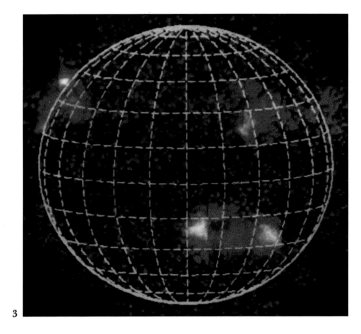

3

Orbiting Observatories

Beyond the Sun is the realm of the stars and galaxies, and this was one of the first areas of research from space suggested by the early spaceflight theorists; Hermann Oberth writing about orbiting telescopes as early as the 1920s. In 1946 astronomer Lyman Spitzer Jr. wrote a RAND report entitled *Astronomical Advantages of an Extra-Terrestrial Observatory,* but it would take 44 years before the ultimate expression of his idea would go into orbit.

The first step in this direction was NASA's Orbiting Astronomical Observatory (OAO) program. One of the major technical problems was designing a pointing system that could hold the satellite steady on the target star. These difficulties delayed the first OAO for three years, but OAO-1 was finally launched on April 8, 1966, and its Atlas Agena placed it successfully into orbit. However, when the pointing system was turned on there were electrical discharges which damaged control circuitry; the batteries overheated and exploded, and five years of effort and millions of dollars had been lost.

There was a better performance from OAO-2, which was launched in December 1968 carrying 11 telescopes and two scanning spectrometers. It operated for four-and-a-half years and returned infrared, ultraviolet, X-ray and gamma ray data from stars, galaxies and comets. The next OAO carried a 36in (910mm) ultraviolet telescope, but during launch on November 30, 1970, the shroud on the Atlas Centaur failed to separate and it fell back to Earth. OAO-3, however, brought the series to a successful close. Launched in August 1972, it carried a 32in (813mm) ultraviolet telescope (designed by Lyman Spitzer) to observe the spectra of bright stars and interstellar matter. OAO-3 was renamed Copernicus – in honor of the 500th anniversary of the birth of Nicholas Copernicus – and operated for nine years, both a scientific and a technical success.

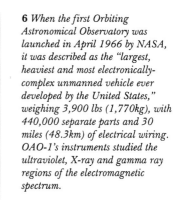

6 *When the first Orbiting Astronomical Observatory was launched in April 1966 by NASA, it was described as the "largest, heaviest and most electronically-complex unmanned vehicle ever developed by the United States," weighing 3,900 lbs (1,770kg), with 440,000 separate parts and 30 miles (48.3km) of electrical wiring. OAO-1's instruments studied the ultraviolet, X-ray and gamma ray regions of the electromagnetic spectrum.*

5

4

Ultraviolet and Infrared

The late 1970s and early 1980s saw the launch of several astronomical satellites. First was the International Ultra-violet Explorer (IUE), launched by a Delta in January 1978 from the Kennedy Space Center. IUE was a highly successful joint project between NASA, ESA and the UK carrying a 17.6in (450mm) ultraviolet telescope. This was followed by the Infrared Astronomical Satellite (IRAS) in January 1983; a joint US/Holland/UK project to map the sky in infrared. Because the telescope detected the heat given off by the stars and other objects in space, it had to be cooled by liquid helium. IRAS operated for 300 days until its coolant supply was exhausted. In that time it observed 250,000 infrared sources, including three rings of dust in the solar system, "infrared cirrus" (graphite particles in interstellar space), five comets and many asteroids. The most significant discovery was a ring of material around the star Vega which could be a planetary system in the process of formation. Several other well known stars, including Fomalhaut, were found

6

1 *Technicians at Perkins-Elmer inspect the Hubble Space Telescope's 94in mirror following the coating of a special reflective surface.*

2 *April 25, 1990: the Hubble telescope is deployed from the cargo bay of the space shuttle Discovery.*

3 *An excellent Hubble example of "gravitational lensing," which occurs when light from a distant source passes through or close to a massive foreground object. Shown here are four images of a very distant quasar created by a relatively nearby galaxy (at the center of the cross) acting as a gravitational lens.*

to have similar rings, the implication being that planetary systems could be more common than earlier thought likely.

The Soviets also orbited an ambitious astronomy satellite about this time. Astron was launched in March 1983 by a D-1e booster which placed it into a highly-elliptical 1,212x124,900 mile (1,950x201,000km) orbit. This took Astron outside the Van Allen radiation belts for much of the time and also greatly reduced orbital decay. The spacecraft carried a Spika ultraviolet telescope. Built together with the French, this was 16.4ft (5m) long with a 31.5in (800mm) mirror; the largest ultraviolet telescope flown up to that date. Astro also carried X-ray detectors and discovered a possible binary star system in the Andromeda constellation.

The Space Telescope

While these satellites were being launched, development in the US was taking place on the Edwin P. Hubble Space Telescope. This was both prolonged and painful. During the late 1960s, a large orbiting telescope had been discussed as a follow-on to the OAOs. The proposal split the astronomical community. The "small science" group wanted an improved OAO, while the "big science" astronomers wanted a revolutionary large telescope able to make sweeping discoveries. There were also technical doubts that a large telescope could be built, but this concern was eased in the mid 1970s by the success of the Big Bird reconnaissance satellites, which used a large mirror as a vital element in their imaging system.

Much of the 1970s was spent building the scientific and political coalition to support the project. Ground-based astronomers saw it as a drain on their funding

After approval was given in 1977, problems continued. The telescope had been underfunded and there was simply not enough money to meet the technical problems. Twice, in 1980 and 1983, the cost overruns threatened to cause the program's cancellation. The situation at Perkin-Elmer, the contractor for the mirrors, was particularly bad, but how bad would not become apparent until after launch. The cost and technical problems delayed the launch date from 1983 until 1986, and then the loss of Challenger in January 1986 delayed it by a further four years.

The Hubble Telescope was finally launched into space in April 1990 aboard the shuttle Discovery. During deployment, one of the solar panels at first failed to unfold, but this was cleared up without the crew having to make a space walk. Over the next several weeks, Hubble's systems were tested: the usual technical problems appeared and these were all too often depicted in the press as major setbacks. Hubble's "first light" (reception of the first images) took place on May 20, and the early pictures were of a better quality than those of a ground-based telescope. It seemed Hubble's problems were over, but in fact they were just beginning.

As the engineers tried to fine-focus the telescope, it became apparent there was a major problem. Due to an error in a test instrument back on Earth, the main mirror had been ground to the wrong shape or "figure." The resulting optical condition, known as "spherical aberration," meant that star images could not be focused to a pinpoint; each star instead showing a "halo." When news of the problem was released on June 27, 1990, there was the predictable storm of press and Congressional ridicule and condemnation.

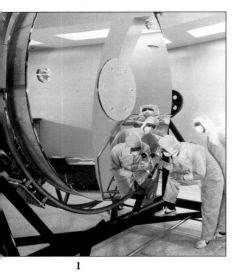

1

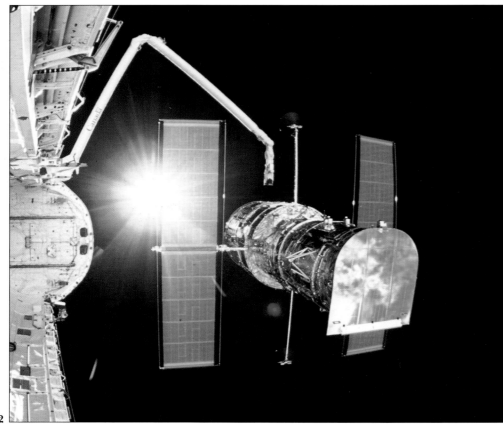

2

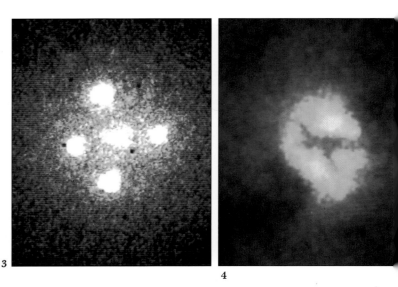

3

4

It was a bad summer for NASA. Two shuttle flights had to be postponed due to hydrogen fuel leaks, and after the second was found the shuttle fleet was grounded, pending solutions. NASA's technical competence was questioned and respect for the agency, slowly rebuilt after the Challenger tragedy, was again weakened. Some astronomers went farther, claiming that ground-based telescopes, equipped with "adaptive optics," could equal Hubble's resolution. (Adaptive optics involve manipulation of the main mirror to compensate for the turbulence of the atmosphere.) This claim, however, ignored several factors. Obviously, any clouds would render an Earth-bound telescope useless, and, more important, Hubble was designed to cover the spectrum from infrared to ultraviolet. Many of these wavelengths are extensively absorbed by the Earth's atmosphere, so no ground-based telescope could hope to rival its research in these areas.

As summer turned to fall in 1990, the Hubble situation became better understood. Computer processing could

4 *A Hubble image of the core of the nearby spiral galaxy M51 showing a dark "X" silhouetted against the galaxy's nucleus. The feature is due to absorption by dust and marks the exact position of a black hole, which may have a mass equivalent to one million stars like the Earth's Sun.*

and NASA's Jet Propulsion Laboratory felt it was a competitor to planetary space missions. In Congress, some seemed to have an emotional opposition to the project and NASA was forced to reduce its cost estimates and the size of the main mirror to 94in (240cm). It was a balancing act between what was desired, what was the minimum size that was scientifically useful, and what Congress would accept.

remove some of the distortions, particularly in the case of bright objects. In early August, a cluster of stars 160,000 light years away was imaged. Ground photos showed 27 stars, but Hubble showed 60. Later in the month Hubble recorded the core of galaxy NGC 7457, revealing stars crowded together at least 30,000 times denser than those in our own Milky Way galaxy. Earlier ground-based photos, including one taken by Dr. Edwin P. Hubble, for whom the telescope was named, gave no hint of this crowding. A second set of images showed a ring of gas around the supernova that exploded in 1987. The ring was 1.3 light years away from the supernova remnant and was composed of hydrogen-rich material emitted from the star as it expanded during the 10,000 years before the explosion. Clearly, Hubble was not the "fiasco" it had been called during the summer, and computer processing was allowing it to conduct a large proportion of the planned research.

Nonetheless, it was true that the faulty mirror prevented Hubble from effectively observing dim objects at the edge of the universe. These were galaxies formed soon after the "big bang," and this was the type of "cutting edge" astrophysical research that the telescope was intended to perform. A new Wide Field/Planetary Camera which takes account of the aberration is being built, and this will be installed during a shuttle mission due to take place in 1993. Unfortunately, ESA does not have the funds to replace its Faint Object Camera, but additional system modifications to the telescope are planned and, assuming no further problems develop, the overall efforts should restore the telescope to close to its planned capability. (From the start, it had been planned that the shuttle would make regular servicing

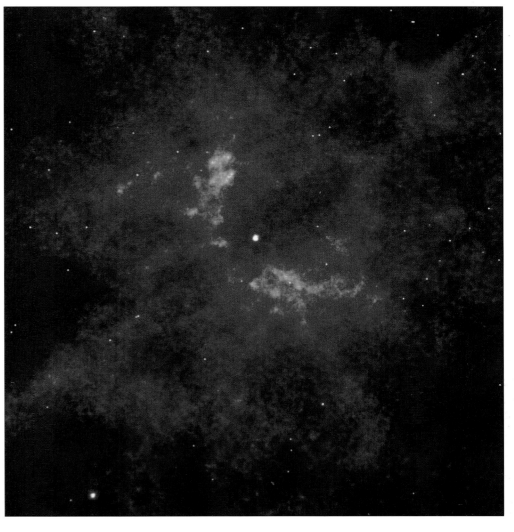

5

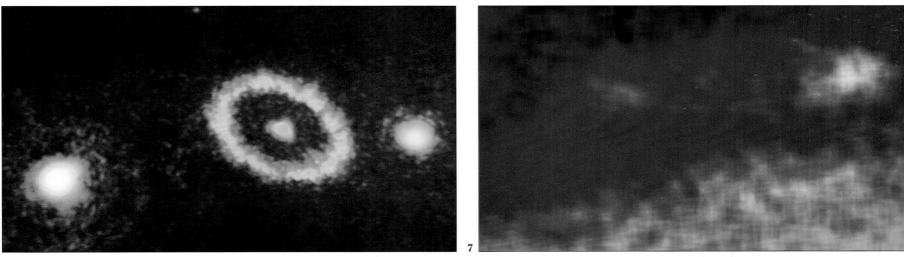

6   7

and, if necessary, repair flights during the telescope's 15-year lifetime.)

Future Missions
Astronomy has been part of manned spaceflight since the beginning. As far back as Gemini 10 in 1966, for example, ultraviolet photos of stars were taken during a spacewalk and on the Apollo 16 mission in 1972 an ultraviolet camera was set up on the lunar surface. The Skylab 4 astronauts observed comet Kohoutek in December 1973 and January 1974, and Salyut and Mir crews have been kept busy with astronomical studies. The most complex manned astronomical mission so far flown was that of Astro-1 aboard the shuttle Columbia in December 1990.

For the future, there is the SIRTF (Space Infrared Telescope Facility), which NASA intends to be a follow on to IRAS, but which is not yet fully funded. ESA is also planning an infrared satellite called ISO (Infrared

Satellite Observatory) and, if funds allow, in 1994 the CIS plans to launch Aelita: a cryogenically-cooled telescope similar to IRAS. At the other end of the spectrum is NASA's EUVE (Extreme Ultraviolet Explorer), which is to make an all-sky survey in extreme ultraviolet wavelengths, with the hope that some 1,000 new sources might be found: the spacecraft was successfully launched in June 1992.

Gamma and X-rays
Another sector of space astronomy is the study of cosmic rays, X-rays and gamma radiation. Cosmic rays were the first type of space radiation studied. At the end of the 19th century detectors were picking up radiation from an unknown source, and in the early years of this century it was found that this radiation increased with altitude. Between 1923 and 1926, Dr. Robert A. Millikan of Cal Tech proved that these "cosmic rays" came from space, and during the following

Shown here are three more images taken by the Hubble Space Telescope in the period from launch to spring 1992.

5 The first clear view of one of the hottest known stars: the central star (the point of light at center) of the nebula known as NGC 2440.

6 A "mysterious" elliptical ring of material around the remnants of Supernova 1987A.

7 ESA's Faint Object Camera obtained this image on February 8, 1992. It is the first direct image of an aurora on Jupiter taken in ultraviolet light – and rated as the best picture of the phenomena at Jupiter ever recorded.

The first deep space views of the Earth in color were obtained by a US Navy satellite in July of 1967 from a distance of 18,000 miles (almost 29,000km). The satellite and instrumentation were developed in the Applied Physics Laboratory of the Johns Hopkins University.

1 *The designers of the DODGE TV camera system are examining a one-inch-diameter vidicon that obtained and stored the images.*

3 *Northwest Africa can be seen in the image obtained on July 25, 1967. This is a composite of separate red, blue and green images transmitted back from the satellite: the multicolored disc seen at top is a calibration device.*

1

2

2 *The Soviet Union when glasnost or openness was introduced adopted an uneven approach to the information released on spaceflight operations. This picture of a Cosmos satellite being prepared for launch at the Plesetsk "spaceport" was an advance in that the site was regarded as top secret for many years, but the release about the launch of Cosmos 2026 on June 7, 1989, merely described it as being "equipped with up-to-date space exploration facilities" – and nothing more.*

lite up to that time, with a weight of 26,900lb (12,200kg), and the first use of the D-1 booster, which the Soviets called the "Proton booster" thereafter. It carried instruments to determine the energy spectrum and chemical composition of cosmic rays, gamma rays and the galactic electron flux. Proton 2 followed in November 1965 and Proton 3 in July of the following year. Proton 4, a heavier, improved satellite for the study of high energy cosmic rays and electrons, was launched in November 1968.

Ironically, X-ray and gamma ray astronomy was first conducted by military satellites. Between October

3

1963 and April 1970, pairs of Vela nuclear detection satellites were launched by the US. They were placed into 68,000 mile (109,412km) high orbits with each pair placed 180° apart. From this lofty perch, the Vela satellites watched for secret Soviet nuclear tests in space or the atmosphere; such tests having been banned by treaty. The X-ray and gamma ray detectors were sensitive enough to pick up a nuclear test of as small as 10 kilotons and (in theory) as far away as the orbits of Venus and Mars. They also picked up emissions from the stars. Vela 10 (launched with Vela 9 by a Titan IIIC in May 1969) provided a decade of X-ray data on the star Cygnus X-1, which is believed to be a black hole: a star 10 to 15 times the mass of the Sun with a diameter of only 5 miles (8km) across. Its surface gravity is so great that not even light can escape. Cygnus X-1 orbits a blue giant star from which it draws off hydrogen gas. As the gas spirals into the black hole it is heated and emits X-rays.

The Velas also detected two unexpected phenomena: X-ray and gamma ray bursters. The first occur in binary star systems composed of a normal star and a neutron star, the latter being the remnant of a supernova explosion. The neutron star pulls hydrogen gas from its companion; this builds up until the temperature and pressure become so high a thermonuclear explosion occurs. The energy is akin to an object 100,000 times brighter than the Sun appearing one second and vanishing the next, and the phenomena reoccur every few hours or days. The first was discovered in 1967, and by the early 1980s about 30 X-ray bursters were known. The source of the more energetic gamma ray bursters is less clear. It may be from a binary star system (similar to the X-ray bursters), a "starquake" on a neutron star, or the impact of an asteroid on a neutron star. They are more numerous than X-ray bursters but never repeat. The most powerful known gamma ray burster occurred on March 5, 1979. It was observed by three

decade a number of manned balloon flights were made to study the rays. Soon after World War II, unmanned plastic Skyhook balloons carried cosmic ray packages to 100,000ft (30,480m). When V-2 and Viking sounding rockets were launched in the late 1940s and early 1950s, they carried similar payloads.

The quest for altitude was prompted by the interaction between cosmic rays and the atmosphere: when a cosmic ray particle hits an oxygen atom it shatters into secondary particles. However, detectors on a satellite above the atmosphere would provide data on largely unchanged cosmic rays for periods of months or years. Of the many early satellites that carried such detectors, the largest were the Soviet Proton spacecraft. Proton 1 was launched in July 1965. It was both the largest satel-

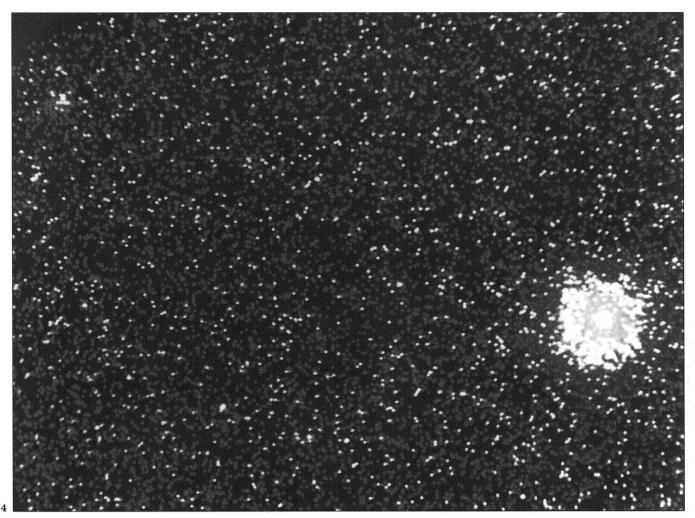

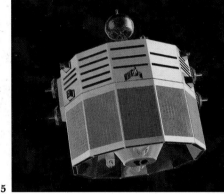

4 *This X-ray image, taken by the second High Energy Astronomy Observatory (HEAO 2 – later renamed Einstein) revealed a newly-discovered object at upper left which it was thought was the most distant and the brightest quasar yet to have been observed to emit X-rays. The light from the object was calculated to have begun its journey more than 10 billion years ago. The bright object at lower right in the picture was another quasar, designated 3C273, which had been discovered about fifteen years before HEAO 2 was launched in November 1978.*
5 *An artist's impression of the COBE satellite. NASA's Cosmic Background Explorer was launched in November 1989 with the objective of obtaining data that could answer some fundamental cosmological questions.*

4   5

Velas, Helios 2 and the Soviet spacecraft Prognoz 7, as well as by Venera 11 and 12. It was traced to a neutron star in the Large Magellanic Cloud, a satellite galaxy of the Milky Way some 186,000 light years away.

International Studies

Scientists followed up the Vela discoveries and X-ray detectors were flown on a number of satellites. Most were to study the Sun. From early sounding-rocket flights it was known that the Sun emitted X-rays, but it was only slowly realized that other stars also emitted them. Because of this it was not until December 1970 that a specialized X-ray astronomy satellite was orbited, when Explorer 42 was launched by a Scout booster from a platform off the coast of Kenya. Because the launch date was Kenyan Independence Day, it was renamed *Uhuru* (Swahili for "freedom"). Within two years, a catalog of 150 X-ray sources had been assembled from Uhuru's data. By 1974 about 35 of these sources had been identified with objects visible on astronomical photos. Uhuru's success resulted in the launch of two more Small Astronomical Satellites (SAS) in 1972 and 1975 (Explorers 48 and 53). These were followed, in turn, by OSO 7 and 8, Britain's Ariel 5, Holland's ANS and the European COS-B and Exosat. In April 1975, Ariel 5 discovered an X-ray nova which, for several months, was the brightest X-ray source in the sky.

The results encouraged NASA to begin development of the High Energy Astronomical Observatory (HEAO). This was one of NASA's highest priority programs in the early 1970s. HEAO 1 was launched in August 1977 by an Atlas Centaur, and in due course it provided a map of 1,500 X-ray sources. HEAO 2 (later renamed Einstein) followed in November 1978 and HEAO 3, which was concerned with cosmic and gamma rays, completed the series in September 1979. HEAO was followed, after a gap of more than a decade caused partly

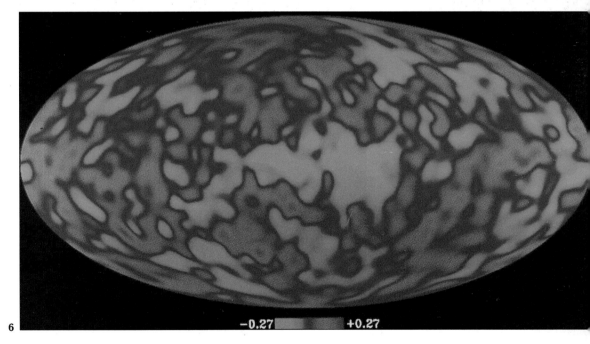

−0.27   +0.27

6

by the Challenger tragedy, by the NASA/West German/UK ROSAT X-ray satellite, which was launched by a Delta II in June 1990. ROSAT also carried a British extreme ultraviolet telescope. The Soviet equivalent is the Gamma-1 high-energy gamma ray observatory based on the Progress spacecraft. Originally planned for a 1984 launch, problems delayed this until July 1990. The most recent systems have an angular resolution many times better than the earlier satellites, and the CIS/French Granat satellite will carry two gamma ray telescopes. Like Gamma-1, Granat has also been delayed for several years.

The US follow-ons to HEAO and ROSAT are the Gamma Ray Observatory and the Advanced X-ray Astrophysics Facility (AXAF); both regarded as being

6 *In spring 1992, it was announced with great excitement that COBE had found evidence of temperature variations in background cosmic radiation which were consistent with predicted "relic" signals resulting from the Big Bang that occurred around 15 billion years ago. This image was released to illustrate the eagerly awaited findings, but in fact it was stated that the color coding revealed mainly instrument noise and that it was computer analyses which had revealed the faint cosmic signals beneath the noise.*

in the series of "great observatories" begun by the launch of the Hubble Space Telescope. Delayed both by the loss of Challenger and by the further shuttle problems in 1990, the GRO was finally launched aboard the orbiter Atlantis in April 1991. At 34,440lb (15,620kg) the heaviest science satellite ever launched, GRO is expected to expand the volume of the observable gamma-ray universe by up to 300 times. AXAF is similar in design to Hubble and also has a planned life of 15 years. Launch from the shuttle is planned for 1996. Both the GRO and AXAF are, again like Hubble, designed for in-orbit servicing. The Japanese are also planning to launch an X-ray satellite: Astro D.

Manned high-energy astronomy studies have also been made. (Studies of the biological effects of cosmic rays and other radiation on crews were important duties during the early manned flights.) Salyut 1 carried the Anna 3 gamma ray telescope. Other Salyut missions carried the FEK-7 photo emulsion camera to detect cosmic rays in the search for "monopoles" (particles with only one magnetic pole, either north or south); the Filin-2 and RT-4 X-ray telescopes; the Yelena gamma ray telescope; and Sitya 4 cosmic ray detectors. The U.S. Spacelab 2 (STS-51F) shuttle mission carried a cosmic ray detector and two X-ray telescopes in addition to

antenna and a 675ft (200m) stabilization boom. These antennas were designed to pick up radio signals from stellar and planetary radio sources. The first RAE was Explorer 38, which was launched into Earth orbit in July 1968 and discovered that the Earth emitted radio signals much like Jupiter. The second RAE, Explorer 49, was launched in June 1973 into a lunar orbit. The Moon blocked radio interference (both natural and man made) and although two antennas did not extend to their full length the spacecraft successfully observed the Sun, Jupiter and objects in deep space. Explorer 49 operated until August 1977 and was destined to be the last US lunar mission for the rest of the present century.

During the Salyut 6/Soyuz 32 mission in 1979, cosmonauts Lyakhov and Ryumin set up the KRT-10 radio telescope. It operated in conjunction with a 229ft (70m) ground telescope to observe radio sources in the Milky Way and the Sun's radio emissions. By combining the two sets of signals (called interferometry), the equivalent of a high resolution, 6,214 mile (10,000km) telescope dish could be formed.

It is only in the decade of the 1990s that the CIS plans to follow up this beginning. At one time, the launch of a number of Radioastrons was planned: 33ft (10m) dish antennas that were to be placed into 435,000 mile

1 *July 1962: the launch of Telstar by a Delta booster. The satellite was privately owned by AT&T in the US and was the first true communications satellite. It was placed in an elliptical orbit and, while its place in history was assured, demonstrated the advantages of a geostationary orbit, where the satellite appeared to be fixed over a given area of the surface from which it was therefore continually in view.*

2 *A Soviet Molniya (Lightning) 1 communications satellite. Much of the Soviet Union was in high northern latitudes so the orbit adopted was a greatly extended ellipse in which the satellite was in view for eight hours at a time – a small network of Molniyas yielding continual coverage.*

the solar telescopes, but the most extensive manned payload is Mir's Kvant-1 module.

Noise from the Galaxy
Radio signals from the Sun, stars and other celestial objects were first discovered in the 1930s, and they began to be studied on a regular basis once World War II had ended. Orbital radio astronomy, however, has been limited. Some early satellites carried radio receivers. The British satellites Ariel 2, 3 and 4 measured radio noise from both the galaxy and very low frequency radio waves from the Earth, but the first dedicated satellites were the Radio Astronomy Explorers. The RAEs carried four 750ft (228m) long antennas, a 120ft (37m) dipole

(700,000km) high orbits. By working in tandem, the antennas would form the equivalent of a dish hundreds of thousands of miles across. Budgetary cuts have now reduced the plan to one Radioastron only, and that in 1994-95. Radio astronomy in space has several advantages over ground-based antennas. As with optical astronomy, an orbital dish can operate in frequencies studied only with difficulty on Earth. Moreover, as indicated above, by placing a telescope in deep space it is possible to create the equivalent of a dish as big as the distance between it and another telescope on Earth. (The distance between two Earth-based telescopes, of course, can be no greater than the Earth's own diameter.) Finally, the size of the dish is not limited by the Earth's gravity. In 1979, the

While most orbital science has been directed outwards, other basic scientific research has been directed to the home planet below. An example of this is geodesy: the study of the exact shape and movements of the Earth's surface. Analysis of Vanguard 1's orbit showed the Earth was very slightly pear-shaped, and to map the slow movements of the Earth's surface NASA launched the Lageos satellite in 1976. It is only 23.6in (60cm) in diameter but weighs 906lb (411kg). In its 3,729 mile (6,000km) high orbit Lageos is not affected by atmospheric drag, and because of the stability of its orbit (known to one metre a year in advance) it is possible to measure the movements of the continents and any changes caused by earthquakes. This is done by bouncing laser beams off the reflectors which cover the outer surface of the satellite. Other, similar satellites are the French Starlette, the Japanese Experimental Geodetic Satellite and the Soviets' Etalon. The launch from a shuttle orbiter of an Italian built Lageos-2 satellite took place in October 1992.

4 *In a clean room at NASA's Goddard SFC, a technician monitors spacecraft response to a laser beam during optical tests on the Lageos-2 spacecraft (Laser Geodynamic Satellite) a joint US-Italian project. The spacecraft was orbited from the shuttle 16 years after the launch of its predecessor in 1976. Both satellites reflect back laser pulses fired from laser ranging stations around the world. By comparing the roundtrip pulse time from one of these ground stations with that of another, scientists are able to determine the amount of movement taking place between the stations concerned with great precision – to less than one inch (about 1 cm). From this, the rates and direction of movements in the Earth's crust can be calculated over a period of time, with important implications for understanding the immense forces at work on and beneath it.*

5

Soviets proposed a possible super radio telescope in space for the 21st century. It would be built from 656ft (200m) diameter modules. Once assembled, the dish could be 0.6 to 6 miles (1 to 10km) across, far larger than any Earth-based dish, and several satellites would be positioned to act as antenna feeds. With two such super dishes, one in Earth orbit and the other at the orbit of Saturn, it would be possible in effect to have a dish 932m miles (1,500m km) across. With such an instrument, it would be possible to study the planets of other stars.

Search for Extraterrestrials
Potentially the most difficult and rewarding task for a radio telescope is SETI: the Search for Extraterrestrial Intelligence. This involves listening for radio signals from non-human civilizations. The task is daunting: the radio spectrum is wide and neutron stars, quasars and other natural radio sources pour out noise. It has been likened to "searching the equivalent of the Encyclopedia Britannica every second to find the part that says 'Hi, we're the aliens'." If such a signal were to be discovered, it would have an impact on every part of human society.

Not surprisingly, SETI has had funding problems. In the late 1970s and early 1980s, Congress removed all SETI funding from NASA's budget. In fact, at one point the agency was actually banned from spending any money on it. Although funding was later permitted, in 1990 the $12m sought was again removed by the House of Representatives, congressmen waving clippings of flying saucer sightings which they linked with SETI. However, full funding for the SETI project was later restored in the House/Senate conference committee.

Communications Satellites
While scientific data from satellites have opened mankind's eyes to the wonders of the universe, it is in the area of applications that space has had the greatest impact on everyday life, and nowhere more so than in the case of communications satellites, which have fundamentally changed business, politics and leisure.

The modern concept of the communications satellite was first proposed by Arthur C. Clarke, the British science fiction writer. In the October 1945 issue of *Wireless World* magazine he proposed placing satellites into a circular orbit 22,300 miles (35,880km) high. At this altitude a satellite takes one day, that is, one Earth revolution, to orbit the Earth. The result is that it seems to stay fixed at the same point in the sky as seen from the Earth's surface. With three such satellites in a "geosynchronous" or "geostationary" or "Clarke orbit," it is possible to cover the entire world. It is worth remembering what the telecommunications situation was like in the 1940s and 1950s. A transatlantic telephone call was made over cables and a wait of hours might be required before a circuit became available. World wide radio was possible using shortwave, but solar activity could disrupt it. Television pictures could be transmitted "live" across a country via relay towers but, because TV signals only traveled "line of sight," it was not possible to transmit, for example, across the Atlantic. Film of an overseas event, such as the Olympics, had to be flown to a recipient country.

The early communications satellites tests were all in low orbit. (A geosynchronous orbit requires a larger booster and is more technically demanding.) The first was Score, launched on December 18, 1958. It was the core stage of an Atlas ICBM equipped with a tape

3 *Echo 1 and 2 were passive communications satellites, launched in 1960 and 1964 respectively, which were targets off which signals were bounced. The future lay with active satellites, but in the 1960s the enormous balloons were prominent objects in the night sky.*

5 *Technicians work on Telstar 2, which was launched in May 1963. Power was supplied from 3,000 solar cells on its surface.*

6 *Syncom 1 – built by Hughes Aircraft – was designed to hover over a fixed longitude, but its orbit was inclined so that it moved 33° north and south of the equator. It failed almost immediately, but two later satellites were more successful.*

recorder and transmitter, and carrying a recorded Christmas message from President Dwight Eisenhower (one of the few such space gestures he made). The Score recorder was also used to store messages for retransmission later when the satellite passed over a ground station. This is called "storage-dump," and a similar procedure was used by the Courier 1B satellite launched in October 1960.

The US also tested passive satellites: Echo 1 and 2. These were large, aluminized balloons (Echo 2 was 135ft – 41m – across) and the signals were simply bounced off the skin. Although bright objects in the night skies during the 1960s, the Echos showed the need for "active" satellites which could re-transmit signals.

The first true communications satellite was ATT's Telstar 1 launched on July 10, 1962. Later that day, the first test TV transmissions were made from the US to France and Britain: an American flag waving in the breeze, with the Andover, Maine ground station in the background. It was "Live via Satellite," and the world would never be the same again. Although Telstar 1 and 2 (launched a year later) and the Relay 1 and 2 test satellites were successful, they showed the limitations of a low-orbit system. ATT at one time envisioned a 30 to 50 satellite network, which would be costly to build and operate. It was soon realized that a single geosynchronous test satellite could also provide an operational system.

The first geosynchronous satellite was NASA's Syncom 1, which was launched on February 14, 1963. The Delta booster placed Syncom 1 into an elliptical transfer orbit up to the required 22,300 miles (35,880km) high, but when the on-board engine fired to circularize the orbit – the standard method of achieving the geosynchronous station – Syncom 1 fell silent. Syncom 2, launched in July of the same year, was more successful and was positioned over the Atlantic. Syncom 3, launched just over a year later, was used for the live broadcast of the opening ceremonies of the Olympic Games from Japan.

Establishment of Intelsat
To operate the communications satellites, the US Congress passed legislation in 1962 establishing the Communication Satellite Corporation (COMSAT). In August 1964, the International Telecommunications Satellite Organization (Intelsat) was established. Intelsat acted as an international coordinating body, assigning orbital "slots" in geosynchronous orbit and radio frequencies. With both COMSAT and Intelsat established, it was time to move to commercial operations. (Inmarsat is a similar international organization which has established a maritime satellite communication network.)

Intelsat 1, better known as Early Bird, was launched on April 6, 1965, and provided 240 telephone circuits or one television channel. With its success, the Intelsat 2 series began. The first Intelsat 2 (F-1) was launched in October 1966 but failed to achieve final orbit due to a failure of its on-board engine. Three more launches followed in 1967: two satellites were placed over the Pacific and one over the Atlantic. Although similar in size and capability to Early Bird, the Intelsat 2s represented a considerable economic advance: a single Early Bird circuit cost $30,000, but on Intelsat 2 satellites this fell to $10,000.

The Intelsat 3 series followed with launches from September 1968 to July 1970. Despite problems (of the eight satellites launched, three failed to reach geosynchronous orbit and another failed because of an antenna malfunction), each Intelsat 3 provided 1,500 phone circuits or four television channels, or a combination of the two. Costs dropped to $2,000 per circuit. Appropriately, it was the Intelsat 3 satellites that relayed television images of the Apollo 11 Moon walks to a world audience estimated at 600m. The Pacific satellites also relayed films of Vietnam combat, a major reason why the war became the first to be "fought" in America's living rooms.

Next was the Intelsat 4 series, with 6,000 telephone circuits or 12 television channels. Seven were launched by Atlas Centaurs between January 1971 and May 1975. Their antenna system was found to be the limiting factor: any more circuits and the calls would interfere with each other. So a new antenna system was built for the Intelsat 4A, giving 6,250 telephone circuits and two television channels. The first was launched in September 1975 and the sixth (and last) in March 1978.

Molniya's Unique Orbit
The Soviets were also launching communications satellites. The first Molniya 1 (Lightning) launch took place in April 1965: the booster was an A-2e. The Molniya satellites used a unique orbit: 334x24,420 miles (538x39,300km) with a period of 12 hours. This high, looping orbit meant the satellite would be visible from the Soviet Union for eight hours at a time, so complete coverage could be achieved with a network of three or more Molniyas. Because of Russia's northerly location, the highly-inclined and elliptical orbit gave better coverage. The first of an improved version, the Molniya 2 series, followed, beginning on November 24, 1971. Nine days before this, the Soviet Union, Cuba, Bulgaria, Hungary, Romania, Mongolia, Poland, East Germany and Czechoslovakia had signed an agreement setting up the Intersputnik group. The first of the Molniya 3 series was orbited in November 1974.

Military Comsats
During the 1960s and 1970s the US was also developing a separate military communications satellite system. The first attempt was the West Ford project: 400m copper "needles," 0.7in (1.8cm) long, were to form a ring around the Earth. Although it worked, astronomers opposed the project and passive satellites were seen as outmoded.

The first operational US military communications satellite was the Initial Defense Communication Satellite System (IDCSS), later renamed Defense Satellite Communication System 1 (DSCS 1). These were launched eight at a time aboard Titan IIICs into sub-synchronous orbits. The satellites seemed to drift slowly across the sky, taking four days to go from one horizon to the other. The first of four sets was launched on June 16, 1966. They were replaced by the DSCS 2 satellites: the first pair being launched in November 1971 by a Titan IIIC. These were larger, more versatile satellites placed into a geosynchronous orbit, but there were early technical problems. Of the first four, three soon failed and the third launch ended in a booster failure. Subsequent DSCS 2s faired better. The replacement was the DSCS 3. The first launch took place on October 30, 1982, which was also the first Titan 34D launch, and two of the later DSCS 3s were deployed from the shuttle during the first mission of Atlantis in October 1985. For some years the Satellite Data System (SDS) was used to relay data from the KH-11 reconnaissance satellites but, when SDS was phased out in the late 1980s, the relay function was switched to the DSCS 3s.

The US Navy uses the Fleet Satellite Communication System (FLTSATCOM), the first of which was launched in February 1978 by an Atlas Centaur. The program was successful but encountered difficulties at its end. The seventh FLTSATCOM was lost on March 26, 1987,

1 *The first weather satellite – Tiros 1 (Television Infrared Observation Satellite) – went into orbit in the spring of 1960. With many more to follow, the satellite was small, as can be seen in this picture.*

2 *As time went on, it seemed that satellites invariably became bigger and bigger: compare Tiros with this GOES (Geostationary Operational Environmental Satellite) launched 26 years later: 14ft 7in (445cm) high and 7ft 1in (216cm) in diameter.*

3 *This particular GOES – G – had a short life, its Delta booster being destroyed by the range safety officer just over one minute into the flight.*

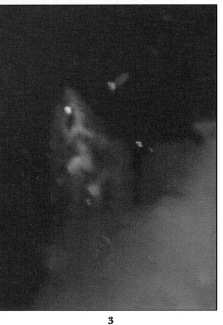

when the booster was struck by lightning after launch, and the final launch was delayed for a considerable period of time when the Centaur stage was damaged during processing by a service platform, launch finally taking place in September 1989. By that time, the Navy had begun using the Leasat. The first, Leasat 2, was launched from the shuttle in August 1984 and Leasat 3 followed in April 1985. Leasat 3 was left stranded in a low orbit by an engine failure, but the satellite was repaired by the crew of the orbiter Discovery in August 1985, when Leasat 4 was deployed. January 1990 saw the launch of Leasat 5 from Columbia.

Soviet military communications satellites are more straightforward. From the 1960s through the 1980s, the Molniya 1 satellites were used for military communications, a storage dump type satellite also being operated. Originally these were launched eight at a time on a C-1 booster, but in the late 1980s they were switched to the F-2 booster and are now launched in a "six pack." Large storage dump satellites were launched singly on C-1 boosters from 1967 through the mid-1980s and were believed to be used for radio transmissions to and from Soviet agents in the West.

Russian Geostationary Comsats

Technically, communications satellites went in several directions in the 1970s and 1980s. One was the continued development of larger, more capable satellites. The first Intelsat 5 was launched in December 1980 by an Atlas Centaur. Unlike earlier Intelsats, it was box-shaped with two solar "wings." In all, nine Intelsat 5s were launched between 1980 and 1984; the seventh and eighth by Ariane boosters from French Guiana. Each had 12,000 telephone circuits and two television channels. The Intelsat 6 satellite reverted to the classic, cylindrical, spin-stablilized form. Some 39ft (11.7m) tall, it has 120,000 telephone circuits and three television channels. Originally, the series was designed for launch aboard the space shuttle as well as Ariane 4, but after the loss of Challenger the shuttle was replaced by the commercial Titan III. The first Titan III launch of an Intelsat 6, however, left the satellite in a low orbit and a rescue mission by the crew of the replacement orbiter Endeavour was carried out in 1992. The first launch of an Intelsat 7 is scheduled for the following year.

In due course the Soviets made the move to geosynchronous orbit. The initial satellite was the Raduga series, commencing in December 1975. This was followed by Ekran satellites, the first being launched in October 1976, which relayed Moscow television transmissions to small ground stations. Next, from December 1978 onwards, was the Gorizont series. All three types used the D-1e launcher. During this time, the Molniya 2 series was phased out, and when in 1980 the Moscow Olympics had a world-wide audience of up to 2,500m people the transmissions were relayed via the Raduga/Ekran/Gorizont satellites.

Domestic Uses

A further, important development was the start of domestic satellite communications. Canada was the first western country to have such a system. This was provided by the Anik satellites launched between 1972 and 1985, the last three on the space shuttle. Indonesia, India, China, Saudi Arabia, Brazil, Mexico, Australia and Japan soon followed.

In the US, domestic phone and television services were provided by a host of satellites: Westar, Comstar, Satellite Business System, GSTAR and Spacenet. The bulk of the payloads carried aboard the shuttle between 1981 and 1986 were communications satellites. When these satellites were removed from the shuttle, commercial launch services the Atlas II, Titan III, Delta II, Pegasus, Proton (USSR) and Long March (China) boosters sprang up, in addition to the highly successful Ariane. The major share of their payloads are also communications satellites reflecting what now appears to be a basic human need for information and entertainment, which has in turn introduced a significant degree of space commercialization. Domestic communications satellites have changed the face of US television. Advances in receiver technology have resulted in a small dish in the backyard being able to pick up the satellite transmissions directly. Because of communications satellites, half the Earth's population can watch the same

4

news or sporting event. When an international crisis breaks, it can be covered live from the scene. It does not take days or weeks for news to spread; it now takes fractions of a second.

The Coming of Tiros 1

Another area where space applications have had a great human impact is weather forecasting. In 1900 the town of Galveston, Texas, was destroyed by a hurricane that hit without warning, and in 1938 another hurricane took the US East Coast by surprise.

The use of satellites to observe weather, to track storms and prevent such surprises, had been proposed early on, and the first weather satellite in the US was Tiros 1 (for Television Infrared Observation Satellite), launched on April 1, 1960, by a Thor Able II. Nine more launches, all by Deltas, followed in the period to 1965. Television cameras photographed clouds, while infrared instruments on some of the satellites determined their altitude and temperature. Tiros 8, launched in 1963,

5

4 *In August 1969 Hurricane Camille hit the headlines in the US, and it is shown here in false color at the time it was crossing the Gulf coast. At that time the digital manipulation of data from satellites was relatively new and pictures such as this, taken from the Nimbus 3 spacecraft's high resolution infrared radiometer (HRIR), looked strange and exotic. In fact, this was a way in which the computer was used to highlight subtle differences in the density of visual signals to make them easier to see and interpret. The image was "sliced" and various colors arbitrarily given to the different slices.*

5 *This is a simulated natural color image of a black and white original from a meteorological satellite.*

was the first to carry Automatic Picture Transmission (APT) equipment which, as soon as they had been taken, transmitted images that could be picked up by simple, home-built receivers below. Almost 900 APT stations were built in 123 countries.

The shift to an operational system came in 1966 with the ESSA (Environmental Science Services Administration) program. The satellites were based on the earlier Tiros series and two were operational at any given time. The even-numbered satellites carried the APT equipment while odd-numbered satellites carried the Advanced Vidicon Camera System (AVCS). ESSA-1 was launched from Cape Kennedy on February 3, 1966, but the rest were launched from Vandenberg AFB in California. The last, ESSA-9, was orbited on February 26, 1969.

The NOAA series (for National Oceanic and Atmospheric Administration) followed and is still current. The earlier satellites in the series were also called ITOS (for Improved Tiros Operational System) and were box shaped, with an array of solar panels. The first was launched in January 1970 by a Delta booster from Vandenberg. The satellite combined the AVCS and APT flown separately on the ESSA satellites, and also carried a two-channel infrared scanner, together with a radiometer to measure the Earth's atmospheric heat. NOAA-2 was an improved satellite with equipment to

environmental phenomena such as ozone depletion over Antarctica.

The Soviets were slow to fly weather satellites. The first full test was Cosmos 122 in June 1966. It was followed by five more Cosmos test missions in 1967 and 1968. The operational system was based on the Meteor 1 satellite, the first of which was orbited in March 1969. Two Meteor satellites were operational at a given time, and in 1971 they began carrying APT equipment compatible with the Western system. The Meteor 2 series was an improved system that provided data on cloud cover and height, snow cover, and sea and air temperatures. The first was launched in July 1975. The follow-on Meteor 3 series was introduced ten years later with a first launch in October 1985. There were problems, however, and it was not until 1988 that the second was launched.

DMSP and Tiros N

The US Air Force had also flown its own weather satellites. The reason for dual civilian and military systems was incompatible data needs: civilian weather forecasters wanted the broad picture, whereas the Air Force needed to know if a small area would be clear of clouds: information required for planning photo reconnaissance coverage. Called the Defense Meteorological Satellite Program (DMSP), launches began in early 1965 and have continued to the present. Two are operational at a time. The latest version, the DMSP Block 5D, provides cloud images in both visible and infrared light, and data on temperature, moisture and auroral activity.

The DMSP Block 5D served as the basis for the civilian Tiros N weather satellite. This carried the Advanced Very High Resolution Radiometer (AVHRR) for securing visible and infrared cloud images. Other instruments yielded data on air temperature, water vapor, ozone content and snow coverage. The satellite could also receive transmissions from buoys, balloons and ground weather stations. Tiros N, a dual test/operational satellite, was launched in October 1978. NOAA-6 followed in June 1979 and three more between 1981 and 1984. NOAA-8 carried an international search and rescue package that could pick up signals from aircraft and ship's emergency locator beacons, and soon after launch it began racking up an impressive number of "saves." The current operational satellites are NOAA-11 and 12, with further launches planned throughout the present decade.

Geostationary Metsats

At the same time, geosynchronous weather satellites were being introduced that could cover much of an entire hemisphere throughout the day. This was the SMS/GOES system (Synchronous Meteorological Satellite/Geostationary Operational Environmental Satellite). SMS-1 was launched in May 1974, SMS-2 followed in February 1975, with operational GOES spacecraft launched in 1975, 1977 and 1978. The GOES concept was also pursued by the European Meteosat satellite, the Japanese Geostationary Meteorological Satellite (GMS) and the Indian INSAT, which also made use of geosynchronous orbit.

In 1978-79, the Global Atmospheric Research Program (GARP) was undertaken. This was an intensive, international weather study using ground observations, ships, buoys, balloons, sounding rockets and satellites. The goal was to determine the practical limits of weather forecasting. The space segment of GARP consisted of three GOES, Meteosat and GMS in geosynchronous orbit, with Tiros N and the Soviet Meteor satellites in low polar orbit. GARP also saw a dramatic demonstration

1

1 Meteorological satellites quickly became an invaluable tool of the forecaster and scientist. However, to the public the images sent back to Earth can appear unexciting. This is where manned missions are valuable because, quite apart from the scientific or technical merit of pictures of the Earth's weather systems, color photographs have an unrivaled impact. This STS-35 image taken in December 1990 shows thunderstorms over the Indian Ocean.

take images both day and night and to monitor atmospheric temperature. Eight ITOS/NOAA satellites in all were launched, two of which suffered booster failures, and the last launch took place in July 1976.

At this same time NASA's Nimbus series was under way. These were research satellites which carried test instruments later flown on operational satellites. Nimbus-1 was launched in August 1964 and Nimbus-7, the last of the series, was orbited on October 24, 1978. A number of instruments aboard Nimbus-7 were still operating at the end of 1990 and obtaining valuable data on critical of the value of satellites. In August and September of 1979, hurricanes *David* and *Frederic* hit the Caribbean and US East and Gulf coasts. Before satellites, the death toll could have been in the thousands: with satellites operating, good warning could be given.

Following GARP, an improved GOES was developed. GOES-4 was launched in September 1980 with GOES-5 and 6 following in 1981 and 1983. One GOES was lost in a 1986 Delta failure, but GOES-7 was orbited in 1987. Extensive improvements are planned for the GOES-Next satellites to be launched in the 1990s, including a search and rescue function which will overcome the weakness of polar orbiting satellites in not always being in line-of-sight of an emergency.

REMOTE SENSING OF EARTH RESOURCES

Remote sensing is the science of obtaining information about an object via some device without being in direct contact with the object. Our eyes and ears could be considered as remote sensing devices, but the term is most commonly associated with observations of the Earth's surface obtained by sensors mounted on aircraft and satellites.

Different materials such as sand, ice and water reflect or emit energy to a varying degree due to their different

2

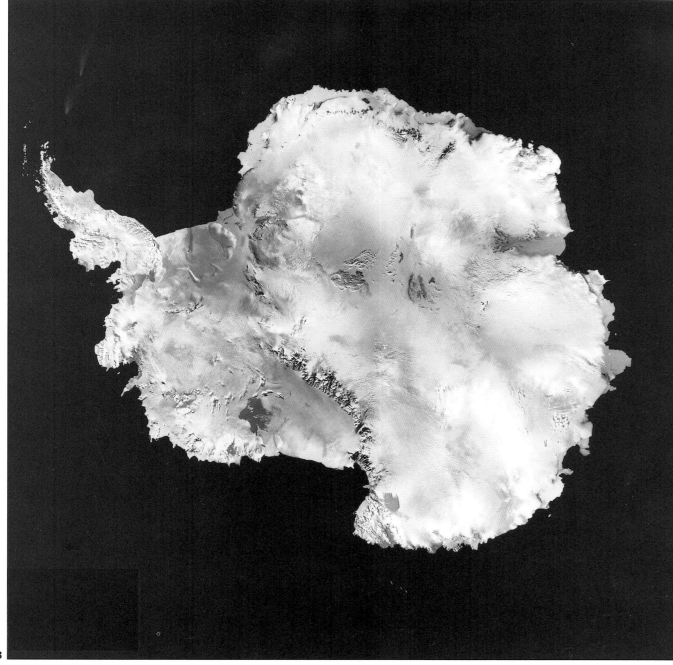

4

5 *For well over a decade the ozone "hole" over the Antarctic has been monitored by NASA's Nimbus 7 research satellite, which has a dedicated instrument aboard called the Total Ozone Mapping Spectrometer (TOMS). This picture plots the hole in October 1990: reds and oranges indicate high ozone concentrations and blues and purples low concentrations. In the fall of 1992 it was reported that TOMS showed the hole to be the largest on record. The shape of Antarctica is overlaid.*

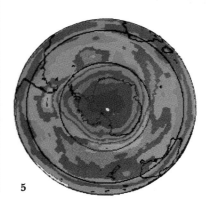

3

5

The value of manned spaceflight is man's ability to discriminate what is studied and photographed; nowhere is this more important than in monitoring the environment. Both of these pictures were taken on the resumption of shuttle flights in September 1988 during the STS-26 Discovery mission.

1 *The Amazon region is obscured by smoke as forest clearing and burning continues apace. The smoke cloud was the largest seen by astronauts to that date and, if it had been over the US, would have covered an area three times the size of Texas.*

2 *Desertification in Africa has been constantly monitored, and Lake Chad in the fall of 1988 was significantly smaller than when photographed from the shuttle in January 1986.*

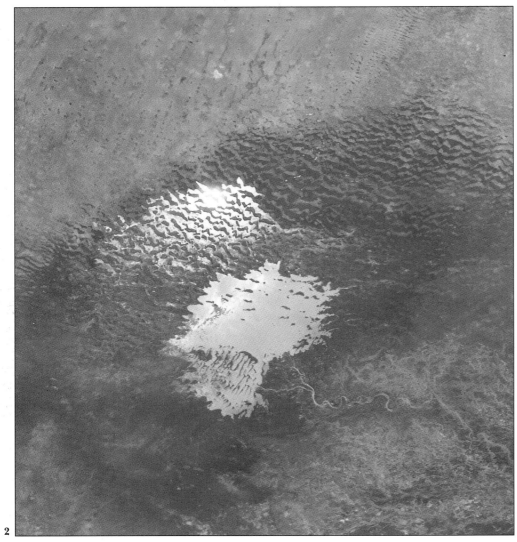

3 *Geostationary satellites afford excellent views of large regions of the Earth's surface but, particularly for the higher latitudes, images suffer from perspective distortion. This superb mosaic of Europe and the Mediterranean region by the UK's DRA was built up from around 25 different images taken by NOAA's polar orbiting Tiros-N satellites between 1979 and 1985. Three wavebands from the satellites' advanced, high-resolution radiometer were used and the resolution in the simulated natural color scene is 1km.*

wavelengths of sunlight reflected back from the Earth's surface and also the Earth's naturally emitted energy: heat. They record the amount of energy received at these wavelengths. "Active" sensors, such as radar, transmit their own energy signal and record changes in the nature and amount of that signal when it is reflected back from the Earth. The information from both types is recorded as a set of digital numbers for a given area on the Earth and these can be used to generate computer images of that area, which are often presented in the at-first-sight bizarre form known as "false color."

Remote sensing satellites operate in both geo-synchronous and polar orbits. The considerable distance of the former, at some 22,300 miles (around 35,880km) from the surface, limits the detail or ground resolution that can be obtained, but they maintain a constant watch on somewhat under one hemisphere of the Earth. Greater resolution is available from polar orbit at around 435 miles (700km) high, in which the satellites follow a fixed path while the Earth rotates below, with the same point being observed at intervals of several days or weeks.

These spaceborne sensors offer an excellent vantage point for observing, measuring and monitoring the Earth's phenomena. Large areas of the Earth can be monitored at one time, rapidly and on a regular basis. Also, because the image information is digital, it can be transmitted directly to receiving stations positioned all around the world and then processed by computer. Observations of the atmosphere, the oceans and land surfaces are regularly obtained and used in a wide variety of applications.

Experience in the Air

Although satellites belong to the space age, remote sensing is not such a new concept. As early as 1840, not long after photography had been publicly demonstrated, the use of airborne photography for mapping was advocated. Indeed, the use of photography from balloons and kites flourished in the latter half of the nineteenth and on into the early years of the present century.

The development of aircraft in the 1900s made aerial photography a more practical proposition, and the onset of World War I ensured a greater interest in its application. The need and ability to obtain reconnaissance information from such remotely-sensed images became highly significant, particularly during World War II. Improvements during this period in camera technology and the use of film sensitive to different regions of the electromagnetic spectrum led to many of the developments in sensors and image interpretation techniques used today.

Of course, aerial photography is still widely used in many military and civilian applications, but the move to space has given remote sensing a totally new and productive vantage point. Manned space-flights using mainly film cameras played an important role in establishing the value of remote sensing from space and still continue to contribute much of value today, but it is fair to say that unmanned satellites, together with developments in sensors, computer processing power and communications technology, currently form the "cutting edge" of the new applications.

Weather satellites were the first operational remote sensing satellites, but they have had a value greater than their direct contribution to weather forecasting and research. For example, NOAA images are used to map seasonal changes in vegetation and thus monitor desertification and pest habitats such as locust breeding grounds. Snow and ice mapping are used to estimate potential flood risks and dangers to shipping, while oil

physical properties. Also, the amount of reflectance or emittance from a particular surface varies depending on the wavelength of the energy that acts on it. Thus, materials can appear quite different when exposed to the blue, green and red light to which our eyes are sensitive, or energy from the infrared part of the spectrum which is invisible to human eyes. By observing energy responses in regions of the spectrum where these differences are distinct, measurements can be made which enable surface materials to be identified and monitored. This "multispectral" information provides a unique way of observing the surface features of the Earth.

There are two main types of sensor carried aboard satellites. "Passive" sensors are sensitive to particular

1 *This radar image from an early flight of the shuttle has a radar image in black and white superimposed on the Landsat scene of the area. The wind-blown sand has been penetrated to reveal previously unknown subsurface drainage features in detail.*

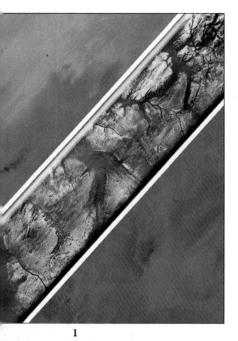

1

slicks and gas flares can be identified and their progress monitored. The last in the Nimbus series of satellites carried a Coastal Zone Color Scanner (CZCS), which was the first purely oceanographic system put into orbit. The instrument measured sea temperature and concentrations of plankton and sediments in estuaries, bays and ocean areas; measurements particularly useful for establishing pollution levels and water quality, evaluating potential fisheries and furthering ocean circulation studies such as movements of the Gulf Stream.

The Coming of Landsat

But the major breakthrough in the remote sensing of Earth resources from space came with NASA's inauguration of the Landsat series of satellites. ("Landsat" was a later name change: the first satellite was called ERTS-1: for Earth Resources Technology Satellite.) These were the first to be designed specifically for the collection of high definition information about the Earth's surface and natural resources, and five Landsats have been launched so far, the first in 1972. Originally designed to record multispectral data passively in four regions of the reflected spectrum, with a ground resolution of about 260ft (80m), the series has developed into a sophisticated Earth observation system. Measurements from seven specific regions of the spectrum (including temperature) and a ground resolution of 100ft (30m) are standard on the Landsat satellites that were launched in the 1980s.

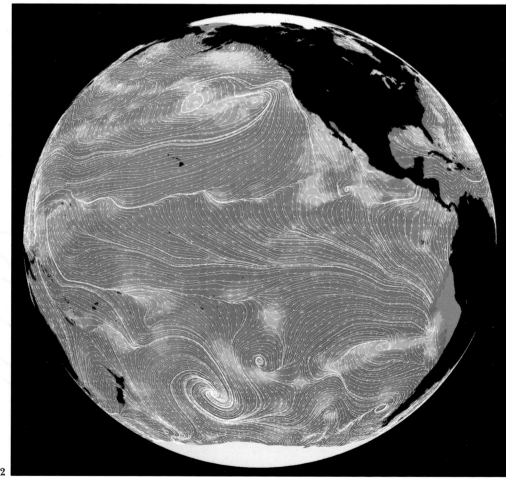

2

2 *The direction of surface winds over the oceans is represented here by white lines with arrows, and wind speed by colors: blue for lowest and yellow for highest. Over 100,000 measurements taken by the Seasat satellite over a 12-hour period were used in this computer-generated image.*

Landsat data are widely applied in many fields. Reproduced here is an image showing the potential for geological mapping around Mount Vesuvius, the only active volcano on the European mainland. Mapping such terrain features in certain spectral regions is important in oil and mineral prospecting, but other land features can also be identified in the image. The port of Naples, its wharves, suburbs and other towns are clearly picked out in light blue, which enables the

mapping of urban spread and change. Forest areas, upland vegetation, fields of different crops, roads, airports and industrial sites are also observable due to their different spectral responses. The image can therefore provide an excellent means of producing land use maps, land resource inventories, estimates of crop yields and information for planning, monitoring and evaluating environmental changes. Sea, lakes and rivers are all easily distinguishable, and coastal sediments are revealed along the beaches. Coastal management, circulation studies and pollution and water quality monitoring are therefore possible.

The Landsat series is expected to continue into the 1990s, providing regular updates of information for these and other future applications. Further improvements including a 50ft (15m) resolution sensor for black and white images are planned for the next generation Landsats 6 and 7. (In the summer of 1992 there were reports of a five-metre resolution "goal" for Landsat 7).

Radar over the Oceans

Seasat – NASA's first satellite radar system – was launched in 1978. Although crippled by a power failure after only three months operation, it allowed the first analysis of information from a spaceborne active remote sensing instrument. Such systems have the unique advantage of being able to penetrate cloud and to make observations in all weathers as well as at night. Seasat had a resolution of over 80ft (25m), and in its short life produced images for a wide range of applications. Oceanographic studies of wave and tidal patterns have enabled the mapping of sandbanks and ocean currents, while measurements of the mountains and trenches of the sea bed terrain have even been made from the data. High definition maps of geological structures on land can be produced which are especially useful in poorly mapped areas, where cloud cover inhibits other sensors. Such maps identify faults and folds in the surface and allow studies of drainage networks and watersheds. Ice mapping despite clouds is useful for surveying purposes but also for locating icepacks and icebergs that might pose a threat to shipping. (Ships themselves can be observed.) In addition, agriculture and hydrology agencies can make use of vegetation mapping and estimates of soil moisture.

The 1980s saw much consolidation of remote sensing as a science, with satellite data and techniques becoming more sophisticated and widely used. Also, remote sensing technology has moved more into the commercial domain from the research sector in recent years. Consequently, the late 1980s and early 1990s has been a period when new initiatives, particularly European but also Asian, have come to the fore.

Developments in France

SPOT – *System Pour l'Observation de la Terre* – is a series of French satellites which brings new features and concepts to remote sensing. SPOT-1 and 2 (launched in 1986 and 1990 respectively) record reflected sunlight in spectral regions similar to the early Landsats, with a ground resolution of 65ft (20m). They can also produce black and white images with 32ft (10m) resolution and have the unique ability of pointing their scanners to view the same point on the Earth from different orbits. This ability provides images similar to stereo aerial photographs and facilitates measurement of the height of the land surface. Combined with the best ground resolution currently available from civilian satellites, this makes SPOT images very useful for land cover mapping, geological and hydrological studies, urban mapping and even topographic map preparation and updating.

ERS-1 is the first of a new generation of European satellites for the 1990s. Due for launch in mid-1991, this system is intended primarily for oceanographic and ice mapping applications. It will carry two active radar instruments designed to provide information on ocean circulation, ice sheets, oil detection and relationships between wind and waves. Some geological and vegetation applications are also planned, while two passive devices will provide images of surface temperatures and atmospheric water vapor. Somewhat later in the decade, Canada plans to launch its Radarsat.

Soviet Activity

The Soviet Union has been active in space remote sensing over the same period as western countries but has only recently started providing data for the civilian and commercial sector. Operational flights of Meteor-Priroda (Meteor-Nature) satellites, broadly similar to Landsat, began in 1981 and the Soviets launched their first multispectral/radar oceanographic spacecraft (Cosmos 1076, also called Okean or Ocean) in 1979. The first civilian remote sensing satellite was launched as recently as 1988. Resurs-O (Resource) provides images of comparatively large areas from two passive sensors with ground resolutions of about 150ft (45m) and 560ft (170m) respectively. The former measures in spectral regions similar to the early Landsats, while the latter is similar to SPOT. Photographic images with 16ft (5m) resolution are available from the military Resurs-F series, which typically only fly for two or more weeks at a time although the lifetime can be prolonged by the addition

operational remote sensing to be conducted. The pressing global environmental issues of today make it safe to assume that remote sensing technology will continue to develop and improve to provide unique information about the Earth well into the 21st century.

More Accurate Navigation

Navigation is yet another space application. The US Navy needed a world-wide system that could provide all-weather navigation for Polaris missile submarines. The first navigation satellite was Transit 1, in a program that became operational with the launch of Transit 5A1 in December 1962. The satellite, in a known orbit, transmitted signals and a ship could calculate its location by using the doppler shift of the radio frequency. (This is the change caused by the satellite's orbital motion.) It was accurate to 200 to 600ft (61-183m) and the system was declassified for civilian use in 1967.

The Soviets were also building missile submarines in the late 1950s and early 1960s: the Hotel, Golf and Yankee classes. They also needed navigation satellites, the first of which was Cosmos 192 launched in November 1967 by a C-1 booster. The system went through four generations using the same procedures as Transit and was released for civilian use in 1978. The Soviets call it Cicada, because the radio beacon sounds like a chorus of insects. At least three of the satellites carried search and rescue equipment.

During the 1960s and 1970s, a number of test navigation satellites were launched in the US: Lofti, TRAAC, 3

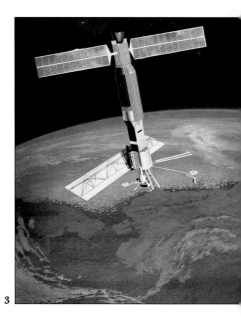

3 An artist's impression of the Seasat satellite. Oceans cannot be monitored over wide areas and at frequent intervals by any means other than from satellites in space: Seasat demonstrated convincingly in its short life of less than four months in 1978 the value of a satellite dedicated to the remote sensing of the oceans with a battery of microwave sensors.

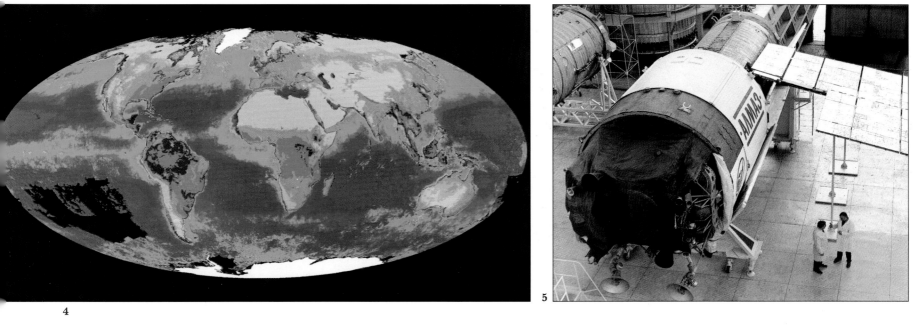

4

5

of solar panels. (Four of the satellites were launched in 1991.) In the late 1980s radar images with 100ft (30m) resolution became available from military spacecraft, and a massive, unmanned radar remote sensing spacecraft based on a Salyut space station was launched in March 1991. Called Almaz (Diamond) it is dedicated to the commercial exploitation of all-weather images of the Earth's surface both at home and abroad.

Japan and India are also increasingly active in remote sensing. Both have launched polar orbiting systems for marine and land monitoring during the 1980s, the latter having similar capabilities to the early Landsats, and more are planned in the year ahead. Internationally, future plans for remote sensing are linked to the development of the wide-ranging program "Mission to Planet Earth." The instruments to be flown are still under development and studies are currently proceeding to determine the best options for the experimental and

Timation, Tip/Triad/Nova and Transat, to develop a system that could locate a ship within tens of feet and was usable by aircraft. This finally emerged as the Navstar/Global Positioning System (GPS), in which location is determined using time signals from a network of 18 satellites. The civilian system provides an accuracy of 98ft (30m) and the military version 30ft (10m). The first Navstar was launched in February 1978, and an improved version followed in 1982. After the Challenger accident, most future Navstar launches were shifted to the Delta II booster. In October 1982, a Soviet D-1e booster launched the first three Glonass (Global Navigation Satellite System) satellites: Cosmos 1413, 1414 and 1415, which use the same procedures as Navstar. Western space analysts call them "Navstarskys." In late summer 1992 it was thought likely that the Glonass system would not be fully operational until 1996-97.

4 NASA's Goddard SFC claims this to be the first image of the global biosphere, produced by combining data from Nimbus 7 for the oceans and from the NOAA-7 weather satellite for the land. The color coding of Nimbus 7 data basically shows the concentrations of marine phytoplankton – red and orange being the highest. NOAA data represents chlorophyll and leaf mass – the dark green areas (rain forests) revealing the highest potential for the production of chlorophyll.

5 March 1991: At Baikonur in Kazakhstan the Soviet Almaz radar remote sensing satellite is prepared for launch. It dwarfs the technicians standing nearby.

16

THE MILITARY IN SPACE

One of the most important yet least understood space activities is that of military reconnaissance. Today, the policies, actions and safety of nations are shaped by images and electronic signals that pour down from "spy satellites." The roots of today's reconnaissance satellites extend back to the years just after World War II. As relations with the Soviet Union worsened, the US felt a desperate need for information on the Russian military to prevent another Pearl Harbor. Several approaches were tried.

Agents were parachuted into Russia but they were captured as soon as they hit the ground. Aerial reconnaissance was also used. US Air Force and Navy aircraft flew along the Soviet border carrying receivers to pick up Soviet radio and radar signals. Starting in 1950, a program of short-range overflights of the Soviet Union and the People's Republic of China covered ports, coastal areas, islands and border areas. The US Air Force also equipped large Skyhook balloons with cameras. During January and February 1956, 448 balloons were launched from bases in Western Europe and drifted across Russia. When they reached Japan, C-119 recovery aircraft attempted to catch the gondolas in mid-air as they descended under parachutes. The results were, at best, mixed; only 44 (about 10 percent) of the gondolas being recovered. The 13,813 photos covered some eight percent of the Sino-Soviet land mass and the high loss rate soon brought an early end to the project.

On July 4, 1956, the first U-2 overflight was made of the Soviet Union. Designed to fly at 70,000 to 80,000ft (21,336 to 24,384m), the aircraft was beyond the reach of Soviet MiGs and anti-aircraft guns. The overflights proved to be a tremendous advance in intelligence gathering and around 90 percent of the intelligence the US had on the Soviet Union came from the U-2, but there were problems. The U-2 had a limited range so some areas could not be covered. More importantly, President Eisenhower was reluctant to authorize U-2 overflights since he feared they might cause a war with Russia, and only about 30 flights took place before Francis Gary Power's U-2 was shot down on May 1, 1960. This meant the US could not solve the foremost intelligence question facing it in the late-1950s: the Missile Gap, or the number of SS-6 ICBMs the Soviets had deployed. To answer this question, the US would have to use satellites.

The US had looked at reconnaissance satellites in the late 1940s, but the cost and technical risks were seen as too high. With the launch of Sputnik and the apparent emergence of the Missile Gap, however, President Eisenhower authorized both an increase in funding and a reorientation. Two separate photo reconnaissance satellites were under development. The first was the Discoverer, from which film was to be returned to Earth by a small capsule. Once re-entry was completed, the capsule would be caught in mid-air by C-119s. The booster was a Thor IRBM with a small upper stage called Agena. The second reconnaissance program was SAMOS (Satellite and Missile Observation System). Unlike Discoverer, SAMOS developed its images on board, following which they were scanned and transmitted to Earth. SAMOS used the Atlas Agena booster.

2

Discoverer 1 was launched on February 28, 1959, only a year after President Eisenhower gave his approval. It reached orbit but began tumbling; the first of an incredible string of failures. Every one of the first 12 Discoverers failed; the boosters exploding during launch, the satellites tumbling in orbit, the capsules being sent into higher orbits rather than re-entering, and the capsules coming down far from the recovery forces. Not until Discoverer 13, launched into a polar orbit from the Vandenberg AFB on August 10, 1960, did everything work. The following day the capsule separated and the retro-rocket fired. The capsule splashed down in the Pacific Ocean and was recovered by frogmen.

1

1 *Launch of an Atlas Agena of the type used for the second generation of US high resolution satellites between 1962-67.*

2 *August 13, 1960: a photograph of the recovered Discoverer 13 capsule taken at Andrews AFB, Maryland. This was the first successful flight in the Discoverer series.*

Discoverer 14 was launched eight days later and was the first to carry a camera, which had a resolution of 10 to 15ft (3 to 5m). The photo targets included a suspected ICBM site at Plesetsk, in northwest Russia. The next day the capsule re-entered and was caught in mid-air by a C-119. The photos were dark and of poor quality, but they did show rail lines that had not been on World War II maps of the area.

Discoverer 17, launched on November 12, 1960, is believed to have photographed the aftermath of the worst space-related accident ever: the Nedelin catastrophe. A prototype SS-7 ICBM was being prepared for launch at Tyuratam on October 24, 1960. Marshal Mitrofan Nedelin, commander of the Soviet Strategic Rocket Forces, was under intense pressure from Nikita Khrushchev to hold to the development schedule. The SS-7 was being counted down but a problem occurred which prevented launch. Rather than wait to drain the fuel from the tanks, Nedelin ordered the engineers to work on the vehicle as it stood on the pad. Nedelin and his staff also left the blockhouse to watch the work from near the base of the SS-7.

At a little after 6:45 pm, as the engineers worked on the rocket, one of them plugged the first stage umbilical cable into the second stage receptacle. This caused the second stage engines to ignite, rupturing the first stage fuel tanks and sending a wave of flame across the pad. Some, like Nedelin, were killed instantly, others tried to run but became entangled in the barbed wire around the pad area and died. In all, 165 people perished in the explosion and following fire. It was not until 1989 that the Soviets admitted the accident had occurred, and that it had been an SS-7. On October 24, 1990, 30 years after the event, they gave the final death toll.

World Crisis

The other reconnaissance satellite program, SAMOS, had a similar run of problems; SAMOS 1 was launched on October 11, 1960, but failed to reach orbit. Of the five SAMOSs launched, only two reached orbit and their images were useless.

The next year saw intensive launch activities. There were a number of failures and problems with poor photos, but between early June and the end of September 1961, against a background of Khrushchev's missile rattling, the Berlin Crisis and the building of the Berlin Wall, the US successfully recovered four Discoverer capsules with photos of suspected ICBM sites. Only Discoverer 29's photos of the Plesetsk area showed any missiles. There were two above-ground launch pads, each having a single SS-6 plus a re-load missile. It was now clear that the Soviet boasts of great missile strength had been a bluff, and Khrushchev quickly withdrew his demands over Berlin.

With their success in finding the truth about the Missile Gap, a curtain of secrecy descended over US reconnaissance satellites. By early 1962, the only information released was a brief launch announcement, and the US government officially denied it was engaged in satellite reconnaissance.

But the Soviets were also developing similar systems. Work began in 1959, when the Vostok was used as the basic spacecraft, the cosmonaut's ejector seat being removed and replaced by a camera system. The modifications were carried out by the Korolev Design Bureau and the project was called Zenit ("Zenith," but not to be confused with the Soviet J-1 booster). The first Zenit launch took place in late 1961, when the A-1 booster failed. The first success came with Cosmos 4, which was launched on April 26, 1962, and successfully recovered three days later.

During 1962, the US reconnaissance satellite program moved to operational status. Two different types of satellite were used. The first was for area surveillance: the spotting of new targets and activities. These satellites were fitted with low-resolution, wide-angle cameras and the Discoverer was used, although that name had been dropped and reference was now made to "Program 162." Once the new targets had been found, the second type – the high-resolution satellite – would make a close examination. The booster was the Atlas Agena. As with the Discoverer/Program 162, film was returned to Earth in a capsule.

The first high resolution satellite was launched on March 7, 1962. The normal lifetime was two days before the Agena's engine was restarted and returned the capsule to Earth. In all, six of the high-resolution satellites were launched in 1962, and there were 22 Discoverer/Program 162 launches in the same year, the latters'

3

4

capsules being recovered after four days in orbit. The intensive launch rate in the early 1960s indicates that the US was conducting a survey of Russia's military and industrial facilities.

Discoverer – the Mainstay

During 1963-66, US reconnaissance satellites were launched at the rate of one high-resolution and one or two area surveillance satellites per month. A number of improvements were introduced in the area surveillance satellites. In 1963 the booster was changed to a Thrust Augmented Thor Agena (TAT Agena): a Thor with three small, solid fuel strap-on rockets, which increased the payload (film, camera etc.) that could be carried. It was not until early 1964 that the Discoverer/Program 162 satellites were finally retired, the 78th and last launch taking place on April 27, 1964. The Discoverer, despite its initial problems, had become the mainstay of the early US reconnaissance satellite program, and it was replaced with a second generation area surveillance satellite with a better camera system and other improvements.

It was also during this period that Soviet reconnaissance satellites became operational. In 1963-64,

5

3 The view from inside a C-119 aircraft as it attempts to recover the Discoverer XIV capsule in August 1960. This first mid-air recovery succeeded on the third attempt, at the end of a 27-hour mission in which Discoverer had covered 450,000 miles (724,205km).

4 A U2 image of an SS-6 launch pad at Tyuratam (Baikonur), taken in 1959.

5 A U2 aircraft in flight. Between 1956 and 1960 U2s flew around 30 missions over the Soviet Union.

1

2

3

launches were made in the spring, summer and fall, but during the harsh Russian winter the program went into "hibernation," presumably because both launch and capsule recovery would have been too difficult. It was not until 1965 that year-round operations began, with the Russian satellites staying in orbit for eight days, which allowed complete coverage of the US.

Like those of the US, the Soviet reconnaissance satellites showed a steady improvement between 1963 and 1966. The second generation high-resolution satellite was introduced with the launch of Cosmos 22 on November 16, 1963. The second generation low-resolution satellite followed in 1966, the first being Cosmos 120, which was launched on June 8. Both of the new satellites used the A-2 booster, which allowed a larger payload than the original A-1. The year 1966 also saw the first use of the old SS-6 pads at Plesetsk for satellite launches. Use of the man-rated Vostok spacecraft meant that the development and the move to operational status of the Soviet program had been smooth. But when there was an in-orbit failure, as with Cosmos 50 in April 1964, the Soviets blew the satellite up to prevent any debris being recovered when it decayed from orbit.

New Generation of Satellies

In the late 1960s, both the US and Soviets introduced third generation reconnaissance satellites. The new US high-resolution satellite used the Titan IIIB Agena D as its booster, and it could be fitted with one of three different imaging packages: a high-resolution camera which reputedly could photograph objects as small as 2ft (61cm) across; a multispectral unit to detect camouflage; or a mapping camera to facilitate locating targets inside Russia.

The new area surveillance satellites were launched by a Long Tank Thrust Augmented Thor Agena D (LTTAT Agena). The first stage was lengthened and changes were made in the strap-ons and the first stage engine. This increased the LTTAT's payload by 20 percent. Like SAMOS, the LTTAT satellite used a radio transmission system to relay images, the signals being sent through a 5ft (1.5m) dish antenna to ground stations or tracking ships. The satellite also carried an infrared scanner to pick up heat sources at night. The initial third generation Atlas Agena-launched, high-resolution satellite lifted off on July 29, 1966, and was immediately phased into operation. The new LTTAT-launched area surveillance satellite had a rocky start. The first was launched on August 9, 1966, but it was nine months before a second followed. This implied there were problems.

The Soviet third generation reconnaissance satellites were significantly improved. The Vostok service module had a cylindrical mid-section added that allowed more batteries and fuel to be carried. A supplementary payload was carried atop the spherical capsule. On high-resolution satellites, this payload was a maneuvering engine which was used to change the orbit so the satellite passed over the same area on the ground each day. On the low-resolution satellites, the supplementary payload was a cylindrical experiment package – either scientific or military – which was separated several days before re-entry. The first of the Soviet third generation satellites (which were the responsibility of the Chelomei Design Bureau, which had taken over control of Zenit in 1965) was Cosmos 208, a low-resolution satellite launched in March 1968. Cosmos 251 (the initial third generation high-resolution satellite) followed in October 1968. The booster for both was the A-2.

The operational philosophies of the US and Soviet reconnaissance programs differed. The standard life-time of the Soviet third generation satellites was 13 days. While the US high-resolution satellites remained in orbit for nine to 13 days, the area surveillance satellites had life-times of between 15 and 64 days before decaying from orbit. And, whereas the Soviets used large numbers of satellites per year, eventually reaching around 30 to 35 during the 1960s, the US used fewer but longer-lived satellites; the number of its launches went down but coverage per year went up.

Big Bird in Service

The fourth generation US satellite combined both high resolution and area surveillance. Popularly called Big Bird (the satellite was so large it required the Titan IIID booster, which in the mid-1980s was replaced with the Titan 34D) it carried a telescope claimed to be able to resolve objects as small as 1ft (30cm) across with camera film being returned via several capsules. Big Bird also carried area surveillance cameras whose images were transmitted back to the surface via a 20ft (6.1m) dish antenna. The first Big Bird was launched on June 15, 1971, and operated for 52 days. Over the program's duration the lifetime greatly increased, from 150 days in the mid-1970s to 190 days in the late 1970s, and up to 275 days at the program's end in the mid-1980s.

The Titan IIIB high-resolution satellites also showed a similar improvement. During the late 1960s they orbited for about two weeks. By the early 1970s this had been raised to 32 days; by the late-1970s it was 76 to 90 days, and when the program ended in the mid-1980s it was up to 128 days. These satellites were believed to carry two capsules and to achieve a resolution of 6in (15cm) across, which was near the theoretical maximum.

The Soviet's fourth generation satellite was based on the Soyuz and carried high-resolution optics. The

Soyuz Orbital Module is believed to have been replaced by two capsules (called "buckets"), and the Descent Module by a camera module. The Service Module's solar panels supplied power while its rocket engine made any orbital maneuvers needed. The first of the new satellites was Cosmos 758, launched on September 5, 1975. The initial lifetime was 29.5 days, which was raised to 44 days by 1983-84 and 50 days by the end of that decade.

The Soviets were also launching improved third generation reconnaissance satellites referred to as "medium resolution," since they shared features of both the high and low resolution satellites. They had a maneuvering engine to correct atmospheric drag and their orbit was higher than that of high-resolution satellites so as to cover a wider area on the ground. Their lifetime was 13 days, and the first to be launched was Cosmos 867, on November 24, 1976.

US Launch Reduction

The introduction of both satellite types was slow. Only one or two were launched per year during the late 1970s. By the early 1980s, however, the Soyuz reconnaissance satellites made up 25 to 30 percent of the total and the medium-resolution satellites about 25 percent, the remainder being third generation, high-resolution satellites. Only one or two of the old third generation low-resolution satellites were launched per year on military mapping missions. The annual total stayed around 30 to 35 per year.

The US, by contrast, continued to reduce the number of launches per year due to the introduction of the fifth generation area surveillance satellite. The first was launched in December 1976 by a Titan IIID from Vandenberg AFB. Although commonly called the KH- (for Key Hole) 11, this was actually the codename for the camera system. Unlike earlier satellites, the KH-11 did not use film. The light from the 92in (2.34m) main mirror was reflected on to an array of several thousand charge-coupled devices (CCDs), where it was converted by each CCD into an electrical current and then into a numerical value: 0 (pure black) to 255 (pure white), depending on the brightness. Once transmitted to Earth, the picture elements (pixels) were reassembled into images by computers. The advantage of no film being used was that images could be obtained as long as the satellite operated. The first KH-11 operated for 770 days and subsequent flights lasted over 1,000 days. The disadvantage of the CCD system was its lower resolution, yielding several feet in the case of the KH-11.

The KH-11 camera was the result of a quarter century of development. The Discoverer/Program 162 had carried the KH-1, 2, 3, 4 and 4A cameras. The TAT Agena area surveillance satellites were equipped with the KH-4A or 4B and the LTTAT Agena also carried the KH-4B camera. The high resolution cameras were the KH-7 (Atlas Agena), KH-8 (Titan IIIB Agena) and KH-9 (Big Bird).

The Soviets also used a CCD camera on their fifth generation satellite, which was again based on Soyuz. The first was Cosmos 1426 launched in December 1982. It remained in orbit for 67 days but by the late 1980s these satellites were operating for 230 to 259 days, with two being launched annually. The Soviet's total remained about 30 reconnaissance satellites per year.

The US reconnaissance effort consisted of two KH-11 satellites in orbit at one time – providing day-to-day, world-wide coverage – while the Big Birds and Titan IIIB satellites covered new targets. The launch rate fell to one Big Bird and/or Titan IIIB per year and a KH-11 was launched as a replacement every one or two years. There was a risk inherent in this policy; any problems

would leave the US in a critical situation, and this became apparent in 1985-86. On August 28, 1985 a KH-11 was lost when its Titan 34D booster failed: the first ever Titan IIID/34D failure. The January 28, 1986, Challenger tragedy grounded the shuttle fleet for almost three years and disrupted the launch of a number of military satellites. Then, on April 18, 1986, the last Big Bird was lost when its Titan 34D booster exploded seconds after lift-off.

The US was now reliant on a single KH-11, the sixth of the series which had been launched in December 1984 and which had therefore been in orbit for sixteen months. If it should fail there was a threat that the US would be left with no space-borne reconnaissance capability. This fear lasted 18 months, until October 26, 1987, when a new KH-11 was successfully launched, thus reestablishing the normal two-satellite "constellation."

4

5

DOD Shuttle Missions

When the shuttle returned to flight in 1988 it began an intensive program of military launches to clear up the backlog. On the second post-Challenger flight (STS-27), which was launched in December 1988, Atlantis carried a sixth generation reconnaissance satellite. Called Lacrosse, it used radar to penetrate cloud cover and darkness. As Russia and Eastern Europe are cloudy much of the year the advantage is obvious. Lacrosse has a round core section with the rectangular radar antenna along the side. Two solar panels, spanning 150ft (46m), provide power for the 10,000-watt radar and a communications antenna points upwards; the Lacrosse, like the KH-11, relaying its data to the ground via communications satellites.

4 December 2, 1988: the shuttle Atlantis lifts off on a DOD mission. It was reported to have launched a satellite called Lacrosse – a radar reconnaissance system which could obtain data through cloud and at night.

5 A 1975 view of an SR-71 in flight. Although "retired" in 1989, the Lockheed-designed and built aircraft was still the fastest in the world. It had no armament, its protection resulting from its speed, stealth and altitude capabilities.

1 *The UK's Skynet 4B military communications satellite launched at the end of 1988. The satellite was hardened against electromagnetic radiation and used signal processing and anti-jamming techniques to provide strong resistance to electronic attack.*

The next step in the reconstruction of the US reconnaissance program was the launch of two advanced KH-11s. These had an improved resolution of 1ft (30cm) and the ability to operate in the infrared, thus equalling the resolution of the Big Birds but also being able to image the surface during darkness. The first was launched in August 1989 aboard the orbiter Columbia (STS-28), and the second followed in February 1990 on Atlantis (STS-36). At first, it was thought that this second advanced KH-11 had broken up after a maneuver following deployment. However, in October 1990 observers in the US and Europe spotted the satellite: it had been boosted into a higher orbit with a greater inclination, and the pieces of "debris" were probably camera covers and shrouds. Another KH-11 was launched on June 8, 1990, by a Titan 4. Thus, as of the late summer of 1990, the US had four KH-11s operating as well as the Lacrosse satellite. One of the KH-11s was the sixth, upon which so much had depended in 1986-87: it was nearly six years old.

With the rebuilding of the US satellite effort, the mach 3 SR-71 Blackbird was retired. This was not due

to obsolescence, for the aircraft was still the fastest in the world, but rather due to cost: $260m to operate a handful of aircraft. There were many inside and outside the government who thought the retirement was most unwise.

Electronic Intelligence

The second major type of space reconnaissance is the gathering of electronic intelligence (ELINT). In modern warfare, awareness of radio and radar signals is of critical importance. For example, one of the first indications that Iraq was about to invade Kuwait in August 1990 was the reactivation of a radar site.

As with photo reconnaissance, in the early years the US used two different types of ELINT satellites. For detection of new radar sites, small piggy-back satellites were launched aboard photo reconnaissance satellites but were then separated and boosted into a higher orbit by an on-board rocket engine. The first was launched on August 29, 1963, aboard a Program 162 flight. Once a new radar site was found, its exact characteristics would be determined by a heavy ELINT satellite, the receiving equipment being carried on an Agena. The first of these was launched on June 18, 1962, by a Thor Agena.

During the 1960s, the US ELINT satellite program followed this pattern. The ELINT sub-satellites were launched aboard Thor and TAT Agenas in 1963 and between 1964 and 1966 on the Atlas Agena high-resolution satellites. In 1967 the sub-satellites were again switched to the LTTAT Agena area surveillance satellites but, when this type was retired in 1972, they were shifted to the Big Bird. On some occasions, two were carried on a single Big Bird. These ELINT sub-satellites were placed into different orbits. A 300 mile (483km) circular orbit was used to observe air defense radars, but Soviet ABM radars, which were very powerful, with beams

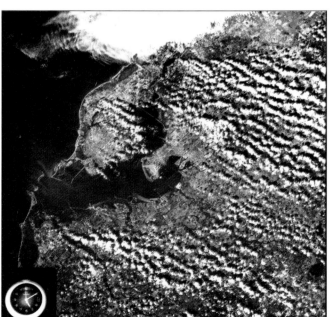

2, 3 *High quality systems aboard civil imaging platforms can suggest the potential in the military area – examples of latter images being released very rarely. Picture **3** here reproduces the full frame of a picture taken by the Skylab 4 crew in February 1974 over Tampa Bay, in Florida. Discernible at lower center are Interbay Peninsula and the runways of MacDill Air Force Base. An enlargement of this small area in the first frame appears as **2** here. The prints are some stages removed from the original film but aircraft were clearly visible in them. Military systems have a much better claimed resolution of 6in – 1ft.*

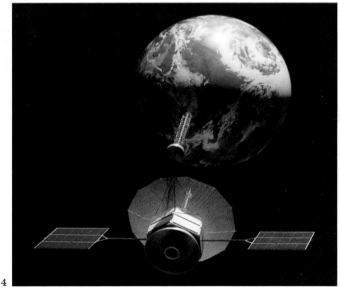

extending far out into space, were monitored from other sub-satellites in a 900 mile (1,448km) high orbit.

The US heavy ELINT satellites were launched at a rate of one or two per year during the mid and late-1960s. After the initial two Thor Agena launches the booster was changed to the TAT Agena D, which carried a larger payload. In 1968 the booster was again switched to the LTTAT Agena, the program ending with a final launch in July 1971.

The Soviets also flew a similar pattern of ELINT satellites. The small satellites were launched on the C-1 booster, and the first was Cosmos 103 in December 1965. As with other Soviet space systems, introduction was slow. There were no launches in 1966 and only two

only receive signals when it had a direct line of sight with the transmitter. Thus, it was easy for the Soviets to schedule sensitive activities, such as ICBM tests, to avoid their brief passes. The Rhyolite satellite was designed to overcome this limitation because, perched in geosynchronous orbit, it could listen in on an entire hemisphere of the Earth. To pick out the weak signal against the noisy background, Rhyolite carried a 70ft (21m) dish antenna. Its primary target was the telemetry from Soviet ICBM tests; analysis of the signals allowing Western intelligence to determine the design and capabilities of the missiles.

The first Rhyolite was launched on March 6, 1973. Its Atlas Agena placed the 606lb (275kg) payload into a geosynchronous orbit above the horn of Africa, from

4 *An artist's impression of a FLTSATCOM communications satellite, a series of which were launched between 1978 and 1989. Use of them was shared by the US Navy, the Air Force and the Department of Defense for UHF communications.*

5 *SS-9 missiles are shown here on parade in Red Square in November 1972. The liquid-fueled, three-stage ICBM was first included in the traditional parade in 1967. Its range was unknown at the time but it was known to be capable of carrying a 20-25 megaton warhead. It was also used as the booster for the Soviet Union's Fractional Orbital Bombardment System (FOBS). In this, the missile placed a nuclear warhead in orbit, but before the orbit was complete the warhead retro-fired and plunged toward its target. To combat the threat the US endeavored to develop anti-missile systems.*

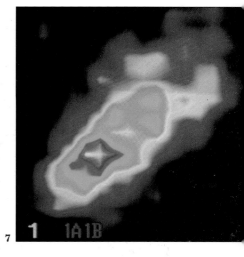

6

7

6 *Both sides in the Cold War watched each other constantly by all available means. More military information was generally available in the West because of its tradition of media freedom, but occasionally the Soviet Union lifted the curtain a fraction. This picture was released in Moscow in November 1974 (on the occasion of Rocket Forces and Artillery Day). The caption states that the missile is to be "serviced" and that day and night and under any weather conditions the Soviet strategic rocket troops are "on guard for the peaceful Soviet people." Reportedly the picture showed an SS-5 missile.*

7 *A false color ultraviolet image from a US Navy satellite showing the plume of a rocket 280 miles (450km) away. The rocket was moving toward the left with the white cross being the hottest part of the plume. Early warning systems are normally based on infrared technology, but experiments to establish the value of other wavelengths are a natural area for research.*

or three in 1967-70. Starting in 1972, the small ELINT satellites were placed into an orbital pattern in which four satellites were positioned 45 degrees apart. The Soviet heavy ELINT satellite (Cosmos 389) made its debut in December 1970. The A-1 booster, which could carry double the payload of the C-1 rocket, was used, but flights were sporadic during the mid-1970s, with only one or two per year.

In the late 1970s, the Soviet ELINT program underwent major changes. The C-1-launched small ELINT satellite had shown poor reliability and was phased out in 1979. The heavy ELINT satellite was moved to the F-1 booster starting in 1978, and the last A-1 launch was in 1983. As with the C-1 satellites, these were placed into an orbital network to provide coverage.

Geosynchronous Orbit
When the US retired the heavy ELINT satellite, it seemed to leave a major gap. This was filled by a revolutionary new system. An ELINT satellite could

where it could monitor Tyuratam ICBM tests from the moment of ignition. It quickly proved a great success. An advanced Rhyolite system, called Argus, was debated during the Ford Administration. This would have had capabilities like those of ELINT ground stations in Turkey and Iran, but it was seen as a costly duplication and Congress did not appropriate the necessary funds.

In the late 1970s, three more Rhyolites were launched. The satellite orbited in May 1977 was located above Borneo to monitor ICBM tests from Plesetsk, and the two that followed in December 1977 and April 1978 were located nearby to act as on-orbit spares.

Another US ELINT satellite was also launched during this period. It was called Jumpseat and was designed to monitor transmissions up to the Soviet Molniya communications satellites. They were put into highly elliptical orbits similar to the Molniyas, and the level of transmissions from each station would give clues to its importance; a ground station for a headquarters would have more traffic than a supply dump, for instance. This

1 *Other military powers besides the US and Russia have satellite reconnaissance programs, or are planning them. France has enjoyed considerable success with its commercial SPOT remote sensing satellites and is now proposing a military version, called Helios. Other European powers have expressed interest and the proposed resolution of the satellite is three feet (1m). This artist's impression shows the satellite over the western Mediterranean.*

is called "traffic analysis" and is an important part of intelligence gathering.

The first Jumpseat was launched on March 21, 1971, and the booster for the early satellites was the Titan IIIB Agena, although, in the early-1980s it was switched to the Titan 33 or 34B Agena. A second launch followed in August 1973 and a third in March 1975. This established the full three-satellite network needed to keep the Soviet Union under constant surveillance. Another satellite also uses this highly-elliptical orbit: the Satellite Data System (SDS), which relays images from KH-11 and Lacrosse satellites. The first SDS launch was in June 1976, and because of the similarity in orbits it is difficult to tell the two satellite systems apart.

1

3

2

2 *A prelaunch photograph of a Program 647/Defense Support Program early warning satellite. The operational program commenced in 1970.*

3 *Launch of an Atlas missile with a MIDAS satellite on board from the Cape Canaveral Air Force Station. The MIDAS (Missile Defence Alarm System) network of early warning satellites were only partially successful and were replaced in due course by DSP satellites, which were placed into a geosynchronous orbit.*

Once the Jumpseat network was established, replacement launches were made. The fourth launch did not take place until February 1978 and the fifth was orbited in December 1980. Between mid-1983 and early 1986 only one Jumpseat and three SDS launches took place. The reason for the phasing out of the Jumpseat was the changing pattern of Soviet space activity, as when Soviet military radio traffic was shifted away from the Molniyas to geosynchronous satellites such as Raduga, Gorizont and Ekran.

New ELINT Satellite

The folly of not building the Argus ELINT satellite became apparent in late 1978, when the revolution in Iran closed US ELINT ground stations there. After the signing of the SALT II Treaty on June 18, 1979, the debate centered on whether it could be verified. To fill the "verification gap" caused by the loss of the Iranian stations, the US developed advanced geosynchronous ELINT satellites. According to some reports, a series of such satellites was flown in the early and mid-1980s under the code name Chalet (later Vortex). The validity of these reports is not clear, but the spacecraft were apparently placed over the Soviet Union and may have been only Early Warning satellites, possibly with piggyback ELINT payloads.

The definitive geosynchronous ELINT satellite made its first flight aboard the space shuttle Discovery in January 1985 during the 51C mission. The Aquacade ELINT satellite (later press reports called it Magnum) had two large dish antennas: one to pick up telemetry, radio, radar and phone calls, and the other to relay the intercepted signals to a ground station in Australia.

When its antennas and solar panels were fully unfurled they spanned 100ft (30m). After shuttle flights resumed, a second Magnum was orbited, again from Discovery, during the STS-33 mission in November 1989.

The Soviets also built their own advanced, very heavy ELINT satellites. The first was Cosmos 1603, launched by a D-1 on September 28, 1984. The second, Cosmos 1656, was also launched on a D-1, but after this the booster was switched to a J-1. As with other Soviet ELINT satellites, the very heavy satellites formed a network in low Earth orbit.

Developments in China and France

Up to the mid-1970s, reconnaissance satellites were solely the domain of the superpowers. By the last decade of the 20th century, however, a number of other countries either had flown or were actively planning them. China's first reconnaissance satellite, China 4, was launched in November 1975. It was a two-part spacecraft with a capsule resembling an enlarged Discoverer capsule and a Service Module carrying the camera which looked out of a 3ft (1m) port. The satellite remained in orbit six days before the capsule separated and retro-fired. Due to limited resources, Chinese launches of this kind are sporadic.

France is the prime mover in a European reconnaissance satellite program in which Italy and Spain are also participating. Called Helios, it is based on the SPOT Earth resources satellite and will have a CCD imaging system yielding a 3ft (1m) resolution. (Helios has also been suggested as the basis for a United Nations-sponsored reconnaissance agency.) US news organizations have thought about a Mediasat, but the extremely high development costs have limited them to using SPOT images of such subjects as Libya's chemical weapons factory and the Chernobyl nuclear power plant. The Israeli and Iraqi space programs are both believed to have reconnaissance satellites as their goal.

Independent ELINT satellites are rarer. The British began secret development of a geosynchronous ELINT satellite called Zircon in the early 1980s. It was based on Rhyolite, Argus and Magnum technology, but the program was reportedly cancelled in 1986 after a BBC TV documentary revealed its existence. It is now planned that Helios will carry ELINT equipment.

Detecting Launches

In the late 1950s and early 1960s, satellites with infrared scanners (to detect the heat of missile plumes) were seen as a way to provide early warning of a Soviet ICBM attack within a minute or two of launch, whereas surface radars in Alaska, Greenland and England could not detect an attack until the missile climbed above the horizon, some 15 minutes after launch. The first US attempt was MIDAS (MIssile Defense Alarm System). Eight MIDAS satellites would be placed in two orbital networks. Orbiting at 2,300 miles (3,704km), the network would have an 86 to 100 percent probability of one satellite detecting a single ICBM launch. However, from the first launch attempt in February 1960 to the conclusion of the program in the mid-1960s, MIDAS was beset by problems. The infrared scanners designed to pick up the hot ICBM exhaust were both unreliable and subject to false alarms. Although they showed satellite detection of missile launches was possible, a new approach was needed.

This new approach was to place a single early warning satellite into a geosynchronous orbit, where it could keep watch over the whole Soviet land mass, thereby eliminating the need for a large and costly orbital network. The first test mission was launched in August 1968, and several more launches provided an interim satellite detection system against submarine-launched missiles.

The operational system, called Program 647, made its debut on November 6, 1970, but the Titan IIIC booster's upper stage malfunctioned and left it in a useless orbit. However, subsequent launches proved the design was a complete success: the infrared telescope was not only able to track missile launches but could even pick up jets. The satellite was later renamed the Defense Support Program (DSP). An advanced version was built for launch on the space shuttle or Titan 4, and both advanced satellite and Titan launcher made their bow on June 14, 1989.

The Soviets were slow to fly early warning satellites and used a different approach. The first Soviet satellite of this kind was Cosmos 520, which was launched on September 19, 1972, a full 12 years after MIDAS 1. Its A-2e booster put it into a highly elliptical orbit similar to that of the Molniya communications satellites, the two high, looping orbits a satellite makes each day allowing it to observe US ICBM sites for a total of 15 hours.

Development was slow, and it was not until 1977 that the three-satellite network became operational. In 1980-82, this was expanded from three to nine satellites, but reliability of the system was believed to be poor and there were reports of the infrared scanners having problems similar to those of MIDAS. The Soviets had to replace almost the entire network in 1984 and 1985. By the late 1980s, three launches annually were needed, and this meant replacement of the network every three years.

4 *A TRW artist's impression of an early 1990s DSP satellite in orbit. The role of the spacecraft was described thus: "Using infrared detectors that sense the heat from missile plumes ... [they] ... detect, characterize and report ballistic missile launches."*

Ocean Surveillance

Another military mission is ocean surveillance, in which the Soviets have been particularly active. In the late 1950s and early 1960s, the major naval threat was US aircraft carriers, and to counter them the Soviets built submarines and surface ships armed with anti-shipping cruise missiles. To locate the target ships the Soviets developed a radar-equipped satellite. The first was Cosmos 198, which was launched in December 1967 by an F-1 booster. To supply power to the radar a small nuclear reactor was used, but this posed a problem. The satellite's orbit was low enough for it to decay in a few months. Therefore, to prevent radioactive debris from reaching Earth the Soviets planned for the reactor to be separated and boosted into a higher orbit, which would take 500 or more years to decay.

The early years of the Soviet program were marked by problems and failures: Cosmos 198 in 1967, for example, operating for only a few days. It was not until 1974 that the Soviets were able to operate the two-satellite network for the designed 75-day lifetime. It was in that same year that the second type of Soviet ocean surveillance satellite was introduced. Cosmos 699 was launched on December 24 using ELINT to spot Western ships, since a carrier task force puts out a large volume of radar and radio transmissions. During the late 1970s a single ELINT ocean surveillance satellite was launched between pairs of nuclear-powered satellites.

5 *Of course, the missile watch was not all on one side. The Soviets would have been monitoring western launches, such as this US Trident C-4 missile booster test from a submerged submarine.*

1 *A Vela satellite which was developed by the US to monitor any violations of the 1963 nuclear test ban treaty. (The satellites could also monitor solar flares and measure other sources of natural radiation.) Twelve Velas were launched between 1963 and 1970 and accumulated more than 40 years of operating life between them.*

1

Given the low reliability of Soviet space systems, failures of the nuclear-powered satellites were inevitable. The first was Cosmos 954, which re-entered on January 24, 1978, with the reactor still attached to the satellite. Radioactive debris was scattered over northern Canada and it took several months to clear it up. The second was Cosmos 1402 in January 1983. When the boost maneuver failed, the reactor core was separated (a post-Cosmos 954 modification), which was intended to cause it to burn up on re-entry. When Cosmos 1402 and the core re-entered any surviving debris came down in the ocean. The third was Cosmos 1900. A command to separate the core failed in mid-April 1988, but the spacecraft had a back-up system which was programmed to boost the reactor if it lost pressurization, attitude control or re-entry heating began. On October 1, 1988, the automatic system was triggered and the reactor was sent into a higher orbit.

The nuclear satellite's replacement is the heavy, ocean surveillance satellite. The first was Cosmos 1870, launched by a D-1 on July 25, 1987, and carrying both radar and visual sensors. The spacecraft was developed from the military Salyut and power comes from solar panels rather than a reactor. The Soviets call it Almaz ("Diamond") and Cosmos 1870 operated for two years.

In contrast, the US Navy was slow to launch ocean surveillance satellites. Although radar satellites were studied, only an ELINT system was actually flown. This was the "White Cloud," the first of which was launched in April 1976. After the main satellite separated from the Atlas F booster it deployed three sub-satellites. The relative spacing between them remained fairly constant, akin to flying in formation, and the sub-satellites picked up transmissions from Soviet ships, the differing arrival times at each satellite being used to find the ships' position.

The threat from Soviet orbital nuclear weapons caused the US to deploy anti-satellite (ASAT) systems in the early 1960s. The first was the Program 505 Nike Zeus ASAT. It was based at Kwajalein Atoll in the Pacific and could intercept satellites up to 200 miles (321km) high. A nuclear warhead was used to "kill" the target satellite, and Nike Zeus was operational from the summer of 1963 until May 1966. The Program 437 Thor ASAT system was based on Johnston Island. The Thor's nuclear warhead could destroy a satellite as high as 805 miles (1,296km) or as distant as 1,726 miles (2,778km). Four unarmed test launches were made in early 1964 before it was declared operational.

Although the two systems were upgraded, and training launches were made, all attempts to develop a non-nuclear ASAT ended in cancellation. In late 1970 the Program 437 Thor ASAT was taken off alert status: it would now require 30 days to ready the Thor for launch. In August 1972 the launch facilities were damaged by Hurricane Celeste: they were never fully repaired and closed down in April 1975. Originally designed to attack orbital nuclear weapons, the ASAT nuclear warhead was seen as too inflexible to deal with new Soviet space threats.

For their part, the Soviets developed an orbital interceptor which makes a high speed pass of the target satellite. As the interceptor closes on the target, a high-explosive warhead is detonated and the target is riddled by shrapnel. The first test interception took place on October 20, 1968. Cosmos 249 maneuvered for two orbits before making a fast fly-by of the Cosmos 248 target satellite. In 1972, additional interceptions were flown which tested various attack profiles and showed an ability to destroy satellites as high as 600 miles (965km).

A second test series was conducted in 1976-78, when a new attack profile appeared. The interceptor would be put into a low orbit by its F-1 booster. As it passed below the target satellite, an on-board engine fired and the interceptor made a "pop-up" maneuver, flying past the target. The process took less than one orbit and gave little warning that an attack was under way. The tests were halted to allow talks on an ASAT ban, but these soon broke down over Soviet demands that the US shuttle be cancelled.

As US/Soviet relations worsened in 1980, Soviet ASAT training exercise flights resumed. On June 18, 1982, Cosmos 1379 intercepted the Cosmos 1375 target. This was part of a seven-hour simulated nuclear attack, involving the firing of two SS-11 ICBMs, an SS-20 mobile IRBM, a submarine launched missile and two ABM-X-3 missiles. After this the Soviets announced a halt of ASAT flights, but the F-1 boosters have remained operational and ready for launch ever since.

In contrast, the US ASAT program has suffered from political problems. A small two-stage rocket was developed for launch from an F-15 fighter. The interceptor was an infrared-guided vehicle which collided with the target and, although several flight tests took place, only one actual interception was made, on September 13, 1985, when the US Solwind satellite was blasted into a cloud of fragments. Congressional opposition then caused the program to be cancelled.

SDI Research

Similar difficulties have faced the US Strategic Defense Initiative ("Star Wars"). Started with the goal of rendering nuclear weapons obsolete, it faced both technical and political problems. It would have to cope with some 2,000 Soviet missiles and 10,000 warheads, the possible weapons including railguns (which magnetically

2

2 *A variety of orbital weapons were researched for the Strategic Defense Initiative – "Star Wars." This Martin Marietta concept dating from 1987 shows a laser demonstrator spacecraft. The size of the spacecraft was such that it would have required a new booster with a thrust of 10m lb to lift it into orbit.*

Threat from Orbit

There are also other, destructive, systems. Early in the space age, the possibilities of orbital nuclear weapons were studied. The Soviets built a system – the Fractional Orbital Bombardment System (FOBS) – which consisted of an SS-9/F-1 ICBM that could place a warhead into orbit. Before completing a full orbit, the warhead was retro-fired (thus the term "Fractional Orbit") to de-orbit and plunge to its target.

accelerated projectiles), lasers, particle beams and ground-based ABM missiles with non-nuclear warheads. The cost raised problems, and many in the press and academic circles viewed any strategic defense of the US as doomed to failure, destabilizing and immoral. With the collapse of the Warsaw Pact and the pressure of the US deficit, support was even more limited.

To some degree, SDI has come to be seen as a means of protecting against accidental or unauthorized launches and third world missile programs. These are more limited threats and could be dealt with by ground-based interceptors; the costly, orbiting laser battle stations no longer being necessary. The irony of the last decade of the 20th century is that several third world countries now have both nuclear weapons and ballistic missiles, weapons that were once the sole province of the superpowers.

The Soviets, however, with ambitious anti-missile programs stretching back for three decades, have deployed 100 ABM missiles around Moscow. Technical shortcomings have foiled many of the Soviet plans and for them, too, an effective SDI-type defensive shield would be as difficult to construct as it would be costly.

Keeping the Peace

Military satellites have critical roles in both peace and war. For thirty years, they have allowed the US to keep watch on the Soviet Union, enabling it to act, not out of fear or suspicion, but on the basis of what the Soviets were actually doing. But this does not guarantee a sound response: for example, there was a period when US intelligence clung to the belief that the Soviets would deploy only a limited number of ICBMs, even as the US missile force was being matched and surpassed. Images and other data are only first steps; humans must interpret them correctly.

Reconnaissance satellites have also made arms control between the superpowers possible, by employing them to count the number of ICBMs, bombers and submarines possessed by each side without the intrusion of on-site inspection. Both the SALT I and II treaties refer to "national technical means of verification," meaning satellites. One specific type of mission, the Vela nuclear detection satellite, was built specifically to police the Limited Test Ban Treaty, which prohibited nuclear explosions in space, in the atmosphere, or underwater. When the Vela satellites were retired in the mid-1970s the nuclear detection instruments were transferred to the DSP early warning satellites and Navstar navigation satellites.

With the apparent end of the Cold War, it would be expected that the satellites' targets would be shifted. In fact, this had long since been the case. In the mid and late 1960s, US satellites were targeted on Russia. If the US needed information on China, an aircraft overflight would be made. With the arrival of Big Bird, coverage of non-Soviet bloc targets went from 10 to 20 percent of the total, and by the mid 1980s, a full 50 percent of KH-11 targets were in the third world.

The Soviets have long used satellites to cover targets outside the US, including the 1969 border disputes with China, the 1971 war between Pakistan and India and the 1973 Middle East war. In the last case, the Soviets launched seven reconnaissance satellites to cover battlefield operations. Examples during the 1980s included fighting in Chad and the Lebanon, as well as the invasion of Grenada.

With US/Soviet tensions easing, such regional wars will be the primary targets for reconnaissance satellites in the 1990s. At the start of the crisis over the Iraqi invasion of Kuwait in 1990, the US had four KH-11s and one Lacrosse operating. The satellites watched Iraqi troop movements, supply dumps, air defenses, missile sites and nuclear and chemical weapons facilities. One or more DSP satellites also kept watch for the Iraqi missile launches, while ELINT satellites monitored communications. Given clear weather, there could have been few hiding places for Iraqi forces, and even in adverse conditions there was the Lacrosse radar satellite which was originally developed to spot Soviet armored forces under cloud cover or at night.

With the threat of regional conflicts developing into chemical or even nuclear warfare, these combat roles of reconnaissance satellites will become even more important.

Indeed, the Gulf War was the first where space systems played a critical role and, in retrospect, it is clear that tactical considerations will play an increasing role

3

in their design and operation. Previously, they were oriented towards strategic intelligence goals. However, a fast changing wartime situation places far different requirements on an intelligence system – particularly on the speed at which the information is processed and sent to field units.

Iraq's use of Scud missiles highlighted several space aspects of the conflict in the Gulf. Early warning satellites could pick up the exhaust plume of the missiles within seconds of launch with the result that warnings could be given to the civilian populations of Israel and Saudi Arabia quickly enough for them to take cover – something Londoners lacked with the V-2s in the Second World War.

The Scuds also showed that SDI was not a theoretical matter. The Patriot TABM (Tactical Anti-Ballistic Missile) proved that it was possible to intercept incoming missiles. With increasing numbers of Third World countries possessing ballistic missiles and developing nuclear weapons, such considerations will become more important in the future.

The political importance of the ABM systems was also highlighted by the conflict in the Gulf. It must be remembered that there were great fears in the early days of the war that Israel would retaliate in response to the Scud attacks. Such retaliation would have run the risk of breaking up the Coalition – but the Patriot TABM, along with "The Great Scud Hunt" and intense diplomatic activity, kept Israel out of the war.

The orbital watch goes on.

3 This view of the Iranian Gulf at night – taken during the STS-35 mission in December 1990 – sets the scene for the war fought between Iraq and the allies seeking to reverse the invasion of Kuwait. Urban lights and oil well flare-off characterize the view, but the war had significance in that it provided practical experience of the success achieved by early warning systems in orbit and anti-missile defenses.

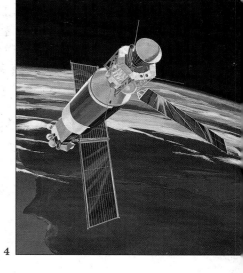

4

4 The reconnaissance satellites orbited by the two superpowers over past decades have remained almost totally secret. Despite media pressure, no illustrations of what the vehicles look like have been released, but an idea of that appearance might be derived from this artist's impression of a USAF "Agena Scientific Set" that Lockheed released in 1971.

17

THINGS TO COME

THINGS TO COME

1 *Early concepts of the space station were a mixture of the fanciful and the practical. This 1976 concept published from the Johnson Space Center proposed the use of shuttle external tanks as part of a station's structure, rather along the lines of Skylab. With flights of the space shuttle years away, the caption cautioned "There are no plans for the construction of such hardware at this time."*

An outline of events in space over the next decade or so has been sketched in previous sections: by and large, the spacecraft that will transmit their findings back from the depths of the solar system or from Earth orbit in that period are either already on the way or at least are funded and being built. But what of further into the future? Such forecasting seems to be expected but is fraught with dangers. The case of the newly-appointed Astronomer Royal in Britain who, in January 1956 (the proximity to the launch of Sputnik 1 is significant),

2 *This concept represented what might be called the mid-term view of the space station; published ten years after the 1976 impression. The most noticeable feature is the twin tower construction with a transverse boom bearing the main solar arrays and modules. The station's servicing facility (teleoperator arms can be seen clearly) was relocated to below the boom and the solar arrays were enlarged from previous proposals.*

referred to the possibility of spaceflight as "utter bilge" serves as a warning – as did an essay in futurology in the 1930s which failed to forecast radar, the jet engine and penicillin. Nonetheless, in broad terms the background to events in the space age so far does give some insights into possible developments.

The visionaries of the pre-spaceflight era usually saw future developments taking place in a very orderly and logical sequence. First of all, Earth orbit would be obtained. This would be followed by the construction of space stations, which would enable humans to be conditioned to the physiological and mental demands of spaceflight, and where missions into deeper space could be prepared. Then would come flights to the Moon and beyond. But it did not happen in that orderly,

logical manner. Earth orbital flights were indeed the first step, but the next significant step was taken on the Moon as a direct result of the Kennedy imperative. Then came space stations (briefly on the part of the US, and far more comprehensively and consistently on the part of the Soviet Union) – a concept beloved of the space-flight fraternity which in the US had to explain their value to an audience that, via the medium of television, had already experienced the excitement of walking on the Moon. This was a difficult task, and the argument was heard that, for all its achievements, Apollo was an aberration. What happened was that, once man had landed on the Moon, and during a politically stormy time in the USA, many politicians and members of the public tended to feel that space had been "done." As a result, NASA encountered great difficulty in securing support for its ensuing plans.

There is some validity in the argument that the Moon landing was a high point, which created difficulties for those charged with formulating subsequent space-flight policies. But the claim that it was an aberration implies that in its absence President and public in the US would have followed the proposals of the visionaries, and proceeded to the next significant step, after orbit was obtained, of building ever more complex space stations. That may have been the case, but equally it may well not have been. One possible alternative result might have been a far greater attachment to unmanned scientific exploration in a comprehensive manner, which would have delighted the fast-growing, new school of space scientists.

allowed the shuttle to proceed in 1972, but NASA's proposals were already a compromise and there was no real commitment to the project from the chief executive's office. President Reagan approved the space station in 1984, but again there was no firm commitment, and NASA for its part, having gathered together a group of potential users whose objectives were in some cases incompatible, deliberately left the details of the station hardware out of its conceptual proposals. This invited active participation – to use a neutral word – from members of the Congress, with the politicians in effect having a direct and detailed control over the project. Over the years, as a result, the station has been (in the jargon of Washington) redesigned, scoped, rescoped, phased, rephased and restructured. For the fiscal year 1991 budget, NASA requested over $2,450m for the

3 *Freedom's central module section as envisaged in the spring of 1991 after repeated modifications. The twin tower has been dispensed with. The US habitation and laboratory modules are at bottom, with the Japanese Experiment Module (JEM) at top right and the ESA laboratory at top left. Shown at the node between the US and ESA modules is an Assured Crew Return Vehicle (ACRV) – an essential vehicle – the construction of which NASA was discussing with Russia and ESA as 1992 closed, there being a possibility that for the short term a modified Soyuz might be chosen.*

4 *Envisaged as part of Freedom sometime in the 1990s is the Circumstellar Imaging Telescope, dedicated to the search for planetary systems around nearby stars and to studying any faint material near bright astronomical bodies. The concept is being developed and studied at NASA's Jet Propulsion Laboratory. The telescope would include a coronagraph to block light from nearby stars and a "super smooth mirror" to facilitate the search. The telescope is shown here attached to the station and being serviced.*

The importance of the Kennedy decision to go to the Moon was its startling firmness and clarity – to land man on the Earth's satellite in a period of time that was short enough to maintain a sense of urgency, and for those participating at the beginning to believe they would be on hand to see the job through. Because of its previous studies, NASA could present the President with the technical options for achieving the goal, and was given the go-ahead. Inevitably, many problems developed, but the target set the pace and forced the answers.

It is important to note that at no time since has such a situation existed in the USA, and that fact affected what happened after 1969, what is happening now, and what will happen for at least some years to come.

Space Station Changes
In the absence of long-term, clearly delineated and adequately funded goals, firmly backed by the President, NASA has been subjected to shifts of opinion and compromises which have changed over time. Richard Nixon

station and received $1,900m. In addition, the Congress told the Agency that there would a ceiling (of $2,500m-$2,600m annually) for the next five years, and that it should resubmit a revised space station plan within 90 days. (In the meantime, the international partners in the station who are expected to meet around one-third of the cost waited on the sidelines with, to put it mildly, considerable concern.)

After a request for additional time, NASA presented its revised proposals in the spring of 1991: they seemed sound and evoked a good response. The size of the station was reduced, with emphasis on prefabricating sections on the ground, thereby reducing the periods of EVA required. The target was for a man-tended capability to be achieved in 1996 after six shuttle flights, and for a permanently-manned station (with a crew of four astronauts) in 1999 after 17 shuttle missions. Vice-President Dan Quayle signalled strong support for the station from the administration – "President Bush and I are prepared to make a commitment to build a permanently-manned space station in this decade" –

In the second half of the 1980s a team of scientists and engineers at the Johnson Space Center began studying a return to the Moon. These two artist's impressions are drawn from the results of the studies.

1 An inflatable habitat with support installations: forerunners of a permanently-inhabited lunar base.

2 A lunar lander separating from an Orbital Transfer Vehicle (OTV) in low orbit over the crater Copernicus.

while the President's chief scientific adviser rebutted criticisms of the station from national science bodies in a forthright manner. "Science [is not] the only or even necessarily a particularly important consideration in decisions involving space …. A vastly more compelling rationale for the space station is that it is the first step in the great adventure that will take our species away from the home planet and eventually out towards the stars." For good measure, the international partners in the station project professed themselves pleased by NASA's preparedness, during this last major review, to pay attention to their interests.

Unfortunately, assaults on the space station in the

responded to a request by the NASA Administrator by presenting a report on *Leadership and America's Future in Space*. This highlighted major possible initiatives as (1) a Mission to Planet Earth; (2) the robotic exploration of the Solar System; (3) the creation of an outpost on the Moon and (4) the human exploration of Mars. (One result of these reports was the setting up of an Office of Exploration in NASA – subsequently redesignated Office of Aeronautics, Exploration and Technology.) At the end of 1990 another report requested by the NASA Administrator dealt with the future of the US space program yet again, recommending reduced reliance on the space shuttle and the redesign of the space station

1

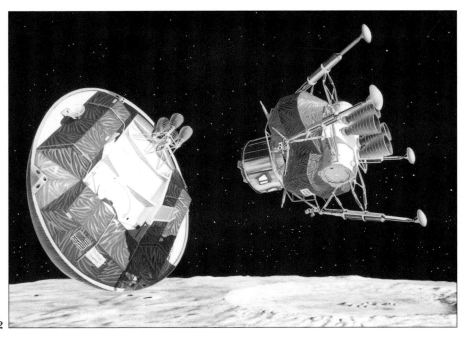

2

3

3 A Russian concept of a solar sail probe. It is proposed to launch a solar sail race to the Moon in the near future.

4 A Boeing concept of a 360ft- (110m) long Mars transfer vehicle with a nuclear thermal rocket engine located at the far end of the boom. A lander is docked at the near end of the vehicle with its aerobrake shield prominent.

Congress (and particularly the House of Representatives) were repeated in 1992 and at one time there were serious doubts about whether the project would survive. The international partners once again expressed grave concern about the uncertainty and while by early September 1992 the station had survived it seemed certain that its funding for fiscal year 1993 would be considerably below the $2,250m requested by NASA.

New NASA Administrator

These difficulties occurred almost at the same time as former astronaut Richard Truly, who was appointed Administrator in 1989, was forced to resign by the President – reportedly because of disagreements with Vice-President Dan Quayle in his role as chairman of the National Space Council. An experienced aerospace industry leader – Daniel Goldin, vice president and general manager of the TRW Space and Technology Group – was quickly appointed to the post but a dramatic change of this kind could have done nothing for morale within NASA, at least in the short term. To be fair, over the years NASA has shown awareness of the value of long term goals and has not been slow to generate such proposals or to receive them from others. Thus, in 1983 and 1986 the Solar System Exploration Committee of NASA's Advisory Council presented proposals for the rest of the century which were balanced and showed full awareness of economic realities. The Space Sciences Board of the National Research Council later presented its views on the shape of the space sciences in the 21st century. (It was decidedly cool on manned spaceflight.) In 1986, the year of the Challenger tragedy, a National Commission on Space appointed by the President published its report *Pioneering the Space Frontier*, and in the next year a group led by astronaut Sally Ride

Freedom, though with different emphasis from that being recommended by Congress. Seeking the views of others is in moderation a sound policy, but even a sympathetic observer must be tempted to conclude that, if words recommending paths to the future could have been replaced by action, the US by now could have had Freedom well under way. Nonetheless, the events of spring 1991, so far as the station was concerned, promised tangible progress at long last.

Back to the Moon

In 1989 (on the 20th anniversary of the Apollo 11 landing) President Bush showed willing by announcing a "long-range continuing commitment." While the task for the 1990s was the space station, it was to be followed by a return to the Moon "to stay" after which was to come "a journey into tomorrow" – a manned mission to Mars. The proposals were referred to as the *Space Exploration Initiative*, and a National Space Council under the Vice-President was to act as the point of focus for developing the national space policy. In May 1990 President Bush set the major target as raising the American flag on Mars before July 2019, although how this could be squared with the concept of the "international participation" in SEI that was to be invited was not immediately apparent. Richard Truly, the NASA Administrator, promptly set up another committee called the Synthesis Group (under Gemini and Apollo veteran Tom Stafford) to examine the ideas submitted from across the nation in response to an appeal for "innovative concepts and technologies … to carry out SEI." The Group reported in the summer of 1991 with four policy alternatives – all of which looked to manned landings on Mars beginning in the second decade of the next century.

4

Although a program extending for over a quarter of a century (more than six presidential terms of office) would inevitably present problems of pace and application, George Bush's reference to a long-range, continuing commitment was promising, and he has supported increased spending on space. Unfortunately, Congress has not been so enthusiastic. Following the marked lack of enthusiasm the previous year, at the end of summer 1992 it was prepared to grant NASA only about 10 percent of the sum of $32m that had been requested for SEI for the ensuing fiscal year. And there, at the time of writing, the matter rests. There is considerable drift in US space policy and as 1992 drew to a close American space historian and author Curtis Peebles wrote disconsolately:

"Congress seems to believe that the only political gain from space is by cutting NASA's budget .... The general budget situation is also grim. Three of the biggest items are debt service, entitlements (social security, farm subsidies and other special interest funding) and the Savings and Loan restructuring. With these non-productive burdens, there is little room for anything "visionary." The press also seems both nihilistic and ignorant about space. During one landing of the shuttle, the reporter of a national radio network seeing a contrail coming off the wings described it as the smoke trail from the orbiter's afterburners – obviously not knowing that the vehicle lands unpowered. The American public seems withdrawn. Survey after survey indicates how little Americans know about the wider world. They seem to have turned away from that world and drawn inward."

Changes in Russia

The amount of attention devoted to the situation concerning spaceflight in the United States basically reflects the traditional openness of its society. It has only been in recent years that the quite opposite nature of society in the Soviet Union has been modified to reveal any significant details about space achievements and future plans there – a great pity in the homeland of one

5

5 *Another possible scenario from the Space Exploration Initiative. A small excursion vehicle has descended to the surface of Phobos – one of the moons orbiting Mars – and a member of the crew has donned a manned maneuvering unit to extend the distance covered during the initial exploration.*

1 An impression of the space observation platform to be launched in the later 1990s as NASA's first step in the "Mission to Planet Earth." As earlier envisaged, the polar orbiting spacecraft were promising to be among the most complex ever launched. Political pressure was brought to bear, however, and NASA was instructed to limit the instrumentation aboard the spacecraft, which were themselves to be modified.

2 Up until the late 1970s serious attention was being paid to the possibility of constructing solar power satellites in orbit around the Earth – one impression of which appears here. The satellites would have been enormous: the mass could have been 50,000 tons, with the solar array collector area measuring 20 square miles.

of the great visionaries of spaceflight and where spaceflight achievements have been of a high order. Plans for missions to Mars in the 1990s are known and the continued, successful occupation of near Earth space with Mir-type stations will probably continue. But in recent years those in charge of Soviet spaceflight have experienced the hitherto unknown constraints of policies insisting on cost effectiveness and the need to earn a profit, with funding for space taking major cuts. These developments intensified following the break-up of the Soviet Union and the tendency of the previously centralised space program to fragment as the leading republics of the CIS sought to establish separate and independent space programs. The concentration of much of the former Soviet space organisation and infrastructure within the Russian republic has mitigated the effects of the break-up – but all sectors of the space effort within the CIS are suffering from the economic dislocation and high inflation which have intensified since the Soviet Union passed into history.

1

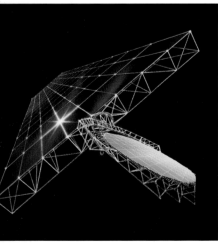

2

Russia has much space experience to offer the world – from proven launchers and the construction and operation of Earth orbiting spacecraft to such applications as the remote sensing of Earth resources and materials processing. That potential is now being energetically marketed world-wide in the drive to earn hard currencies and to finance the future prosperity as well as expansion of the industry. The response abroad has been reasonably enthusiastic – allied with a little caution about the ability to deliver given the present domestic situation in the CIS generally. The next decade and onwards could witness a melding of the (principally) Russian capabilities in space with those of the other leading space powers to the benefit of all. Time will show.

ESA, Japan and China

While ESA's achievements in space have been considerable, there have been second thoughts about the ambitious program adopted in 1987 and – while the situation is far from clear – it seems likely that the future emphasis in both manned and unmanned spaceflight will be on cooperative projects with other space

powers. Given the unfortunate experiences with project cancellations in the USA, it would not be surprising if such cooperation was sought increasingly with Japan and the CIS.

However, the situation seems clearer in other parts of the world. Japan will continue its advance in many areas of space (including manned spaceflight eventually), the only questions being how long it will take to solve the problems being encountered in the development of its H-II launcher and in what area it will choose to offer commercial services. Within the limitations imposed by general economic needs, China seems set on developing its commercial launcher services further, and it would not be surprising to find it developing a manned spaceflight capability of a basic kind once sufficient power and reliability have been secured with launchers.

Mission to Planet Earth

All too often considerations of spaceflight tend to center on dramatic projects, such as unmanned interplanetary fly-bys or manned "firsts," to say nothing of the romance of the far future. In fact much of what has been achieved (and most successfully) has concerned applications technology such as communications and meteorological satellites. Plans for the future, both at national and international level, indicate continual development of such applications which have clearly identified and financially strong user groups – sometimes of a private commercial kind, sometimes largely state owned and sometimes a mixture of both – to stimulate both technological and commercial progress at a fast pace. The remote sensing of Earth resources, begun systematically in space with the launch of ERTS (Landsat) 1 in 1972, has not progressed so convincingly because it does not have such easily identifiable user groups as is the case with other applications, and there have been strong doubts about continuity in the orbiting of satellites. However, the launch of French SPOT remote sensing satellites (backed by a vigorous marketing campaign which increased turnover by one-quarter to almost $38m in 1991 compared with 1990), ESA's development of its ERS-1 satellite (with a second to follow) and NASDA's of both ocean and land sensing spacecraft will provide a considerable boost in the next few years.

This boost will be only part of an inevitably increasing emphasis on the world environment generally. The concept of a Mission to Planet Earth was proposed in the 1987 Ride Report, where it was defined as "an initiative to understand our home planet, how forces shape and affect its environment, how that environment is changing and how those changes will affect us." By obtaining such data and understanding, the report foresaw the possibility of "[eventually developing] the capability to predict changes that might occur – either naturally, or as a result of human activity." NASA embraced the concept with alacrity, and proposed a major, 15-year program called EOS (Earth Orbiting System), in which a battery of geostationary and polar orbiting spacecraft contributed by NASA, ESA, Japan and other countries would provide comprehensive data on the atmosphere, oceans and land. NASA's first polar orbiter was scheduled for a 1998 launch, and it proposed spending on EOS amounting to $17,000m by the year 2000. By agreement the program was also made the principal theme of the International Space Year taking place in 1992, a supporting group EURISY (European Association for the ISY) being dedicated to the "protection of our natural resources and the harmonious evolution of our planet." Of major importance in a project of this type and magnitude is adequate provision for the ground-

based, data handling segment, since an enormous capacity will be required to receive data and process it to a usable form in a reasonable time. (It has been estimated that EOS will generate every 30 days a volume of data equivalent to that in the Library of Congress.) NASA has recognized this fact and is proposing to invest 60 percent of total EOS funding on the ground-based Data and Information System and science. Perhaps predictably NASA's plans for EOS have come under political pressure and during 1991/2 it was instructed to reduce spending to the year 2000 by over one half to $8 billion by restricting the instrumentation to be flown on a modified system of satellites.

Materials Processing – a Query

The point was made above that progress in the remote sensing of Earth resources has not taken place so surely as have other applications. The use of the micro-gravity conditions existing in orbit is at an even earlier stage so far as applications, as distinct from basic research, is concerned. In micro-gravity factors affecting physical processes on the surface of the Earth, such as sedimentation and convection, are virtually eliminated, and for a considerable number of years the potential theoretical and ultimately practical value of what has come to be called MRPS (materials research and processing in space) has attracted researchers in such specialisms as electronics, biotechnology, metals and alloys, polymers and ceramics, as well as glass. The phenomena have been studied in many countries: aircraft flying parabolic trajectories and drop towers provide researchers with short periods of micro-gravity, but for obvious reasons orbital flight provides the opportunity for far more extensive study, provided astronaut movements and other in-flight events such as thruster firings (which can seriously degrade the "quietness" required during micro-gravity work) are kept to a minimum.

The more prolonged materials research and processing conducted on NASA space shuttle missions and Soviet space stations has led to conflicting reports on the value of the work. The portrayal of what amounts to new industries resulting from microgravity conditions in space contrasts with emphasis on the gains made by using far more sophisticated techniques on the surface of the Earth, usually at lower cost. This is an area where time will show what scientific and applications benefits can be derived from micro-gravity work in space. It will continue to be conducted on manned missions, but it may well be that automated systems placed on mantended "free-fliers" will become increasingly popular. Certainly, continued access to orbital space is important for significant progress to be made. While it is impossible to say, we should not be surprised if, over the next decade, major improvements in existing materials such as pharmaceuticals, crystals and semi-conductors, optical fibers and ceramics, as well as high strength alloys and foam metals, in addition to totally new products, were to result from MRPS.

Private Sector Interest

Both remote sensing and micro-gravity applications touch upon another important theme of the future: the extent to which "commercialization" (defined here as participation by the private sector in the West) will develop. Traditionally, the heavy costs of spaceflight could only be borne by governments, although those in the West often used non-state-owned contractors to build launchers, spacecraft and other equipment. In the decade of the 1980s it became a fashion in the US, UK and some other countries to shed state financed activities and to place them in the market place – the logic being that the private sector could usually be relied upon to perform more efficiently (with the implication being that if a particular activity could not be performed efficiently and at a profit then it should be allowed to fail). In the post-Challenger period it was official policy in the USA for aerospace companies to develop expendable launch vehicle (ELV) services, but in a sense the appearance of "commercial" versions of the Atlas, Delta and Titan launchers was a continuation of what had gone before, although in a different guise.

3

The post-Challenger disaster period saw a growing emphasis on the provision of commercial (i.e. non NASA) launcher systems and other initiatives.

3 *One of the most imaginative developments (yet relatively economic for customers wishing to put limited payloads into Earth orbit) has been the Pegasus delivery system, which is launched from an aircraft at high altitude. The system is now operational and is a joint enterprise between Orbital Sciences and Hercules Aerospace in the US.*

4 *February 14, 1990: lift off of a commercial Delta booster with a USAF payload aboard.*

5 *A commercial Titan launcher lifts off in June 1990 on the way to placing an Intelsat comsat in orbit.*

4

5

The future of launching into Earth orbit must lie not with the vertical rocket but with reusable orbiters that can take off and land horizontally on runways.

1 *The British Hotol project (Horizontal Take Off and Landing) is a single stage to orbit vehicle demanding the development of engines capable of working both in the atmosphere and the vacuum of space. A subsequent proposal for an interim stage Hotol envisaged it being launched at high altitude from a Russian Antonov An 225 transport.*

A more significant development was the attempt to privatize the provision of remote sensing data in the US, with Eosat (the Earth Observation Satellite Company) taking over the functions performed previously by the Department of Commerce. (A similar exercise was taking place in terms of data handling in the UK in the winter of 1990-91.) Responsibility for the development of new Landsat spacecraft and securing adequate revenue from data which had previously been supplied by the Government at a much lower cost is not an immediate prescription for success, and Eosat has not had an easy time.

There is a central problem here. Energetic commercial companies are prepared to face risks, but they must be calculated risks, with the possibility of a reasonable return over a time scale which does not result in impossible financial stress. NASA has for some years pursued an energetic policy of encouraging "doing business in space," but it is one thing to invest money in communications satellites or services (which is highly

US, and private industry in both the UK and Federal Germany has over recent years conducted studies aimed at solving the single most important problem in the future development of spaceflight: that of reducing launch costs significantly. While costs using vertical launchers could certainly be trimmed by optimum design of both the vehicle and its servicing infrastructure, by far the greatest gains would result from the development of reusable vehicles operating from airfields. MBB in Germany has proposed the two-stage Sanger. This would be capable of reaching all orbits from take off in Europe: its first stage would be an airbreathing, turbo-ramjet hypersonic aircraft, and its second stage (in both manned and cargo versions) would be a reusable and rocket propelled winged vehicle. The logic of this type of vehicle has been obvious for decades, and a two-stage vehicle launching horizontally was one of the early options proposed for the US shuttle. MBB has envisaged the development of its system some time after funding

1

3

2

2 *US industry has been examining a proposed National Aerospace Plane (NASP or X-30) for several years. One impression of the vehicle appears here: the project's future looked very unsure as 1992 ended, but the attitude of a new administration in Washington might give a fresh stimulus.*

profitable and became established early on with few problems) and quite another to invest large sums of money in the potential of remote sensing and microgravity processing. Clearly, regular and reliable access to space must be a major consideration for an investor, with the cost of that access being probably even more important.

But, even if a little haltingly, private capital outside of the lucrative space communications industry and the traditional US launcher manufacturers is beginning to come forward. In the US, the Orbital Sciences Corporation was first formed in 1982 and, together with Hercules Aerospace, developed the Pegasus orbital delivery system, which is deployed from an aircraft and which is now operational – the first launch from a B-52 having taken place in April 1990. Spacehab Inc. has developed a commercial, mid-deck augmentation module (CMAM) for the shuttle, which will make a pressurized research and development laboratory available, and a major contract has already been signed with NASA for flights beginning in 1993. There is a powerful representation of former senior NASA managers on Spacehab's staff, and the same is the case with Space Industries Inc., a Houston based company offering a wide range of commercial services.

Reducing Costs

The "enterprise culture" is not, of course, limited to the

for ESA's Ariane 5/Hermes tailed off, but the future of Hermes is now very unclear. An early calculation of the cost of development of the MBB concept put it at up to $15,000m, with delivery of payloads into low Earth orbit being about one-quarter the cost of Ariane 5 or the space shuttle.

The British proposal was essentially more advanced. Hotol (for Horizontal Take Off and Landing) would be a single stage to orbit, fully reusable vehicle. This would be powered by a Rolls Royce dual mode rocket motor using atmospheric oxygen and on-board liquid hydrogen for take off, and on-board liquid oxygen and hydrogen for achieving orbit. In 1985 British Aerospace proposed Hotol as an alternative to France's Hermes. A joint government-industry proof of concept investigation was concluded successfully in 1987, but the government refused further funds when it contemplated possible development costs of around $8,000-10,000m. While a manned version could be developed, Hotol (not only by means of the innovative engine and other technology but in terms of its total system design) is essentially dedicated to the task of launching an unmanned vehicle into space and delivering its payload into orbit at the minimum cost, calculated in the mid-1980s at 20 percent of shuttle costs. It was recognized that such an innovative project would have to be international in character, but even so its prospects dimmed after the initial moves, although they received a new boost in the fall of 1990 with the news that joint Soviet-UK studies were to be conducted on an "interim" HOTOL vehicle which would be launched from a Soviet Antonov AN-225 transport aircraft. The studies were continuing in 1992.

The US, too, has been conducting preliminary studies of a National Aero-Space Plane (NASP or the X-30), but the cooling of Department of Defense interest in the project (in which it was originally seen as a replacement for both ICBMs and manned existing bombers) caused the National Space Council to slow the pace of research by contracting aerospace companies, and a decision on whether to build and fly-test an X-30 will not now be taken until 1993. However, in whatever final form it appears, it is evident that the totally reusable vehicle giving low cost access to space is the only way to proceed. The high development funding involved, in which heavy costs earlier will lead to far lower costs later, seems certain to cause this to be a major international project.

In general, the extent of such international co-operation in spaceflight has expanded greatly over the years. Often it is regarded as a way to lower costs, but the additional complications introduced into the planning, design, manufacture, integration and general management of projects can frequently lead to such expectations being less than fulfilled. Not infrequently, cooperation of this kind does not run smoothly, as in the case of the original ESA/NASA joint mission to study the Sun and the development of the space station.

Space Colonies – and Starships

Spaceflight reality often seems to differ greatly from the concepts of the visionaries. The latter, quite rightly, concentrate upon engineering, technological and scientific matters and ignore the vagaries of political considerations. We have already made reference to the logical progression proposed for the move into Earth orbit followed by the development of space stations, which would act as the way stations for journeys to deeper space. So far it has not happened like that, although the sequence may be reestablished in the years ahead.

Ignoring political factors has both advantages and disadvantages. If spaceflight visionaries endeavored to take account of such factors the results of their work would tend to be restricted, lacking in insight and daring. That they do not is an advantage. The dis-advantage is that time scales can be seriously distorted, and compromises resulting from reality lead in unwanted directions. Moreover, the understandable aim that spaceflight should become as routine or as "ordinary" as possible paradoxically runs counter to the inspiring stimulus provided by the grand designs of the visionaries.

Between ten and twenty years ago there was much discussion of massive solar power satellites beaming "clean" energy down to the surface of Earth from orbit. There was an enormous enthusiasm for the concept of space colonies, where the materials found in interplanetary space could be worked economically without the need of continually fighting to overcome the forces of Earth's gravity "well." There is a simple elegance in the concept of using the pressure of sunlight to power spacecraft across interplanetary distances and a constant attraction in designing vehicles to travel to nearby stars using current or likely near-future technology. While such ideas may be far-seeing there is no need for them to be far-fetched. Thus Alan Bond, the leader of the British Interplanetary Society study group which conceived the Project Daedalus star probe system in 1978, was the designer of the revolutionary

4

power plant proposed for use in the Hotol single stage to orbit vehicle.

Not all the ideas of the visionaries will be translated into reality. Equally, new technological discoveries may lead to opportunities which we cannot begin to visualize at the moment. Any disappointment felt at our lack of success in fulfilling the hopes of the space visionaries could well reflect a mis-match in the time scale of developments as much as anything else. Forward move-ment in spaceflight does not seem to occur along a logical and direct path at a steady pace. It eddies like the action of water and, being of humankind, it has some human failings built into it. Thus Wernher von Braun, the builder of the magnificent Saturn Vs, spent a period of his life building rockets that bombarded the south of England and other areas of Europe in World War Two. Again, a complex spacecraft destined for interplanetary space can cost as much as a major hospital (or two) but it is our inadequacies which cause us to fail to see that with wisdom we could have both spacecraft and hospital.

Humankind, as imperfect as it is, strives towards new horizons, both within and without the body and the mind. Space is the ultimate horizon because, paradoxically, it is without end. And in the absence of the possibility of complete fulfilment lies the eternal challenge of the stars.

Despite the post-Apollo malaise in the US, scientists, engineers and dreamers there and elsewhere in the 1970s continued to envisage the far future. The economic realities of the 1990s seem to suggest the conversion of such concepts into something more as being unlikely over the next few decades, but fore-casting the technological future is notoriously difficult. And human-kind can always dream.

3 The Daedalus starship: an artist's impression of the vehicle proposed following a British Interplanetary Society study.

4 A 21st Century space colony as envisaged by the late Gerard O'Neill and his colleagues. Each cylinder is 19 miles (32 km) long and four miles (6,400m) in diameter, accommodating between 200,000 and several million inhabitants. Each cylinder would rotate around its axis once every two minutes to create Earth-like gravity: solar energy would supply power, and lunar or asteroid raw materials would be used for construction.

INDEX

Bootprint on the Moon: undisturbed by humankind, it will remain unchanged for aeons as testimony to the first small step on the journey to the stars.

ACKNOWLEDGEMENTS

A book of this kind owes a great deal to many individuals and organizations. In the time-honored phrase it is impossible to thank all by name, but the authors either individually or jointly would like to thank in particular: Dr. W.D. Compton, James E. Oberg, Phil and Sue Henderson, Roy Gibson, Dr. Simon Mitton, Patrick Moore, Beatrice Lacoste, and L.J. Carter.

Of space agencies, companies and institutions across the world the following were most helpful: the British Interplanetary Society; the National Air and Space Museum, Washington; the Canadian Space Agency; NASDA and ISAS in Japan; CNES, Arianespace and Aerospatiale in France; ISRO in India; INPE in Brazil; the EURISY group; and the various offices and stations of ESA and NASA. In NASA, however, Lee D. Saegesser and the staff of the NASA History Office; Althea Washington at NASA HQ in Washington; Ellen Seufert at the Goddard SFC; and Mike Gentry and Lisa Vazquez at the Johnson Space Center must be singled out by name. Most of the US aerospace companies approached for pictures and information responded generously and are acknowledged in the picture credits. SpotImage and its chief executive G. Brachet were equally helpful – as were Mrs M.H.E. Williams in the Science Information Unit of the Australian Defence Science and Technology Organisation and Frank Chapman in the Photographic Section of DSTO.

Picture Credits
Aerospatiale (France) 190/1,2,3 191/4 224/1; American Science and Engineering 120/2; British Aerospace 187/3 189/3 199/5 222/1 234/1 British Interplanetary Society 13/4; CIA/Curtis Peebles 219/4; CNES (France) 185/4 187/6,7; Defence Research Agency (DRA) U.K., 213/2,3 215/3; Dornier (Germany) 192/3; DSTO – Department of Defence (Australia) 18/1; ESA 189/5 191/5; ESTEC 188/1; GE Astro-Space Division 173/4 (RCA)198/2 (RCA)210/1 232/1 ; Grumman Aerospace 128/3 171/6; Hughes Aircraft 168/2 197/6 210/2; ISAS (Japan) 193/4,5 194/1,2 195/5; ISRO (India) 198/1,3; Johns Hopkins University 206/1,3; Lebedev D.A/Spaceflight 27/2,3,4,5 30/3 35/4; Lockheed Missiles and Space 128/2 219/5 225/5 227/4; Martin Marietta 111/2 131/3 170/2 226/2 ; Max-Planck-Institut für Aeronomie/H.K. Keller 199/6 MBB (Germany) 234/2; NASA/ESA 204/3 205/6,7; NASA/Esther C. Goddard 14/1,2 NASA/Smithsonian Institution/Lockheed 204/2 NASDA (Japan) 193/6 195/3,4; New China Pictures/Eastfoto, N.Y. 196/1,2,3 197/4,5; Naval Research Laboratory (NRL) 22/3 70/3 115/4,5 120/1 147/5,6 202/1,2 223/7; Orbital Sciences Corporation 233/3; Rockwell International 134/1; Rockwell/Rocketdyne Division 234/3; Sovfoto/Eastfoto, N.Y. 11/3 12/1,2 15/4 16/1 17/2,3 18/1,2 19/3 20/1,2 26/1 28/1,2 29/3 30/1 31/2 32/1 34/1,2 36/1,2 37/4,5 38/1,2 95/3,4 96/1,2 97/3 98/1 99/2 100/1 101/2,3,4 102/1,2,3 103/4 104/1,2 105/3,4,5 106/1,3 108/2 109/3 148/1,2 149/3,4,5,6 160/1,2,3 161/4,5,6 174/1,2 175/3,4,5 176/1,2,3 177/4,5,6 178/1,2,3 179/4,5,6,7,8 180/1 181/3,4 182/1,2,3,4 183/5,6 184/1,2,3 185/5 201/5,6 206/2 208/2 217/5 223/5,6 230/3; Space Frontiers/H.J.P. Arnold 88/1 127/4 163/4,7 180/2; Tass/London 29/4,5 32/2 36/3 37/5 38/3 39/4 106/2 107/5,6 108/1 Thiokol 130/1 154/1; TRW 207/4 222/4 225/4 226/1; USAF 233/4,5; USAF/Curtis Peebles 218/1,2,3 220/1,2 224/2,3; USGS 92/2,3 93/4 163/5 199/4,7. All other pictures courtesy US National Aeronautics and Space Administration and supplied from the Space Frontiers Collection.